Biology and Applications of Bacteriophages

Biology and Applications of Bacteriophages

Edited by Emma Richardson

New York

Hayle Medical,
750 Third Avenue, 9th Floor,
New York, NY 10017, USA

Visit us on the World Wide Web at:
www.haylemedical.com

ISBN: 978-1-64647-544-5

Cataloging-in-Publication Data

Biology and applications of bacteriophages / edited by Emma Richardson.
    p. cm.
Includes bibliographical references and index.
ISBN 978-1-64647-544-5
1. Bacteriophages. 2. Bacteriophages--Health aspects. 3. Bacteriophages--Therapeutic use.
I. Richardson, Emma.
QR342 .B333 2023
579.26--dc23

Table of Contents

Preface.. VII

Chapter 1   **Bio-Control of *Salmonella* Enteritidis in Foods Using Bacteriophages**...................................1
Hongduo Bao, Pengyu Zhang, Hui Zhang, Yan Zhou, Lili Zhang and Ran Wang

Chapter 2   **Coordinated DNA Replication by the Bacteriophage T4 Replisome**...19
Erin Noble, Michelle M. Spiering and Stephen J. Benkovic

Chapter 3   **Bacteriophage-Derived Vectors for Targeted Cancer Gene Therapy**.......................................34
Md Zahidul Islam Pranjol and Amin Hajitou

Chapter 4   **Effect of Bacteriophage Infection in Combination with Tobramycin on
the Emergence of Resistance in *Escherichia coli* and
Pseudomonas aeruginosa Biofilms** 51
Lindsey B. Coulter, Robert J. C. McLean, Rodney E. Rohde and Gary M. Aron

Chapter 5   **The Role of the Coat Protein A-Domain in P22 Bacteriophage Maturation**............................60
David S. Morris and Peter E. Prevelige, Jr.

Chapter 6   **Photodynamic Inactivation of Mammalian Viruses and Bacteriophages**................................75
Liliana Costa, Maria Amparo F. Faustino, Maria Graça P. M. S. Neves,
Ângela Cunha and Adelaide Almeida

Chapter 7   **Structural Aspects of the Interaction of Dairy Phages with Their Host Bacteria** 116
Jennifer Mahony and Douwe van Sinderen

Chapter 8   **Understanding Bacteriophage Specificity in Natural Microbial Communities**...................132
Britt Koskella and Sean Meaden

Chapter 9   **Interaction of Bacteriophage λ with Its *E. coli* Receptor, LamB** 151
Sujoy Chatterjee and Eli Rothenberg

Chapter 10   **The Staphylococci Phages Family**...169
Marie Deghorain and Laurence van Melderen

Chapter 11   **Lysogenic Conversion and Phage Resistance Development in Phage Exposed
Escherichia coli Biofilms**..189
Pieter Moons, David Faster and Abram Aertsen

Chapter 12   **Genomic Sequences of two Novel *Levivirus* Single-Stranded RNA Coliphages
(Family *Leviviridae*): Evidence for Recombination in Environmental Strains** 201
Stephanie D. Friedman, Wyatt C. Snellgrove and Fred J. Genthner

Chapter 13   **Function and Regulation of Clustered Regularly Interspaced Short Palindromic Repeats (CRISPR)/CRISPR Associated (Cas) Systems** 222
Corinna Richter, James T. Chang and Peter C. Fineran

Permissions

List of Contributors

Index

Preface

Every book is initially just a concept; it takes months of research and hard work to give it the final shape in which the readers receive it. In its early stages, this book also went through rigorous reviewing. The notable contributions made by experts from across the globe were first molded into patterned chapters and then arranged in a sensibly sequential manner to bring out the best results.

Bacteriophages or phages are the viruses that infect bacterial cells. They replicate within these bacteria. Bacteriophages can be categorized on the basis of their morphological characteristics, nucleic acid content, site where they are usually found, and the bacterial target. Phage therapy is a specialized application of bacteriophages used for treatments wherein bacterial infections are treated using bacteriophages. The genomes of bacteriophage are used to encode highly pathogenic bacterial toxins including cholera toxin in Vibrio cholerae and diphtheria toxin in Corynebacterium diphtheria. Despite its benefits, there are some risks associated with phage therapy. For instance, it may lead to development of antibiotic resistance and reduced activity due to immune system response. This book contains some path-breaking studies on bacteriophage biology. The various advancements in bacteriophage research have been glanced at and their applications as well as ramifications have been analyzed in detail. The book will serve as a valuable source of reference for the graduate and postgraduate students.

It has been my immense pleasure to be a part of this project and to contribute my years of learning in such a meaningful form. I would like to take this opportunity to thank all the people who have been associated with the completion of this book at any step.

Editor

Bio-Control of *Salmonella* Enteritidis in Foods Using Bacteriophages

Hongduo Bao [1,†], Pengyu Zhang [2,†], Hui Zhang [1], Yan Zhou [1], Lili Zhang [1] and Ran Wang [1,*]

[1] Institute of Food Safety, Jiangsu Academy of Agricultural Sciences, State Key Laboratory Cultivation Base of MOST- Jiangsu Key Laboratory of Food Quality and Safety, Nanjing 210014, China; E-Mails: baohongduo@163.com (H.B.); zh851200@163.com (H.Z.); zhou.yan.77@hotmail.com (Y.Z.); lilizhangnj@163.com (L.Z.)

[2] Ginling College, Nanjing Normal University, Nanjing 210097, China; E-Mail: panpan.jump@163.com

[†] These authors contributed equally to this work.

[*] Author to whom correspondence should be addressed; E-Mail: ranwang@jaas.ac.cn

Academic Editor: Abram Aertsen

Abstract: Two lytic phages, vB_SenM-PA13076 (PA13076) and vB_SenM-PC2184 (PC2184), were isolated from chicken sewage and characterized with host strains *Salmonella* Enteritidis (SE) ATCC13076 and CVCC2184, respectively. Transmission electron microscopy revealed that they belonged to the family *Myoviridae*. The lytic abilities of these two phages in liquid culture showed 10^4 multiplicity of infection (MOI) was the best in inhibiting bacteria, with PC2184 exhibiting more activity than PA13076. The two phages exhibited broad host range within the genus *Salmonella*. Phage PA13076 and PC2184 had a lytic effect on 222 (71.4%) and 298 (95.8%) of the 311 epidemic *Salmonella* isolates, respectively. We tested the effectiveness of phage PA13076 and PC2184 as well as a cocktail combination of both in three different foods (chicken breast, pasteurized whole milk and Chinese cabbage) contaminated with SE. Samples were spiked with 1×10^4 CFU individual SE or a mixture of strains (ATCC13076 and CVCC2184), then treated with 1×10^8 PFU individual phage or a two phage cocktail, and incubated at 4 °C or 25 °C for 5 h. In general, the inhibitory effect of phage and phage cocktail was better at 4 °C than that at 25 °C, whereas the opposite result was observed in Chinese cabbage, and phage cocktail was better than either single phage. A significant reduction in bacterial numbers (1.5–4 log CFU/sample,

$p < 0.05$) was observed in all tested foods. The two phages on the three food samples were relatively stable, especially at 4 °C, with the phages exhibiting the greatest stability in milk. Our research shows that our phages have potential effectiveness as a bio-control agent of *Salmonella* in foods.

Keywords: *Salmonella* Enteritidis (SE); phages; phage cocktail; food; bio-control

1. Introduction

Salmonella is a gram-negative bacterium that is one of the principal causes of food-borne diseases. Presently, over 2500 serotypes of *Salmonella* are known [1], and the most common worldwide is *Salmonella* Enteritidis (SE) [2]. SE is a food-borne pathogen that is a significant food safety concern globally. Since the period 1996–1999, the recorded incidence of human SE infection in the Foodborne Diseases Active Surveillance Network (FoodNet) has increased by 44% [3]. In Canada, SE was detected in various commodities, most frequently in chicken (with PT13, PT8 and PT13a predominating) [4]. In China, Ke's study of 1764 clinical *Salmonella enterica* isolates in Guangdong province showed that ~15% of isolates were SE, which was the primary cause of salmonellosis in adults [5]. Despite improved preventive and control strategies in chicken commercial flocks and in the food industry, SE infection still poses a constant problem [6–9]. Moreover, with the misuse of antimicrobials in many farms including disease treatment and growth promotion in domestic livestock, many SE are resistant to several antimicrobial agents.

New environmentally-friendly intervention strategies are needed to treat microbial infection. In recent years, applications of bacteriophages to control bacterial pathogens have received new interest [10]. They have also been identified as a prospective alternative bio-control method for infections and contaminations by antimicrobial resistant pathogens [11]. Bacteriophages, or phages, are abundant in the environment, with an estimated ratio of 10:1 to their bacterial hosts. In many experimental studies, phages have been used to control *Salmonella* contamination in a variety of foods. Goode *et al.* [12] used phages to well-control SE on chicken skin that had been inoculated with commercially relevant numbers of bacteria (*i.e.*, 1 log CFU/cm^2) . In Leverentz's study [13], *Salmonella*-specific phage could reduce *Salmonella* numbers in experimentally contaminated fresh-cut melons and apples stored at various temperatures. They also found that a phage mixture was better in reducing *Salmonella* populations than chemical sanitizers on honeydew melon slices [13]. The *Salmonella* phage (Felix-O1), which has a broad host range within the genus *Salmonella*, demonstrated an approximately about 2 log units reduction in *Salmonella* Typhimurium DT104 inoculated on chicken frankfurters [14]. In addition, phages had successfully controlled the growth of other important pathogens such as *Listeria monocytogenes* [15,16], *Escherichia coli* O157:H7 [17,18], and *Shigella* [19].

Our study involved the isolation from chicken excretion sewage of two new lytic SE phages using SE ATCC13076 and CVCC2184 as hosts. We determined their host ranges and lytic activity against hosts *in vitro*. The major aim of this study was to evaluate the potential of the two individual phages, or a mixture of the two (cocktail), to control SE contamination in three kinds of foods.

2. Materials and Methods

2.1. Salmonella Cultures, Media and Growth Conditions

A total of 311 epidemic *Salmonella* spp. strains were used in this study. Some of the strains were kindly donated by Guoxiang Cao (Chinese Academy of Agricultural Sciences, Yangzhou, China), Guoqiang Zhu (Yangzhou University, Yangzhou, China), Yuqing Liu (Shandong Academy of Agricultural Sciences, Jinan, China), Yanbing Zeng (Jiangxi Academy of Agricultural Sciences, Nanchang, China) and Jiansen Gong (Poultry institute, Chinese Academy of Agricultural Sciences, Yangzhou, China). The others were isolated between 2010 and 2015 from chicken farms and foods by our laboratory and stored in our lab. Of these, *S.* Enteritidis (SE) ATCC13076 and CVCC2184 were used to isolate bacteriophages. They were grown at 37 °C in Luria-Bertani broth (LB, Beijing land bridge technology Co., LTD, Beijing, China) or LB supplemented with 1.5% agar.

2.2. Isolation and Purification of Lytic Salmonella Phages

Fifty chicken excretion sewage samples were collected from 10 chicken farms in Jiangsu Province, China. Then these samples were used to isolate *Salmonella* phage as previously described [20]. Phages were propagated on SE ATCC13076 and CVCC2184 using LB (0.6% agar) soft agar overlays [21]. Incubating the double-layer agar plates for a longer period of time (3–4 days), the host cells that survive in the middle of the plaque were surveyed for whether they are resistant strains or lysogens.

To prepare high titers of phage, crude phage lysate propagating using a single plaque with its hosts was filtered a 0.22 μm filter, then NaCl (0.5 M, final concentration) and polyethylene glycol (PEG 8000) (10%, final concentration) (Amersco, Solon, OH, USA) were added to the supernatant and incubated overnight at 0 °C. Finally, phage particles were precipitated by centrifugation ($10,000\times g$, 15 min, 4 °C), and resuspended in SM buffer (5.8 g/L of NaCl, 2.0 g/L of $MgSO_4$, 50 mL/L of 1 M Tris, pH 7.5, 5 mL/L of presterilized 2% gelatin). The final concentration was 1.0×10^{10} PFU/mL.

2.3. Morphology of the Isolated Phages

Freshly-purified phage suspended pellets in SM buffer were used in this experiment. A small drop of phage was loaded onto a carbon-coated copper mesh grid and excess phage suspension was removed with filter paper. Negative staining of phages with 1% (*w/v*) phosphotungstic acid and phage images were obtained by a Transmission Electron Microscope (TEM) (H-7650, Hitachi High-Technologies Corporation, Tokyo, Japan) at an acceleration voltage of 80 kV.

2.4. Thermal and pH Stability

For thermal-stability testing, tubes with phages were kept in a water bath ranging from 30 to 90 °C for 30 min or 60 min. For pH-stability testing, samples of phages were mixed in a series of tubes containing buffer peptone water (BPW) of different pH (adjusted using NaOH or HCl) and incubated for 2 h at 37 °C. Bacteriophage titers were all determined using the double-layer agar plate method.

2.5. In Vitro Experiment of Phage Mediated Lysis

Overnight cultures of *Salmonella* ATCC13076 and CVCC2184 were diluted to $\sim 10^4$ CFU/mL in fresh medium. A single phage stock (PA13076 or PC2184) was added to give three multiplicity of infection (MOI) of 10^2, 10^3 and 10^4, respectively. The mixtures were then incubated at 37 °C for 5 h with gentle shaking. Phage-free culture (containing only bacteria) and bacteria-free culture (containing only phage) were also included as controls. Bacterial counts were determined at 0, 0.5, 1, 2, 3, 4 and 5 h.

2.6. Host Ranges of Phage

Three hundred eleven strains of epidemic *Salmonella* spp. were used in this study. Host ranges of phage were determined by spot test. The method was spotting 10 μL of phage preparation ($\sim 10^8$ PFU/mL) on lawn cultures of the bacteria strains. Plaque formation was monitored after incubation at 37 °C for 24 h.

2.7. Food Sample Preparation

Three different kinds of foods were purchased at local supermarkets and included: chicken breast, pasteurized whole milk and Chinese cabbage. All food samples were preliminarily analyzed to check for the possible natural contamination by *Salmonella*, according to the standard procedures. Samples of chicken breast were sliced aseptically into 2 cm × 2 cm squares (about 1 g), and placed into Petri dishes. Pasteurized whole milk samples were divided into 1 mL/sample in a biosafety cabinet. Chinese cabbage was washed with water, followed by washing with sterile water and 75% ethanol, prior to its use in the assays. The sanitized cabbage samples were then sliced into 2 cm × 2 cm squares, and placed in Petri dishes.

2.8. Individual Phage Treatment of Their Respective Hosts

SE ATCC13076 and CVCC2184 were grown individually in LB at 37 °C overnight. Phages PA13076 and PC2184 were used individually to challenge their appropriate hosts which were added to the three kinds of foods. Experiments were conducted at two temperatures (4 or 25 °C) to represent refrigeration and room temperatures. For each sample, 25 μL of diluted host strains (4×10^5 CFU/mL) were carefully pipetted onto the surface of the meat and cabbage or into the milk samples and allowed to attach for 15 min at room temperature in a biosafety cabinet. This was followed by adding 25 μL of diluted phages (4×10^9 PFU/mL) per sample. For the controls, the same volume of SM buffer was used instead of the phage suspension. All samples were performed in triplicate. Bacteria and its phage were monitored by viability counting on LB plates after 0, 1, 2, 3, 4 and 5 h of phage treatment. To quantify *Salmonella*, each chicken breast piece and Chinese cabbage was homogenized in 5 mL PBS. *Salmonella* were detected directly in milk samples. *Salmonella* were counted by pouring larger aliquots (1 mL) of diluted or undiluted of milk sample or the homogenates with molten LB agar on 90-mm plates. Simultaneously, the concentration of phage was determined at each of the monitored times. Phage counts were done by the agar-overlay technique. Aliquots of 100 μL of serial 10-fold dilution from the samples were mixed with 100 μL host cells and 4 mL molten LB soft agar (0.6%). The detection limits for bacteria or phage enumeration were 1 CFU/Sample or 10 PFU/Sample in milk samples and 5 CFU/Sample or 50 PFU/Sample in chicken breast and Chinese cabbage samples. Values less than the detection limit for chicken breast and Chinese cabbage samples were replaced with 5 CFU/Sample.

2.9. Phage Cocktail Control of SE Mixture

The efficacy of the phage cocktail (combination of PA13076 and PC2184) was studied on food samples (chicken breast, pasteurized whole milk and Chinese cabbage) that were experimentally contaminated with a mixture of equal numbers of *Salmonella* ATCC13076 and CVCC2184. The purified phage stocks of PA13076 and PC2184 were used to make a phage cocktail in SM buffer with a combined titer of 4×10^9 PFU/mL. Briefly, before adding the 25 μL of phage cocktail, the prepared food samples were pre-incubated with 12.5 μL 4×10^5 CFU/ mL of each SE strains and allowed to attach for 15 min. After phages were added, samples were further incubated at 4 °C or 25 °C for up to 5 h. The numbers of viable *Salmonella* concentration were calculated as described above. The detection limits for numbers of bacteria were also the same as above.

2.10. Determination of Phage Stability on Foods

Twenty-five microliters of diluted phages (1.6×10^{10} PFU/mL) were inoculated directly onto the surface of the meat and cabbage or into the milk sample. These samples were then incubated at 4 °C or 25 °C. The phage titers were determined at 0, 5, 24, 48 and 72 h, separately. The detection limits for phage titers were the same as in Section 2.8.

2.11. Statistical Analysis

Bacteria and phage concentrations of each sample were determined by duplicate plating. Results are shown as mean values of the logarithm of CFU/sample and the standard deviations of the mean are indicated by error bars. Each phage-treated sample was compared to its control counterpart using one-way ANOVA. Significant differences were discriminated using Duncan's test with significance set at $p < 0.05$. All data were analyzed using SPSS 16.0 (SPSS Inc., Chicago, IL, USA).

3. Results

3.1. Lytic Phages Isolation and Purification

Two new SE phages were isolated and designated as vB_SenM-PA13076 (PA13076) and vB_SenM-PC2184 (PC2184) by their ability to propagate on host strains SE ATCC13076 and CVCC2184, respectively. The bacteria present in the middle of the plaque could not form lysogen. These two phages are indeed strictly lytic. In spot tests, phages PA13076 formed clearing zones on lawns of ATCC13076 and phages PC2184 formed clearing zones on lawns of CVCC2184. Both PC2184 and PA13076 could cross-react and lyse to the other's hosts. The two phages formed round and clear zones on their own hosts, and the size of plaques were 0.5 to 1 mm.

3.2. Phages Morphologies

TEM images of phage PA13076 and PC2184 are shown in Figure 1. They all had the characteristics of a *Myoviridae* family (with a contractile tail). The head of PA13076 was oval and 66 nm in diameter. Its tail was 90 nm in length and 18 nm in diameter. PC2184 possessed an icosahedral head (diameter, 65 nm) and a contractile tail (length, 106 nm and diameter, 17 nm).

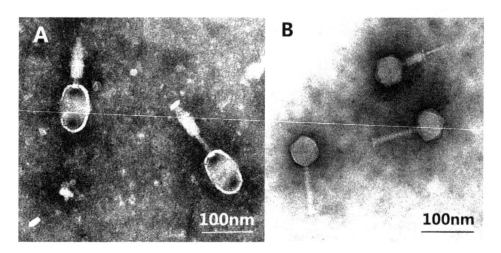

Figure 1.   Transmission electron micrographs (TEM) of phage PA13076 (**A**) and PC2184 (**B**).

3.3. Thermal and pH Stability

Phage PA13076 and PC2184 were stable between 30 °C to 50 °C for 30 min and 60 min. At 60 °C, there was 2-log reduction for PA13076, whereas PC2184 had only a slight reduction. PA13076 were not detectable at 70 °C and PC2184 were not detectable for 60 min at 80 °C (Figure 2A,B). The titers of phage were relatively stable at pH 6 to 11 for PA13076 and at pH 5 to 11 for PC2184. Their titers declined dramatically under lower or higher pH conditions (Figure 2C).

3.4. Lytic Activity of Phages PA13076 and PC2184 on Its Host in Vitro

Figure 3A shows the reduction of SE ATCC13076 growth compared to phage-free control (MOI = 0) when phage were added at MOIs of 10^2, 10^3 and 10^4 to host cells initially present at 1.46×10^4 CFU/mL. PA13076 achieved a peak reduction of 0.9, 1.6, and 1.8 log CFU/mL after 0.5 h, respectively. The number of viable SE ATCC13076 was reduced by about 3.5-log when treated with phage at MOI of 10^2 and 10^3 compared to the phage-free control after 5 h. Moreover, the number of viable SE ATCC13076 was reduced by 5.5-log when treated with PA13076 at MOI of 10^4. Figure 3B describes the same experiment using 1.23×10^4 CFU/mL of SE CVCC2184 and different MOIs of PC2184. When the highest MOI (10^4) of PC2184 was used, SE CVCC2184 numbers were reduced dramatically compared with the control; no viable bacteria were detected by 2 h. The number of viable *Salmonella* remaining at each time was dependent on the MOI used; higher MOI resulted in lower numbers of viable bacteria.

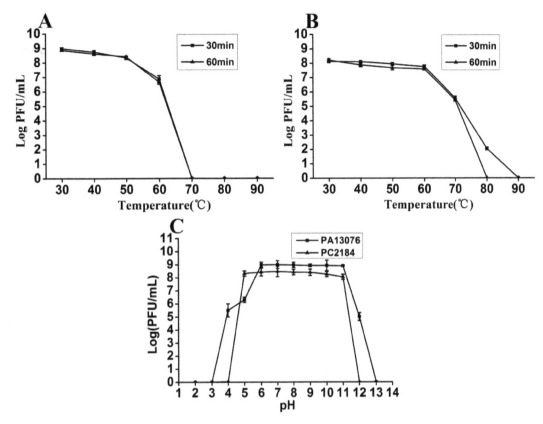

Figure 2. The thermal stability of phages PA13076 (**A**) and PC2184 (**B**); and the pH stability of phages PA13076 and PC2184 (**C**).

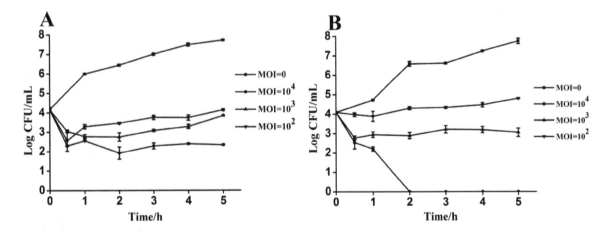

Figure 3. Lytic effects of phage PA13076 and PC2184 against the specified hosts of liquid cultures *in vitro*: (**A**) PA13076 and (**B**) PC2184.

3.5. Host Ranges of Phage PA13076 and Phage PC2184

The phage PA13076 and phage PC2184 both possessed wide host ranges. The results indicated that phage PA13076 had a lytic effect on 222 of the 311 epidemic *Salmonella* isolates (71.4%), whereas PC2184 produced a lytic effect on 298 isolates (95.8%) (Supplementary Table S1).

3.6. Efficacy of Individual Phage in the Bio-Control SE in Contaminated Foods

No host organisms were isolated from uninoculated control samples. The results shown in Figures 4–6 demonstrate the effects of single phage PA13076 and phage PC2184 on SE ATCC13076 and SE CVCC2184 contamination, respectively, at 4 and 25 °C on food samples (chicken breast, pasteurized whole milk and Chinese cabbage) and the concentration of phage on the surfaces of these food products at 4 and 25 °C are also shown. The titers of phage PA13076 and PC2184 remained stable and slightly higher than the initial inoculum over 5 h.

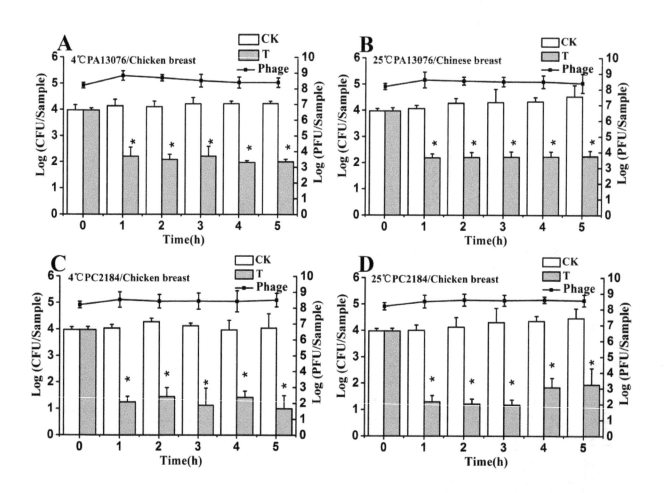

Figure 4. Effects of individual phages on growth of SE ATCC13076 and CVCC2184 on the surface of chicken breast at 4 °C and 25 °C. Each chicken breast sample was inoculated with either SE ATCC13076 (**A,B**) or SE CVCC2184 (**C,D**) (1×10^4 CFU), and phage PA13076 or PC2184 was applied (1×10^8 PFU) later (CK, inoculated control without phage; T, treated with phage). The titers of phage were also detected each sampling time (indicated in a dotted line). Date represent the mean $\pm$ S.D. ($n = 3$), * represents $p < 0.05$ (Duncan's test).

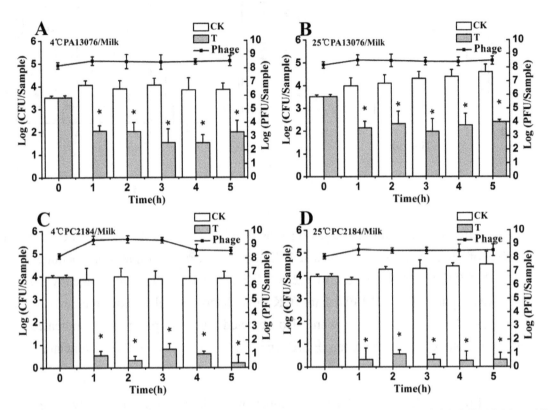

Figure 5. Effects of individual phages on growth of SE ATCC13076 and CVCC2184 in milk samples at 4 °C and 25 °C. Each milk sample was inoculated with either SE ATCC13076 (**A,B**) or SE CVCC2184 (**C,D**) (1 × 10⁴ CFU), and phage PA13076 or PC2184 was applied (1 × 10⁸ PFU) later (CK, inoculated control without phage; T, treated with phage). The titers of phage were also detected each sampling time (indicated in a dotted line). Date represent the mean ± S.D. ($n = 3$), * represents $p < 0.05$ (Duncan's test).

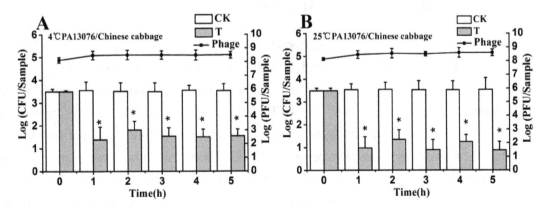

Figure 6. *Cont.*

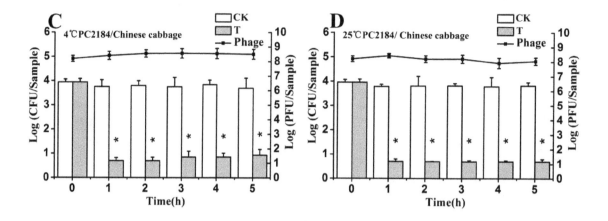

Figure 6. Effects of individual phages on growth of SE ATCC13076 and CVCC2184 on the surface of Chinese cabbage samples at 4 °C and 25 °C. Each sample was inoculated with either SE ATCC13076 (**A,B**) or SE CVCC2184 (**C,D**) (1×10^4 CFU), and phage PA13076 or PC2184 was applied (1×10^8 PFU) later (CK, inoculated control without phage; T, treated with phage). The titers of phage were also detected each sampling time (indicated in a dotted line). Date represent the mean $\pm$ S.D. ($n = 3$), * represents $p < 0.05$ (Duncan's test).

The efficacy of phage PC2184 to reduce the number of viable bacteria was clearly better than phage PA13076 at 4 °C and 25 °C. PA13076 and PC2184 treated groups (T) showed significantly lower ($p < 0.05$) SE counts compared with the positive control (CK) group at 4 °C and 25 °C in each of the three foods types. The most significant reductions in viable SE for the two phages were observed in pasteurized whole milk (Figure 5).

3.7. Efficacy of Phage Cocktail on Reducing SE Mixture

Viable counts increased slightly during incubation of SE mixture (SE ATCC13076 and SE CVCC2184) at 25 °C on chicken breast and in milk, but, at 4 °C they were stable. Bio-control by PA13076 and PC2184 cocktail resulted in a decreasing viable *Salmonella* counts of at least 1.65 log CFU/sample (Figure 7A,B), 3.89 log CFU/sample (Figure 7C,D) and 2.9 log CFU/sample (Figure 7E,F) on the three kinds of foods in the first 1 h, which was followed by a small amount of regrowth during the remaining incubation period at 25 °C. There was almost complete elimination of viable bacteria in pasteurized whole milk at 4 and 25 °C. The effects of the phage cocktail on Chinese cabbage were better than for chicken breast. After 5 h at 4 °C, the SE concentration was reduced by about 3.86 log CFU/sample on Chinese cabbage and 2.5 log CFU/sample on chicken breast relative to the initial concentration of bacteria at 0 h. The reduction in SE counts was somewhat less on Chinese cabbage (3.0 log CFU/sample) at 25 °C after 5 h.

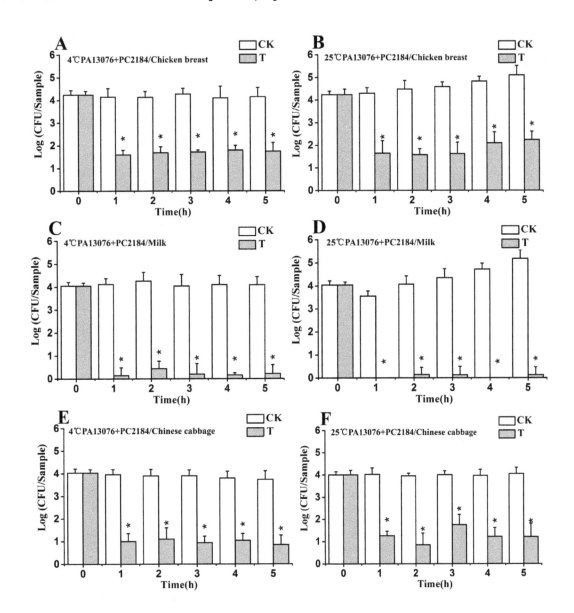

Figure 7. Efficacy of phage cocktail on reducing SE mixture treated food samples at 4 °C and 25 °C: (**A,B**) chicken breast; (**C,D**) pasteurized whole milk; and (**E,F**) Chinese cabbage. Each sample was inoculated with 1×10^4 CFU of SE mix (CK), or 1×10^4 CFU of SE mix and 1×10^8 PFU of phage cocktail (T). Date represent the mean $\pm$ S.D. ($n = 3$), * represents $p < 0.05$ (Duncan's test).

3.8. Stability of Phage on the Treated Foods

The concentration of phage PA13076 and phage PC2184 added to the food samples were monitored over 72 h. In milk samples, we observed no significant loss in titers of phage PA13076 at 4 °C and 25 °C (Figure 8C), however there was a significant loss in titers of phage PC2184 at 4 °C and 25 °C at 72 h (Figure 8D). For phage PA13076 and PC2184 added to chicken breast, titers remained stable up to 48 h at 4 °C and up to 24 h at 25 °C (Figure 8A,B). Phage PA13076 and PC2184 all decreased about 2 log

CFU/sample and 4 log CFU/sample on chicken breast at 48 and 72 h, respectively, at 25 °C. However, in Chinese cabbage (Figure 8E,F), these two phages are relatively stable up to 48 h at 4 °C and up to 5 h at 25 °C.

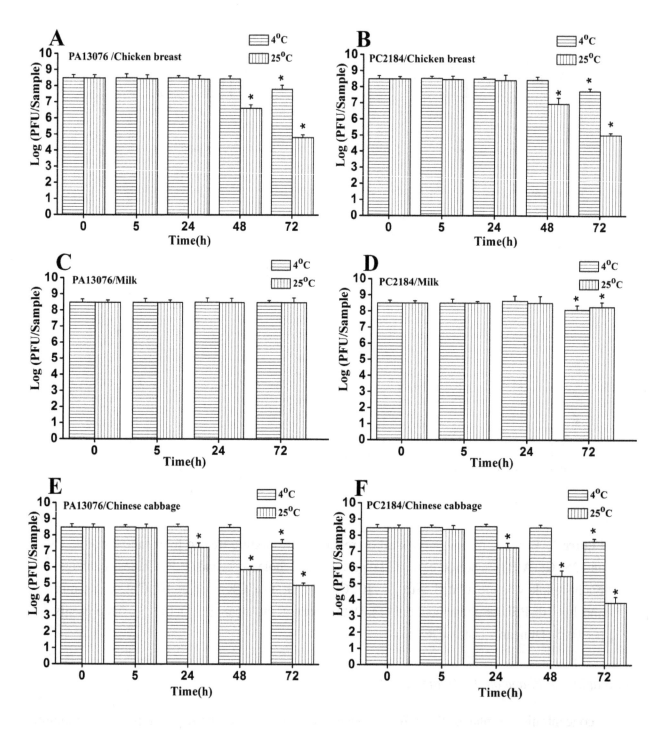

Figure 8. Stability of phage PA13076 and PC2184 over 72 h of incubation on the three kinds of food at 4 °C and 25 °C (the data for the milk sample at 48 h was lacking). Each sample was inoculated with 4×10^8 PFU of phage. Date represent the mean ± S.D. ($n = 3$), * represents $p < 0.05$ compared to the original phage numbers (Duncan's test).

4. Discussion

In this study, two lytic bacteriophages were successfully isolated from chicken sewages and characterized as *Salmonella* Enteritidis phages vB_SenM-PA13076 (PA13076) and vB_SenM-PC2184 (PC2184) by international common method according to their morphologies and their host's serovar. PC2184 was very similar to the previously *Salmonella* phage FGCSSa1 in ultrastructure and size [22]. Although, PA13076 belonged to *Myoviridae*, as does PC2184, but the shape of PA13076's head is very different from other reported *Salmonella* phage *Myoviridae*s [23,24] and *Siphoviruses* [25,26]. Resistance to heat and pH are important for bio-control applications. If these characteristics are too narrow, then phages may be ineffective in actual use. The two phages were both relatively stable between pH 6 and 9, which is compatible with the range (pH 5.5–7.0) of many foods. For thermal tolerance, the two phages were rapidly inactivated if the temperature was higher than 60 °C (for PA13076) or 70 °C (for PC2184), both of the phages were rapidly inactivated. The thermal stability was similar to the *Lactobacillus plantarum* phage PhiJL-1 [27].

As seen in Figure 3, there were differences in the killing ability between the two phages *in vitro*. Phage PC2184 showed impressive killing kinetics in liquid culture, whereas phage PA13076 was not as efficient. In general, the lysis data for the two phages demonstrated that the bactericidal activity was related to the MOI. In the present study, the MOI ratio of 10^4 showed the highest reduction rate of host cell counts (about 5.5-log for PA13076 and 7.0-log for PC2184). The results demonstrated that phage PC2184 killed its hosts almost completely after 2 h with a sufficiently high MOI (10^4). Lytic activities of these two phages *in vitro* showed that a much lower ratio of 10^2 or 10^3 also gives good killing. Theoretically, a low MOI ratio is advantageous for the commercial feasibility of large-scale application, as it would reduce the cost of preparation, purification and application of phage products. In this assay, we also found that there was significant survival of pathogens when SE cultures were inoculated with phage PA13076 or PC2184 at MOI of 10^2 and 10^3. This has previously been shown for the *E. coli* phage Mu^L [28] and the *Salmonella* phage FGCSSa1 [22], which did not completely lyse their hosts. Carey-Smith [22] suggested that only a subpopulation of the host was susceptible to phage infection.

Based on the above *in vitro* results, we designed a comprehensive study to determine whether these two phages would significantly reduce the population of *Salmonella* Enteritidis at high MOI (10^4) grown on foods. Typically, relatively low numbers of *Salmonella* cells are present in foods (including our experiments) [29]. Commercial foods like chicken breast, pasteurized whole milk or Chinese cabbage are usually refrigerated at 4 °C or remain at room temperature (25 °C). Therefore, these two temperatures were chosen to test the activities of individual phages and their combination (cocktail). In the bio-control study, we demonstrated that the individual phage and the cocktail could reduce SE counts on chicken breast surface at 4 °C or 25 °C with a short contact time. In our study, SEs eliminated more counts when the phage cocktail was applied compared to individual phage treatment. Thus, the phage cocktail therapy was more efficacious. These data is consistent with numerous published papers. For example, O'Flynn *et al.*, (2004) confirmed that a phage cocktail with MOI of 10^6 could completely eliminate *E. coli* O157:H7 in seven of nine cases on meat [30]. Hooton *et al.*, (2011) showed that a mixture of phages (PC1) was more effective at controlling *Salmonella* Typhimurium U288 on pig skin at a MOI of 10 or above [31]. Considering physical limitations of the solid matrix for proper dissemination of phage, applying more phage generally resulted in greater inactivation [32]. Our initial phage doses (10^8 PFU), which ensure complete contact with *Salmonella* hosts (10^4 CFU), caused significant reductions of *Salmonella* without the need for phage replicating. These kinds of therapy are called passive, which can reduce the likelihood of development of bacteria resistance [31]. Considering that much lower *Salmonella* concentrations were present, more phage from the beginning of the experiment is necessary to use. Because the growth of SE was greatly suppressed at 4 °C, the ability of phage or phage cocktail gave more favorable results than the experiment at 25 °C.

In experiments on pasteurized whole milk, the ability of the individual phage PC2184 and phage cocktail to control SE was almost indistinguishable. In Guenther's study [32], similar results were obtained. *Listeria* counts rapidly dropped below detectable levels in liquid foods treated with phages, such as chocolate milk and mozzarella cheese brine, and the levels of residual bacteria were much lower than in other foods [32]. As suggested by Guenther, the greater efficacy of phages in liquid foods may the phage particles being freely suspended. Thus, the individual phage and phage cocktail treatment are likely to be a powerful bio-control method to reduce and possibly eliminate SE in the milk industry.

In recent years, vegetables have been implicated as potential vehicles of bacterial pathogens, including *Salmonella* [33]. In Chinese cabbage bio-control experiment, our results suggested that phage PA13076, PC2184 or the cocktail were each more effective at reducing SE on the surface of Chinese cabbage at 25 °C than at 4 °C, which was opposite to the effects in chicken breast and pasteurized whole milk. This indicates that successful phage bio-control depends on the food matrix and temperature. Previous studies also found that the phage activity was sensitive to the physiological state of the host, which can be affected by growth conditions such as temperature, nutrient availability and oxygen tension [34,35]. These same studies showed that phage ECP-100 application significantly reduced the concentration of viable *E. coli* organisms on tomato slices by *ca.* 99% during storage (10 °C) for 24 h [36].

Phage PA13076 and PC2184 used in this study showed greater stable at 4 °C than at 25 °C. Furthermore, they were more stable in liquid samples (pasteurized whole milk) than in chicken breast and Chinese cabbage; they remained at above 50% of the initial phage concentration. More stable phage have been reported when used in spiced chicken over 72 h at 4 °C [19] and phage P100 in raw salmon fillet tissue over 10 days of storage at 4 °C [37] were reported.

The results obtained in this work have shown that bacteriophage or phage cocktail has the potential to be developed as an alternative strategy to combat SE infection in food. However, it is important to ensure that no horizontal gene transduction occurs while using the phage. Moreover, isolations of new phages having a stronger lytic activity are also needed.

5. Conclusions

Our study has shown that bacteriophages or a phage cocktail can reduce or completely eliminate SE inoculated with 10^4 CFU SE *in vitro* and in foods. These results demonstrated that bacteriophage treatment has the potential to be developed as an alternative strategy to prevent SE infection in food safety. However, more work needs to be done to determine whether phages can be used to disinfect food products.

Acknowledgments

We thank Matthew K. Ross (Mississippi State University) for manuscript review and professional English editing. We would like to thank G.X. Cao, G.Q. Zhu, Y.Q. Liu, Y.B. Zeng and J.S. Gong for the gift of epidemic *Salmonella* strains. This work was supported by a grant from the Jiangsu Province Natural Science Foundation (No. BK2012788), the Jiangsu Agricultural Science and Technology Foundation (No. CX(12)5040), and the Natural Sciences Foundation of China (NSFC No. 31402234).

Author Contributions

Hongduo Bao isolated and characterized the virulent phages used in this study. She also designed the study, analyzed the data and wrote the manuscript. Pengyu Zhang, Hui Zhang, Yan Zhou and Lili Zhang performed the bio-control experiments. Ran Wang contributed to the analysis of the results. All authors approved the final version.

References

1. Dunkley, K.D.; Callaway, T.R.; Chalova, V.I.; McReynolds, J.L.; Hume, M.E.; Dunkley, C.S.; Kubena, L.F.; Nisbet, D.J.; Ricke, S.C. Foodborne Salmonella ecology in the avian gastrointestinal tract. *Anaerobe* **2009**, *15*, 26–35. [CrossRef] [PubMed]

2. Galarce, N.E.; Bravo, J.L.; Robeson, J.P.; Borie, C.F. Bacteriophage cocktail reduces Salmonella enterica serovar Enteritidis counts in raw and smoked salmon tissues. *Rev. Argent. Microbiol.* **2014**, *46*, 333–337. [CrossRef]

3. Chai, S.J.; White, P.L.; Lathrop, S.L.; Solghan, S.M.; Medus, C.; McGlinchey, B.M.; Tobin-D'Angelo, M.; Marcus, R.; Mahon, B.E. Salmonella enterica serotype Enteritidis: Increasing incidence of domestically acquired infections. *Clin. Infect. Dis.* **2012**, *54*, S488–S497. [CrossRef] [PubMed]

4. Nesbitt, A.; Ravel, A.; Murray, R.; McCormick, R.; Savelli, C.; Finley, R.; Parmley, J.; Agunos, A.; Majowicz, S.E.; Gilmour, M.; *et al.* Integrated surveillance and potential sources of Salmonella enteritidis in human cases in Canada from 2003 to 2009. *Epidemiol. Infect.* **2012**, *140*, 1757–1772. [CrossRef] [PubMed]

5. Ke, B.; Sun, J.; He, D.; Li, X.; Liang, Z.; Ke, C.W. Serovar distribution, antimicrobial resistance profiles, and PFGE typing of Salmonella enterica strains isolated from 2007–2012 in Guangdong, China. *BMC Infect. Dis.* **2014**, *14*. [CrossRef] [PubMed]

6. Gorissen, B.; Reyns, T.; Devreese, M.; De Backer, P.; Van Loco, J.; Croubels, S. Determination of selected veterinary antimicrobials in poultry excreta by UHPLC-MS/MS, for application in Salmonella control programs. *Anal. Bioanal. Chem.* **2015**, *407*, 4447–4457. [CrossRef] [PubMed]

7. Hotes, S.; Traulsen, I.; Krieter, J. Salmonella control measures with special focus on vaccination and logistic slaughter procedures. *Transbound. Emerg. Dis.* **2011**, *58*, 434–444. [CrossRef] [PubMed]

8. Milbradt, E.L.; Zamae, J.R.; Araujo Junior, J.P.; Mazza, P.; Padovani, C.R.; Carvalho, V.R.; Sanfelice, C.; Rodrigues, D.M.; Okamoto, A.S.; Andreatti Filho, R.L. Control of *Salmonella* Enteritidis in turkeys using organic acids and competitive exclusion product. *J. Appl. Microbiol.* **2014**, *117*, 554–563. [CrossRef] [PubMed]

9. Penha Filho, R.A.; de Paiva, J.B.; da Silva, M.D.; de Almeida, A.M.; Berchieri, A., Jr. Control of *Salmonella* Enteritidis and *Salmonella* Gallinarum in birds by using live vaccine candidate containing attenuated *Salmonella* Gallinarum mutant strain. *Vaccine* **2010**, *28*, 2853–2859. [CrossRef] [PubMed]

10. Kutateladze, M.; Adamia, R. Phage therapy experience at the Eliava Institute. *Med. Mal. Infect.* **2008**, *38*, 426–430. [CrossRef] [PubMed]

11. Mann, N.H. The potential of phages to prevent MRSA infections. *Res. Microbiol.* **2008**, *159*, 400–405. [CrossRef] [PubMed]

12. Goode, D.; Allen, V.M.; Barrow, P.A. Reduction of experimental *Salmonella* and Campylobacter contamination of chicken skin by application of lytic bacteriophages. *Appl. Environ. Microbiol.* **2003**, *69*, 5032–5036. [CrossRef] [PubMed]

13. Leverentz, B.; Conway, W.S.; Alavidze, Z.; Janisiewicz, W.J.; Fuchs, Y.; Camp, M.J.; Chighladze, E.; Sulakvelidze, A. Examination of bacteriophage as a biocontrol method for salmonella on fresh-cut fruit: A model study. *J. Food Prot.* **2001**, *64*, 1116–1121. [PubMed]

14. Whichard, J.M.; Sriranganathan, N.; Pierson, F.W. Suppression of Salmonella growth by wild-type and large-plaque variants of bacteriophage Felix O1 in liquid culture and on chicken frankfurters. *J. Food Prot.* **2003**, *66*, 220–225. [PubMed]

15. Carlton, R.M.; Noordman, W.H.; Biswas, B.; de Meester, E.D.; Loessner, M.J. Bacteriophage P100 for control of Listeria monocytogenes in foods: Genome sequence, bioinformatic analyses, oral toxicity study, and application. *Regul. Toxicol. Pharmacol.* **2005**, *43*, 301–312. [CrossRef] [PubMed]

16. Bigot, B.; Lee, W.J.; McIntyre, L.; Wilson, T.; Hudson, J.A.; Billington, C.; Heinemann, J.A. Control of Listeria monocytogenes growth in a ready-to-eat poultry product using a bacteriophage. *Food Microbiol.* **2011**, *28*, 1448–1452. [CrossRef] [PubMed]

17. Magnone, J.P.; Marek, P.J.; Sulakvelidze, A.; Senecal, A.G. Additive approach for inactivation of *Escherichia coli* O157:H7, Salmonella, and Shigella spp. on contaminated fresh fruits and vegetables using bacteriophage cocktail and produce wash. *J. Food Prot.* **2013**, *76*, 1336–1341. [CrossRef] [PubMed]

18. Patel, J.; Sharma, M.; Millner, P.; Calaway, T.; Singh, M. Inactivation of *Escherichia coli* O157:H7 attached to spinach harvester blade using bacteriophage. *Foodborne Pathog. Dis.* **2011**, *8*, 541–546. [CrossRef] [PubMed]

19. Zhang, H.; Wang, R.; Bao, H. Phage inactivation of foodborne Shigella on ready-to-eat spiced chicken. *Poult. Sci.* **2013**, *92*, 211–217. [CrossRef] [PubMed]

20. Bao, H.; Zhang, H.; Wang, R. Isolation and characterization of bacteriophages of Salmonella enterica serovar Pullorum. *Poult. Sci.* **2011**, *90*, 2370–2377. [CrossRef] [PubMed]

21. Swanstrom, M.; Adams, M.H. Agar layer method for production of high titer phage stocks. Proceedings of the Society for Experimental Biology and Medicine. *Soc. Exp. Biol. Med.* **1951**, *78*, 372–375. [CrossRef]

22. Carey-Smith, G.V.; Billington, C.; Cornelius, A.J.; Hudson, J.A.; Heinemann, J.A. Isolation and characterization of bacteriophages infecting Salmonella spp. *FEMS Microbiol. Lett.* **2006**, *258*, 182–186. [CrossRef] [PubMed]

23. Santos, S.B.; Kropinski, A.M.; Ceyssens, P.J.; Ackermann, H.W.; Villegas, A.; Lavigne, R.; Krylov, V.N.; Carvalho, C.M.; Ferreira, E.C.; Azeredo, J. Genomic and proteomic characterization of the broad-host-range Salmonella phage PVP-SE1: Creation of a new phage genus. *J. Virol.* **2011**, *85*, 11265–11273. [CrossRef] [PubMed]

24. Whichard, J.M.; Weigt, L.A.; Borris, D.J.; Li, L.L.; Zhang, Q.; Kapur, V.; Pierson, F.W.; Lingohr, E.J.; She, Y.M.; Kropinski, A.M.; *et al.* Complete genomic sequence of bacteriophage felix O1. *Viruses* **2010**, *2*, 710–730. [CrossRef] [PubMed]

25. O'Flynn, G.; Coffey, A.; Fitzgerald, G.F.; Ross, R.P. The newly isolated lytic bacteriophages st104a and st104b are highly virulent against Salmonella enterica. *J. Appl. Microbiol.* **2006**, *101*, 251–259. [CrossRef] [PubMed]

26. Kang, H.W.; Kim, J.W.; Jung, T.S.; Woo, G.J. wksl3, a New biocontrol agent for Salmonella enterica serovars enteritidis and typhimurium in foods: Characterization, application, sequence analysis, and oral acute toxicity study. *Appl. Environ. Microbiol.* **2013**, *79*, 1956–1968. [CrossRef] [PubMed]

27. Lu, Z.; Breidt, F., Jr.; Fleming, H.P.; Altermann, E.; Klaenhammer, T.R. Isolation and characterization of a Lactobacillus plantarum bacteriophage, phiJL-1, from a cucumber fermentation. *Int. J. Food Microbiol.* **2003**, *84*, 225–235. [CrossRef]

28. Fischer, C.R.; Yoichi, M.; Unno, H.; Tanji, Y. The coexistence of *Escherichia coli* serotype O157:H7 and its specific bacteriophage in continuous culture. *FEMS Microbiol. Lett.* **2004**, *241*, 171–177. [CrossRef] [PubMed]

29. Guenther, S.; Herzig, O.; Fieseler, L.; Klumpp, J.; Loessner, M.J. Biocontrol of Salmonella Typhimurium in RTE foods with the virulent bacteriophage FO1-E2. *Int. J. Food Microbiol.* **2012**, *154*, 66–72. [CrossRef] [PubMed]

30. O'Flynn, G.; Ross, R.P.; Fitzgerald, G.F.; Coffey, A. Evaluation of a cocktail of three bacteriophages for biocontrol of *Escherichia coli* O157:H7. *Appl. Environ. Microbiol.* **2004**, *70*, 3417–3424. [CrossRef] [PubMed]

31. Hooton, S.P.; Atterbury, R.J.; Connerton, I.F. Application of a bacteriophage cocktail to reduce Salmonella Typhimurium U288 contamination on pig skin. *Int. J. Food Microbiol.* **2011**, *151*, 157–163. [CrossRef] [PubMed]

32. Guenther, S.; Huwyler, D.; Richard, S.; Loessner, M.J. Virulent bacteriophage for efficient biocontrol of Listeria monocytogenes in ready-to-eat foods. *Appl. Environ. Microbiol.* **2009**, *75*, 93–100. [CrossRef] [PubMed]

33. Spricigo, D.A.; Bardina, C.; Cortes, P.; Llagostera, M. Use of a bacteriophage cocktail to control Salmonella in food and the food industry. *Int. J. Food Microbiol.* **2013**, *165*, 169–174. [CrossRef] [PubMed]

34. Cohen, S.S. Growth Requirements of Bacterial Viruses. *Bacteriol. Rev.* **1949**, *13*, 1–24. [PubMed]

35. Hadas, H.; Einav, M.; Fishov, I.; Zaritsky, A. Bacteriophage T4 development depends on the physiology of its host *Escherichia coli*. *Microbiology* **1997**, *143*, 179–185. [CrossRef] [PubMed]

36. Abuladze, T.; Li, M.; Menetrez, M.Y.; Dean, T.; Senecal, A.; Sulakvelidze, A. Bacteriophages reduce experimental contamination of hard surfaces, tomato, spinach, broccoli, and ground beef by *Escherichia coli* O157:H7. *Appl. Environ. Microbiol.* **2008**, *74*, 6230–6238. [CrossRef] [PubMed]

37. Soni, K.A.; Nannapaneni, R. Bacteriophage significantly reduces Listeria monocytogenes on raw salmon fillet tissue. *J. Food Protect.* **2010**, *73*, 32–38.

2

Coordinated DNA Replication by the Bacteriophage T4 Replisome

Erin Noble, Michelle M. Spiering and Stephen J. Benkovic *

Pennsylvania State University, Department of Chemistry, 414 Wartik Laboratory, University Park, PA 16802, USA; E-Mails: eun2@psu.edu (E.N.); mms36@psu.edu (M.M.S.)

* Author to whom correspondence should be addressed; E-Mail: sjb1@psu.edu

Academic Editor: David Boehr

Abstract: The T4 bacteriophage encodes eight proteins, which are sufficient to carry out coordinated leading and lagging strand DNA synthesis. These purified proteins have been used to reconstitute DNA synthesis *in vitro* and are a well-characterized model system. Recent work on the T4 replisome has yielded more detailed insight into the dynamics and coordination of proteins at the replication fork. Since the leading and lagging strands are synthesized in opposite directions, coordination of DNA synthesis as well as priming and unwinding is accomplished by several protein complexes. These protein complexes serve to link catalytic activities and physically tether proteins to the replication fork. Essential to both leading and lagging strand synthesis is the formation of a holoenzyme complex composed of the polymerase and a processivity clamp. The two holoenzymes form a dimer allowing the lagging strand polymerase to be retained within the replisome after completion of each Okazaki fragment. The helicase and primase also form a complex known as the primosome, which unwinds the duplex DNA while also synthesizing primers on the lagging strand. Future studies will likely focus on defining the orientations and architecture of protein complexes at the replication fork.

Keywords: DNA replication; T4 bacteriophage; replisome

1. Introduction

Bacteriophages were first discovered in the early 20th century due to their ability to kill bacteria [1]. Apart from their therapeutic uses, bacteriophages were found to encode proteins that carried out many of the same basic processes that are found in eukaryotic cells. The T4 bacteriophage, which infects *Escherichia coli*, is one of the best-studied viruses in this group. Its double-stranded DNA genome encodes all of the proteins necessary to carry out viral DNA replication in the infected cell. The components of the T4 replisome can be purified and used to reconstitute DNA replication *in vitro*. This system has been well characterized as a model for DNA replication at a fork [2–4]. The T4 replisome consists of eight proteins, which together catalyze coordinated leading and lagging strand synthesis (Figure 1). These proteins are similar in structure and function to their eukaryotic homologues [5]. Studies on the T4 system have contributed greatly to the understanding of DNA replication and paved the way for current studies on human and yeast DNA replication. This review will cover the current understanding of T4 DNA replication and highlight areas where recent research has yielded new mechanistic insight into functioning of the T4 replisome. For more detail on other prokaryotic model systems, see recent reviews highlighting studies of the T7 bacteriophage and *E. coli* replisomes [6–8].

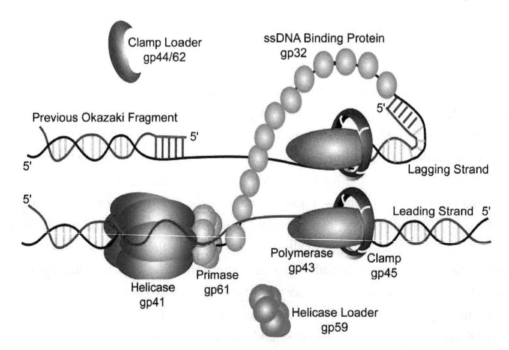

Figure 1. A model of the T4 bacteriophage DNA replisome. Replication of T4 genomic DNA is accomplished by a replication complex composed of eight proteins. The helicase (gp41) and primase (gp61) interact to form the primosome with the assistance of the helicase loader (gp59). The primosome complex encircles the lagging strand DNA, unwinding duplex DNA while synthesizing RNA primers for use by the lagging strand polymerase (gp43). DNA synthesis on both strands is catalyzed by a holoenzyme complex formed by a polymerase (gp43) and a trimeric processivity clamp (gp45). The clamp is loaded onto the DNA by the clamp loader complex (gp44/62). The leading and lagging strand holoenzymes interact to form a dimer. Single-stranded DNA formed by the helicase is coated with single-stranded DNA-binding protein (gp32).

2. T4 Replication Fork Components

T4 replication can be initiated via several different pathways [9]. Two specialized structures, R-loops and D-loops, have been shown to be important. R-loops form at T4 origin sites where an RNA primer is synthesized. D-loops are formed by the recombination machinery and are used to initiate origin-independent DNA synthesis. These two mechanisms of DNA replication initiation of have been reviewed elsewhere [10].

Synthesis of the T4 genomic DNA is accomplished by a holoenzyme complex composed of the gp43 polymerase and the gp45 sliding clamp [11–13]. On the leading strand, DNA synthesis is carried out continuously by one holoenzyme complex. On the lagging strand, DNA is synthesized in the opposite direction of the progression of the replication fork. Multiple priming events allow a second holoenzyme complex to carry out DNA synthesis discontinuously in 1 to 2 kb fragments known as Okazaki fragments. While there is no available crystal structure for the T4 gp43, the structure for the RB69 bacteriophage gp43 has been solved alone and as part of a binary and ternary complex [14–16]. The two proteins are 62% identical and 74% similar and thus, the proteins are likely very similar in topology. The RB69 structure reveals five conserved domains in a configuration similar to that of the eukaryotic B family polymerases. The *N*-terminus contains a 3′ to 5′ exonuclease active site. This truncated exonuclease domain from T4 gp43 has been isolated and the structure solved [12]. The catalytic activity of this domain is independent from the rest of the polymerase, as it retains full exonuclease activity *in vitro* [17]. The *C*-terminus of RB69 gp43 is organized into conserved finger, palm, and thumb domains, which catalyze DNA polymerization 5′ to 3′ [15].

The T4 sliding clamp, gp45, is a ring-shaped, trimeric protein that serves as a processivity factor for the polymerase [18,19]. The inner diameter of the ring is about 35 Å, which is large enough to accommodate duplex DNA. Unlike clamps in other systems, the T4 clamp exists in solution as a partially open ring with one of the three subunit interfaces disrupted [20–22]. Once loaded onto DNA, the interior of the clamp interacts with the DNA phosphate backbone through a number of basic residues and anchors the polymerase to the DNA [19]. gp43 has a C-terminal PIP box domain that mediates the interaction of the polymerase and the sliding clamp [23].

The circular gp45 clamp is loaded onto the DNA by a clamp loader complex. In T4, four gp44 subunits associate with one gp62 subunit forming the gp44/62 clamp loader [24]. Each gp44 subunit binds ATP and the complex has a strong DNA-dependent ATPase activity [25,26]. The clamp loader is a member of the AAA+ family of ATPases, but unlike other enzymes of this type, clamp loaders are pentameric rather than hexameric. This asymmetry results in a gap that allows the clamp loader to specifically recognize the primer-template junction when loading a clamp [27,28].

The T4 helicase, gp41, forms a hexamer upon binding GTP or ATP [29]. This active form of the helicase hydrolyzes GTP/ATP to move along single-stranded DNA [30,31]. Electron microscopy has revealed that there are two forms of the hexameric gp41, a symmetric ring and a gapped asymmetric ring [32]. The "open" ring is thought to be important for the loading of the helicase onto DNA [29]. As part of the replication fork, gp41 unwinds the double stranded DNA by traveling 5′ to 3′, encircling the lagging strand while excluding the leading strand [33]. The preferred substrate for the helicase is a forked DNA with both 5′ and 3′ single-stranded DNA regions, suggesting the protein interacts with both the leading and lagging strands [33,34]. T4 also encodes two other helicases, UvsW and Dda. Both accessory helicases have been suggested to have roles in replication initiation, recombination, and repair (see review [35]).

Priming on the lagging strand is catalyzed by the gp61 primase, which interacts with gp41 to form the primosome [36]. This primosome synthesizes pentaribonucleotides from 5′-GTT-3′ priming sites. The 3′-T is necessary for priming but is not used to template the primer; the resulting primers have the sequence 5′-pppACNNN-3′ [37]. At high concentrations *in vitro*, gp61 alone can synthesize some RNA primers, but they are typically dimers primed from a 5′-GCT-3′ site [37,38]. In the presence of gp41, the rate of primer synthesis increases and shifts to pentaribonucleotide products primed from 5′-GTT-3′ sites, which is the priming site used *in vivo* [38,39]. gp61 alone is monomeric, but in the presence of gp41 and/or DNA, it oligomerizes into a hexameric ring [32,40].

Exposed single-stranded DNA is bound by gp32, which is necessary for DNA replication *in vivo*. It has many functions including preventing the formation of DNA secondary structure, protecting DNA from nuclease digestion, and stimulation of the gp43 synthesis rate and processivity [41–43]. A crystal structure of gp32 in complex with DNA reveals three domains. The N-terminus binds other gp32 monomers allowing for oligomerization, the *C*-terminus mediates interactions with other proteins such as the T4 polymerase, and the core domain binds single-stranded DNA [44].

In vivo a helicase loader, gp59, is required for origin-dependent initiation of replication [45]. In the presence of gp32, the helicase cannot efficiently load onto the DNA fork without the addition of gp59. gp59 interacts with gp41 stoichiometrically and helps to displace gp32, allowing the helicase to load [46]. gp59 is thought to mediate loading by inducing a conformational change in gp41 that promotes DNA binding [47]. It is unclear if gp59 dissociates or remains as part of the replication complex [48,49]. Binding events between gp43 and gp59 have been observed using single-molecule FRET [50].

3. Holoenzyme Formation

The gp43 polymerase alone can only copy short stretches of single-stranded DNA without dissociating [51]. The gp45 sliding clamp is a homotrimeric ring that allows gp43 to catalyze processive DNA synthesis. It is loaded onto DNA by gp44/62 with the clamp loader specifically recognizing the free 3′ end of the primer-template junction. As the clamp is partially open in solution, the function of the T4 clamp loader is to stabilize the open clamp and direct it onto DNA in the correct orientation. Crystal structures of the clamp/clamp loader complex, both with and without DNA, have provided detailed insight into how loading occurs [52]. The clamp loader has a low affinity for the clamp until the binding of ATP through an AAA+ module in each of the gp44 subunits. ATP binding causes the clamp loader subunits to adopt a spiral conformation that can bind to the clamp and open it further, allowing it to be loaded onto DNA. The opening of the clamp occurs in two planes. Movement of $\sim$9 Å in the plane of the ring allows single-stranded DNA to pass through the gap, while an out-of-plane shift of $\sim$23 Å results in a twisted conformation of the clamp, aligning it with the helical structure of the DNA. DNA binding stimulates the ATPase activity of the clamp loader and the hydrolysis of ATP in each of the four gp44 subunits [24,53]. This hydrolysis triggers a change in the conformation of the clamp loader, which closes the clamp around the DNA.

Once the clamp is closed around the DNA, it must be bound by the polymerase to form the holoenzyme. This process has been characterized using a FRET-based assay to monitor clamp loading and holoenzyme assembly. The clamp and clamp loader complex rapidly bind to the DNA after ATP binding. In the absence of the gp43 polymerase, the clamp and clamp loader remain as a complex and dissociate from the DNA together. In the presence of the polymerase, a functional holoenzyme forms in three kinetically distinct steps. The first corresponds to the hydrolysis of ATP and the dissociation of the clamp loader. The subsequent two steps involve slower conformational changes leading to the formation of a stable complex. The dissociation of the clamp in the presence of the polymerase is significantly slower than the clamp alone [54]. This stable holoenzyme complex is then able to efficiently carry out processive DNA synthesis on the leading strand and discontinuous DNA synthesis on the lagging strand.

4. Holoenzyme Processivity

The holoenzyme on the leading strand synthesizes DNA in the same direction as the movement of the replication fork. *In vivo*, the T4 genome can be synthesized within 15 min [55]. The half-life of the holoenzyme complex has been measured as 11 min as part of a moving fork and about 6 min on a small, defined DNA fork structure [56,57]. Given the half-life of the holoenzyme and the speed of synthesis, it is possible that the entire T4 genome could be synthesized by a single holoenzyme on the leading strand. While this highly processive holoenzyme would be advantageous on the leading strand, the lagging strand is synthesized discontinuously and the holoenzyme must repeatedly dissociate and rebind for synthesis of each Okazaki fragment.

A more recent study probing the processivity of the T4 holoenzyme confirmed the long half-life during replication using a standard dilution experiment [58]. However, it was found that an inactive mutant of the polymerase (D408N) was able to rapidly displace the wild-type polymerase and inhibit DNA synthesis. This inhibition occurred on both the leading and lagging strands. These results suggest that although the polymerase will not readily dissociate on its own, it can be actively displaced by a second polymerase without affecting DNA synthesis. The exchange process was termed dynamic processivity and is thought to be mediated through interactions with gp45 [58]. The *C*-terminus of gp43 is essential for polymerase binding to the clamp, but its deletion does not affect DNA polymerization [23]. When polymerase containing this deletion was used as a trap, it could no longer displace the replicating polymerase [58]. As the clamp is trimeric, it is hypothesized that multiple polymerases could bind and facilitate the exchange. This "toolbelt" model for the clamp has been suggested in other systems as well, with numerous proteins involved in DNA replication and repair also containing clamp binding domains [59,60]. In the T7 system, where there is no sliding clamp, the exchange process has been shown to be mediated by an interaction between the polymerase and the helicase [61]. It is thought that the helicase can bind multiple polymerases facilitating exchange on the leading strand and recycling on the lagging strand.

5. Coupling of Helicase and Polymerase for Leading Strand Synthesis

While both gp41 helicase and gp43/gp45 holoenzyme can function independently *in vitro* to unwind duplex DNA, the two enzymes work best when their activities are combined. The helicase alone is significantly slower and less processive than the replication fork, and the holoenzyme is very inefficient at strand displacement synthesis [33,62]. Together, the helicase and holoenzyme are able to efficiently carry out leading strand synthesis [63]. In the presence of a macromolecular crowding reagent, only gp43 and gp41 are needed, indicating the clamp does not play a role [64]. While the functional coupling between the two proteins has been clearly demonstrated, there is no evidence of a physical interaction between gp43 and gp41 [65,66]. One study also found that the T4 polymerase could be replaced with another processive polymerase and still carry out strand displacement synthesis, but could not be replaced with a low processivity polymerase [65]. This suggests that each enzyme is stabilized on the DNA replication fork by the activity of the other, with the helicase providing single-stranded DNA that the polymerase then traps.

In the T7 system, it was reported that nucleotide incorporation by the polymerase provided the driving force to stimulate helicase activity, but a detailed mechanism for helicase-polymerase coupling was not described [67]. A more recent single-molecule study of the coupling in the T4 system used magnetic tweezers to monitor both coupled and uncoupled activity [68]. A DNA hairpin was tethered to a glass slide with a magnetic bead on the other end. Force was applied to destabilize the duplex and assist enzymes in opening the hairpin. At low force, where the duplex of the hairpin is stable, the helicase moved at 6 times slower than its maximal translocation rate and showed sequence dependent pausing. As higher force was applied, the rate of helicase activity increased dramatically. Additionally, at low helicase concentrations, significant helicase slippage was observed involving the reannealing of tens to hundreds of base pairs. This fits with the passive model of helicase activity previously demonstrated, in which the helicase is not efficient in destabilizing duplex DNA and relies on transient fraying of base pairs to move forward [69].

The T4 holoenzyme was found to have very low strand displacement activity at low force and mainly exhibited exonuclease activity [68]. When higher forces destabilized the duplex, the holoenzyme was able to replicate the hairpin at maximal speeds. At moderate forces, the holoenzyme exhibited pausing and stalling. The proportion of holoenzymes observed synthesizing DNA, pausing, or degrading DNA was highly dependent on the force used. This indicates that at higher forces the holoenzyme is able to stay in the polymerization mode, while lower forces shift the holoenzyme to the exonuclease mode. When pausing and exonuclease events were excluded from analysis, the holoenzyme activity fits with a model of a strongly active motor. The basis for collaborative coupling then emerges in a model where the helicase provides the single-stranded DNA for the holoenzyme, but also prevents the fork regression pressure from switching the polymerase into the exonuclease mode. As the holoenzyme is kept in its highly processive polymerization mode, it stimulates the activity of the helicase and prevents slippage backwards [68].

6. Coordination of Helicase and Priming on the Lagging Strand

The leading and lagging strands are thought to be synthesized at the same net rate, despite the need for repeated priming and extension events on the lagging strand [4,70,71]. Priming is catalyzed by a gp61-gp41 complex known as the primosome. Both priming and DNA unwinding activity are stimulated when both proteins are present [34,38,39]. There is strong biochemical evidence for the interaction of the hexameric gp41 helicase and oligomeric gp61 primase [34,36,72,73]. Importantly, a gp61-gp41 fusion protein has been shown to have close to wild-type priming and helicase activity and can successfully catalyze coordinated leading and lagging strand synthesis [74].

This tight coordination of activity is clear, despite the fact that the helicase travels 5′ to 3′ unwinding duplex DNA while the primase synthesizes primers 3′ to 5′ on the same strand. There are three models for how this coupling can occur. The first model suggests that the helicase, and possibly the whole replisome, pauses while the primers are being synthesized. In the second model, primase subunits dissociate from the helicase and are left behind to synthesize primers. In the third model, coupling is accomplished by the formation of priming loops wherein the lagging strand folds back allowing for priming. The loop is then released after the primer is synthesized.

By observing helicase and priming activity on DNA hairpins using magnetic tweezers, the role of the three models in the T4 primosome could be directly observed [74]. In the T7 system, both pausing of the primosome [75] and priming loops have been reported [76]. The T4 study yielded no evidence of pausing of the T4 primosome. However, clear evidence of both primase disassembly and looping were seen in these experiments, indicating that there are two different mechanisms used by T4 to couple the helicase and primase (Figure 2). While primase disassembly was the predominant mode, in the case where the primase and helicase were fused only the looping mechanism was seen.

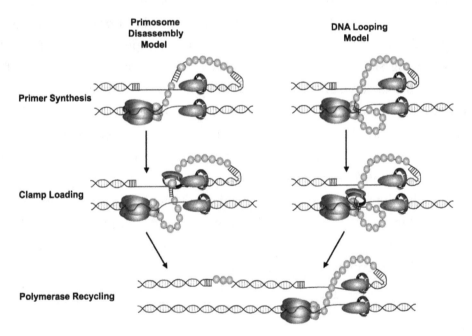

Figure 2. The two models of primosome activity used by T4 to initiate lagging strand synthesis. The helicase (gp41) and primase (gp61) interact as stacked rings encircling the lagging strand. This complex unwinds duplex DNA while synthesizing pentaribonucleotide RNA primers for use by the lagging strand polymerase (gp43). Primer synthesis occurs while the helicase continues to unwind DNA in the opposite direction. Two models have been proposed to accommodate these coupled activities. In the primosome disassembly model (shown left), one of the primase subunits dissociates from the primosome complex and remains with the newly synthesized primer. In the DNA looping model (shown right), the excess DNA unwound by the helicase during primer synthesis loops out allowing the primase to stay intact. In both models, the clamp loader (gp44/62) loads a clamp (gp45) onto the newly synthesized primer. The lagging strand polymerase is then signaled to release and recycle to the new primer.

7. Recycling of the Lagging Strand Polymerase

The trombone model was proposed to explain the coordination of leading and lagging strand synthesis with the two polymerases synthesizing in opposite directions. In this model, the lagging strand DNA loops out during the formation of each Okazaki fragment [4]. These loops have been visualized in electron micrographs of T4 replication products [48]. The lagging strand polymerase is retained as part of the replisome after completing synthesis of each Okazaki fragment [4]. It dissociates from the DNA, but then rapidly binds the next primer to continue synthesis. This recycling of the lagging strand polymerase is supported by numerous studies. While the clamp, clamp loader, primase, and gp32 have all been shown to exchange with proteins in solution during replication, the polymerase is resistant to dilution [77–80]. The size of the Okazaki fragments is also independent of polymerase concentration [4,58]. Importantly, the leading and lagging strand polymerases interact in the presence of DNA, which provides a mechanism for tethering the lagging strand polymerase to the replisome [66].

While the holoenzyme on the leading strand is highly processive, on the lagging strand it must repeatedly dissociate. The trigger for the dissociation of the lagging strand polymerase has not clearly been defined despite a number of studies. Several models have been proposed with two gaining the most support and evidence suggests that both play a role during replication [81]. The collision model proposes that the lagging strand polymerase dissociates after colliding with the end of the previous Okazaki fragment, and this stimulates the primase to synthesize a new primer [62,82]. However, it has been also shown that dissociation of the lagging strand polymerase can occur before reaching the previous Okazaki fragment leaving single-stranded DNA gaps [81]. To account for this observation, the signaling model has been proposed where recycling is triggered by the synthesis of a new primer and the timing controlled by gp61 [80,81,83]. Recently, additional signals have been proposed to regulate this recycling in other replication systems such E. coli and T7. These new triggers include tension induced dissociation of the polymerase [84], primer availability [85], and a third polymerase [86]. While it has been shown that a third T4 polymerase does not seem to play a role in Okazaki fragment synthesis [87], the nature of the signal for recycling is still unknown.

8. Future Directions

The major unanswered questions concerning T4 DNA replication involve understanding the dynamics and organization of the proteins at the replication fork. While the protein complexes involved in replication, the primosome, holoenzyme, and single-stranded DNA binding protein, have been extensively studied, their orientations and spatial juxtapositions at the fork are unclear. According to the trombone model, the two polymerases are thought to interact in opposite orientations, but how these proteins assemble at the fork has not been demonstrated. It is also not known how the polymerases at the replication fork are able to readily exchange with polymerases in solution. It is possible that the polymerase and clamp transiently separate yielding the dynamic processivity that has been observed. Another area of uncertainty is the trigger for recycling of the lagging strand polymerase. A number of possible signals have been suggested but none of these models have been proven. Fluorescence resonance energy transfer (FRET) and single-molecule experiments will likely play an important role in resolving these uncertainties and more clearly defining the organization and coordination of the T4 replisome.

Acknowledgments

This work is supported by National Institutes of Health (NIH) grant GM013306 (Stephen J. Benkovic). Erin Noble is supported by the NIH under Award Number F32GM110857.

Author Contributions

All authors contributed to the conception and editing of this review. Erin Noble wrote the manuscript.

References

1. Sulakvelidze, A.; Alavidze, Z.; Morris, J.G. Bacteriophage therapy. *Antimicrob. Agents Chemother.* **2001**, *45*, 649–659. [CrossRef] [PubMed]
2. Nossal, N.G. Protein-protein interactions at a DNA replication fork: Bacteriophage T4 as a model. *FASEB J.* **1992**, *6*, 871–878. [PubMed]
3. Liu, C.; Burke, R.; Hibner, U.; Barry, J.; Alberts, B. Probing DNA Replication Mechanisms with the T4 Bacteriophage *in Vitro* System. In *Cold Spring Harbor Symposia on Quantitative Biology*; Cold Spring Harbor Laboratory Press: Cold Spring Harbor, NY, USA, 1979; pp. 469–487.
4. Alberts, B.; Barry, J.; Bedinger, P.; Formosa, T.; Jongeneel, C.; Kreuzer, K. Studies on DNA Replication in the Bacteriophage T4 *in Vitro* System. In *Cold Spring Harbor Symposia on Quantitative Biology*; Cold Spring Harbor Laboratory Press: Cold Spring Harbor, NY, USA, 1983; pp. 655–668.
5. Mueser, T.C.; Hinerman, J.M.; Devos, J.M.; Boyer, R.A.; Williams, K.J. Structural analysis of bacteriophage T4 DNA replication: A review in the virology journal series on bacteriophage T4 and its relatives. *Virol. J.* **2010**, *7*. [CrossRef] [PubMed]
6. Lee, S.-J.; Richardson, C.C. Choreography of bacteriophage T7 DNA replication. *Curr. Opin. Chem. Biol.* **2011**, *15*, 580–586. [CrossRef] [PubMed]
7. Hamdan, S.M.; Richardson, C.C. Motors, switches, and contacts in the replisome. *Annu. Rev. Biochem.* **2009**, *78*, 205–243. [CrossRef] [PubMed]
8. Van Oijen, A.M.; Loparo, J.J. Single-molecule studies of the replisome. *Annu. Rev. Biophys.* **2010**, *39*, 429–448. [CrossRef] [PubMed]
9. Mosig, G.; Colowick, N.; Gruidl, M.E.; Chang, A.; Harvey, A.J. Multiple initiation mechanisms adapt phage T4 DNA replication to physiological changes during T4's development. *FEMS Microb. Rev.* **1995**, *17*, 83–98. [CrossRef]
10. Kreuzer, K.N.; Brister, J.R. Initiation of bacteriophage T4 DNA replication and replication fork dynamics: A review in the virology journal series on bacteriophage T4 and its relatives. *Virol. J.* **2010**, *7*. [CrossRef] [PubMed]
11. Benkovic, S.J.; Valentine, A.M.; Salinas, F. Replisome-mediated DNA replication. *Annu. Rev. Biochem.* **2001**, *70*, 181–208. [CrossRef] [PubMed]

12. Reddy, M.K.; Weitzel, S.E.; von Hippel, P.H. Assembly of a functional replication complex without ATP hydrolysis: A direct interaction of bacteriophage T4 gp45 with T4 DNA polymerase. *Proc. Natl. Acad. Sci. USA* **1993**, *90*, 3211–3215. [CrossRef] [PubMed]

13. Sexton, D.J.; Berdis, A.J.; Benkovic, S.J. Assembly and disassembly of DNA polymerase holoenzyme. *Curr. Opin. Chem. Biol.* **1997**, *1*, 316–322. [CrossRef]

14. Franklin, M.C.; Wang, J.; Steitz, T.A. Structure of the replicating complex of a pol α family DNA polymerase. *Cell* **2001**, *105*, 657–667. [CrossRef]

15. Wang, J.; Sattar, A.A.; Wang, C.; Karam, J.; Konigsberg, W.; Steitz, T. Crystal structure of a pol α family replication DNA polymerase from bacteriophage RB69. *Cell* **1997**, *89*, 1087–1099. [CrossRef]

16. Shamoo, Y.; Steitz, T.A. Building a replisome from interacting pieces: Sliding clamp complexed to a peptide from DNA polymerase and a polymerase editing complex. *Cell* **1999**, *99*, 155–166. [CrossRef]

17. Lin, T.-C.; Karam, G.; Konigsberg, W.H. Isolation, characterization, and kinetic properties of truncated forms of T4 DNA polymerase that exhibit 3′–5′exonuclease activity. *J. Biol. Chem.* **1994**, *269*, 19286–19294. [PubMed]

18. Bruck, I.; O'Donnell, M. The ring-type polymerase sliding clamp family. *Genome Biol.* **2001**, *2*. [CrossRef]

19. Moarefi, I.; Jeruzalmi, D.; Turner, J.; O'Donnell, M.; Kuriyan, J. Crystal structure of the DNA polymerase processivity factor of T4 bacteriophage. *J. Mol. Biol.* **2000**, *296*, 1215–1223. [CrossRef] [PubMed]

20. Soumillion, P.; Sexton, D.J.; Benkovic, S.J. Clamp subunit dissociation dictates bacteriophage T4 DNA polymerase holoenzyme disassembly. *Biochemistry* **1998**, *37*, 1819–1827. [CrossRef] [PubMed]

21. Alley, S.C.; Shier, V.K.; Abel-Santos, E.; Sexton, D.J.; Soumillion, P.; Benkovic, S.J. Sliding clamp of the bacteriophage T4 polymerase has open and closed subunit interfaces in solution. *Biochemistry* **1999**, *38*, 7696–7709. [CrossRef] [PubMed]

22. Millar, D.; Trakselis, M.A.; Benkovic, S.J. On the solution structure of the T4 sliding clamp (gp45). *Biochemistry* **2004**, *43*, 12723–12727. [CrossRef] [PubMed]

23. Berdis, A.J.; Soumillion, P.; Benkovic, S.J. The carboxyl terminus of the bacteriophage T4 DNA polymerase is required for holoenzyme complex formation. *Proc. Natl. Acad. Sci. USA* **1996**, *93*, 12822–12827. [CrossRef] [PubMed]

24. Jarvis, T.; Paul, L.; Hockensmith, J.; von Hippel, P. Structural and enzymatic studies of the T4 DNA replication system. II. Atpase properties of the polymerase accessory protein complex. *J. Biol. Chem.* **1989**, *264*, 12717–12729. [PubMed]

25. Jarvis, T.; Newport, J.; von Hippel, P. Stimulation of the processivity of the DNA polymerase of bacteriophage T4 by the polymerase accessory proteins. The role of atp hydrolysis. *J. Biol. Chem.* **1991**, *266*, 1830–1840. [PubMed]

26. Rush, J.; Lin, T.; Quinones, M.; Spicer, E.; Douglas, I.; Williams, K.; Konigsberg, W. The 44p subunit of the T4 DNA polymerase accessory protein complex catalyzes ATP hydrolysis. *J. Biol. Chem.* **1989**, *264*, 10943–10953. [PubMed]

27. Bowman, G.D.; O'Donnell, M.; Kuriyan, J. Structural analysis of a eukaryotic sliding DNA clamp-clamp loader complex. *Nature* **2004**, *429*, 724–730. [CrossRef] [PubMed]

28. Simonetta, K.R.; Kazmirski, S.L.; Goedken, E.R.; Cantor, A.J.; Kelch, B.A.; McNally, R.; Seyedin, S.N.; Makino, D.L.; O'Donnell, M.; Kuriyan, J. The mechanism of ATP-dependent primer-template recognition by a clamp loader complex. *Cell* **2009**, *137*, 659–671. [CrossRef] [PubMed]

29. Dong, F.; Gogol, E.P.; von Hippel, P.H. The phage T4-coded DNA replication helicase (gp41) forms a hexamer upon activation by nucleoside triphosphate. *J. Biol. Chem.* **1995**, *270*, 7462–7473. [PubMed]

30. Young, M.C.; Schultz, D.E.; Ring, D.; von Hippel, P.H. Kinetic parameters of the translocation of bacteriophage T4 gene 41 protein helicase on single-stranded DNA. *J. Mol. Biol.* **1994**, *235*, 1447–1458. [CrossRef] [PubMed]

31. Liu, C.; Alberts, B. Characterization of the DNA-dependent gtpase activity of T4 gene 41 protein, an essential component of the t4 bacteriophage DNA replication apparatus. *J. Biol. Chem.* **1981**, *256*, 2813–2820.

32. Norcum, M.T.; Warrington, J.A.; Spiering, M.M.; Ishmael, F.T.; Trakselis, M.A.; Benkovic, S.J. Architecture of the bacteriophage T4 primosome: Electron microscopy studies of helicase (gp41) and primase (gp61). *Proc. Natl. Acad. Sci. USA* **2005**, *102*, 3623–3626. [CrossRef] [PubMed]

33. Venkatesan, M.; Silver, L.; Nossal, N. Bacteriophage T4 gene 41 protein, required for the synthesis of RNA primers, is also a DNA helicase. *J. Biol. Chem.* **1982**, *257*, 12426–12434. [PubMed]

34. Richardson, R.W.; Nossal, N. Characterization of the bacteriophage T4 gene 41 DNA helicase. *J. Biol. Chem.* **1989**, *264*, 4725–4731. [PubMed]

35. Perumal, S.K.; Raney, K.D.; Benkovic, S.J. Analysis of the DNA translocation and unwinding activities of T4 phage helicases. *Methods* **2010**, *51*, 277–288. [CrossRef] [PubMed]

36. Zhang, Z.; Spiering, M.M.; Trakselis, M.A.; Ishmael, F.T.; Xi, J.; Benkovic, S.J.; Hammes, G.G. Assembly of the bacteriophage T4 primosome: Single-molecule and ensemble studies. *Proc. Natl. Acad. Sci. USA* **2005**, *102*, 3254–3259. [CrossRef] [PubMed]

37. Cha, T.; Alberts, B. Studies of the DNA helicase-RNA primase unit from bacteriophage T4. A trinucleotide sequence on the DNA template starts rna primer synthesis. *J. Biol. Chem.* **1986**, *261*, 7001–7010. [PubMed]

38. Hinton, D.; Nossal, N. Bacteriophage T4 DNA primase-helicase. Characterization of oligomer synthesis by T4 61 protein alone and in conjunction with T4 41 protein. *J. Biol. Chem.* **1987**, *262*, 10873–10878. [PubMed]

39. Cha, T.A.; Alberts, B.M. Effects of the bacteriophage T4 gene 41 and gene 32 proteins on rna primer synthesis: The coupling of leading-and lagging-strand DNA synthesis at a replication fork. *Biochemistry* **1990**, *29*, 1791–1798. [CrossRef] [PubMed]

40. Yang, J.; Xi, J.; Zhuang, Z.; Benkovic, S.J. The oligomeric T4 primase is the functional form during replication. *J. Biol. Chem.* **2005**, *280*, 25416–25423. [CrossRef] [PubMed]

41. Huberman, J.A.; Kornberg, A.; Alberts, B.M. Stimulation of t4 bacteriophage DNA polymerase by the protein product of T4 gene 32. *J. Mol. Biol.* **1971**, *62*, 39–52. [CrossRef]

42. Huang, C.; Hearst, J.; Alberts, B. Two types of replication proteins increase the rate at which T4 DNA polymerase traverses the helical regions in a single-stranded DNA template. *J. Biol. Chem.* **1981**, *256*, 4087–4094. [PubMed]

43. Huang, C.-C.; Hearst, J.E. Pauses at positions of secondary structure during *in vitro* replication of single-stranded fd Bacteriophage DNA by T4 DNA polymerase. *Anal. Biochem.* **1980**, *103*, 127–139. [CrossRef]

44. Shamoo, Y.; Friedman, A.M.; Parsons, M.R.; Konigsberg, W.H.; Steitz, T.A. Crystal structure of a replication fork single-stranded DNA binding protein (T4 gp32) complexed to DNA. *Nature* **1995**, *376*, 362–366. [CrossRef] [PubMed]

45. Dudas, K.C.; Kreuzer, K.N. Bacteriophage T4 helicase loader protein gp59 functions as gatekeeper in origin-dependent replication *in vivo*. *J. Biol. Chem.* **2005**, *280*, 21561–21569. [CrossRef] [PubMed]

46. Ishmael, F.T.; Alley, S.C.; Benkovic, S.J. Assembly of the Bacteriophage T4 helicase architecture and stoichiometry of the gp41-gp59 complex. *J. Biol. Chem.* **2002**, *277*, 20555–20562. [CrossRef] [PubMed]

47. Delagoutte, E.; von Hippel, P.H. Mechanistic studies of the t4 DNA (gp41) replication helicase: Functional interactions of the c-terminal tails of the helicase subunits with the T4 (gp59) helicase loader protein. *J. Mol. Biol.* **2005**, *347*, 257–275. [CrossRef] [PubMed]

48. Chastain, P.D.; Makhov, A.M.; Nossal, N.G.; Griffith, J. Architecture of the replication complex and DNA loops at the fork generated by the bacteriophage t4 proteins. *J. Biol. Chem.* **2003**, *278*, 21276–21285. [CrossRef] [PubMed]

49. Nossal, N.G.; Makhov, A.M.; Chastain, P.D.; Jones, C.E.; Griffith, J.D. Architecture of the Bacteriophage T4 replication complex revealed with nanoscale biopointers. *J. Biol. Chem.* **2007**, *282*, 1098–1108. [CrossRef] [PubMed]

50. Zhao, Y.; Chen, D.; Yue, H.; Spiering, M.M.; Zhao, C.; Benkovic, S.J.; Huang, T.J. Dark-field illumination on zero-mode waveguide/microfluidic hybrid chip reveals T4 replisomal protein interactions. *Nano Lett.* **2014**, *14*, 1952–1960. [CrossRef] [PubMed]

51. Mace, D.C.; Alberts, B.M. T4 DNA polymerase: Rates and processivity on single-stranded DNA templates. *J. Mol. Biol.* **1984**, *177*, 295–311. [CrossRef]

52. Kelch, B.A.; Makino, D.L.; O'Donnell, M.; Kuriyan, J. How a DNA polymerase clamp loader opens a sliding clamp. *Science* **2011**, *334*, 1675–1680. [CrossRef] [PubMed]

53. Berdis, A.J.; Benkovic, S.J. Role of adenosine 5′-triphosphate hydrolysis in the assembly of the bacteriophage T4 DNA replication holoenzyme complex. *Biochemistry* **1996**, *35*, 9253–9265. [CrossRef] [PubMed]

54. Perumal, S.K.; Ren, W.; Lee, T.-H.; Benkovic, S.J. How a holoenzyme for DNA replication is formed. *Proc. Natl. Acad. Sci. USA* **2013**, *110*, 99–104. [CrossRef] [PubMed]

55. Mathews, C.K. *Bacteriophage T4*; Wiley Online Library: Washington, DC, USA, 1983.

56. Kaboord, B.F.; Benkovic, S.J. Accessory proteins function as matchmakers in the assembly of the T4 DNA polymerase holoenzyme. *Curr. Biol.* **1995**, *5*, 149–157. [CrossRef]

57. Schrock, R.D.; Alberts, B. Processivity of the gene 41 DNA helicase at the bacteriophage T4 DNA replication fork. *J. Biol. Chem.* **1996**, *271*, 16678–16682. [PubMed]

58. Yang, J.; Zhuang, Z.; Roccasecca, R.M.; Trakselis, M.A.; Benkovic, S.J. The dynamic processivity of the T4 DNA polymerase during replication. *Proc. Natl. Acad. Sci. USA* **2004**, *101*, 8289–8294. [CrossRef] [PubMed]

59. Maga, G.; Hübscher, U. Proliferating cell nuclear antigen (PCNA): A dancer with many partners. *J. Cell Sci.* **2003**, *116*, 3051–3060. [CrossRef] [PubMed]

60. Maul, R.W.; Scouten Ponticelli, S.K.; Duzen, J.M.; Sutton, M.D. Differential binding of *Escherichia coli* DNA polymerases to the β-sliding clamp. *Mol. Microbial.* **2007**, *65*, 811–827. [CrossRef] [PubMed]

61. Johnson, D.E.; Takahashi, M.; Hamdan, S.M.; Lee, S.-J.; Richardson, C.C. Exchange of DNA polymerases at the replication fork of bacteriophage T7. *Proc. Natl. Acad. Sci. USA* **2007**, *104*, 5312–5317. [CrossRef] [PubMed]

62. Hacker, K.J.; Alberts, B.M. The rapid dissociation of the T4 DNA polymerase holoenzyme when stopped by a DNA hairpin helix. A model for polymerase release following the termination of each okazaki fragment. *J. Biol. Chem.* **1994**, *269*, 24221–24228. [PubMed]

63. Cha, T.-A.; Alberts, B.M. The bacteriophage t4 DNA replication fork. Only DNA helicase is required for leading strand DNA synthesis by the DNA polymerase holoenzyme. *J. Biol. Chem.* **1989**, *264*, 12220–12225. [PubMed]

64. Dong, F.; Weitzel, S.E.; Von Hippel, P.H. A coupled complex of T4 DNA replication helicase (gp41) and polymerase (gp43) can perform rapid and processive DNA strand-displacement synthesis. *Proc. Natl. Acad. Sci. USA* **1996**, *93*, 14456–14461. [CrossRef] [PubMed]

65. Delagoutte, E.; von Hippel, P.H. Molecular mechanisms of the functional coupling of the helicase (gp41) and polymerase (gp43) of bacteriophage T4 within the DNA replication fork. *Biochemistry* **2001**, *40*, 4459–4477. [CrossRef] [PubMed]

66. Ishmael, F.T.; Trakselis, M.A.; Benkovic, S.J. Protein-protein interactions in the bacteriophage T4 replisome the leading strand holoenzyme is physically linked to the lagging strand holoenzyme and the primosome. *J. Biol. Chem.* **2003**, *278*, 3145–3152. [CrossRef] [PubMed]

67. Stano, N.M.; Jeong, Y.-J.; Donmez, I.; Tummalapalli, P.; Levin, M.K.; Patel, S.S. DNA synthesis provides the driving force to accelerate DNA unwinding by a helicase. *Nature* **2005**, *435*, 370–373. [CrossRef] [PubMed]

68. Manosas, M.; Spiering, M.M.; Ding, F.; Croquette, V.; Benkovic, S.J. Collaborative coupling between polymerase and helicase for leading-strand synthesis. *Nucl. Acids Res.* **2012**, *40*, 6187–6198. [CrossRef] [PubMed]

69. Lionnet, T.; Spiering, M.M.; Benkovic, S.J.; Bensimon, D.; Croquette, V. Real-time observation of bacteriophage T4 gp41 helicase reveals an unwinding mechanism. *Proc. Natl. Acad. Sci. USA* **2007**, *104*, 19790–19795. [CrossRef] [PubMed]

70. Salinas, F.; Benkovic, S.J. Characterization of bacteriophage T4-coordinated leading-and lagging-strand synthesis on a minicircle substrate. *Proc. Natl. Acad. Sci. USA* **2000**, *97*, 7196–7201. [CrossRef] [PubMed]

71. Yang, J.; Trakselis, M.A.; Roccasecca, R.M.; Benkovic, S.J. The application of a minicircle substrate in the study of the coordinated T4 DNA replication. *J. Biol. Chem.* **2003**, *278*, 49828–49838. [CrossRef] [PubMed]

72. Jing, D.; Beechem, J.M.; Patton, W.F. The utility of a two-color fluorescence electrophoretic mobility shift assay procedure for the analysis of DNA replication complexes. *Electrophoresis* **2004**, *25*, 2439–2446. [CrossRef] [PubMed]

73. Jing, D.H.; Dong, F.; Latham, G.J.; von Hippel, P.H. Interactions of bacteriophage t4-coded primase (gp61) with the t4 replication helicase (gp41) and DNA in primosome formation. *J. Biol. Chem.* **1999**, *274*, 27287–27298. [CrossRef] [PubMed]

74. Manosas, M.; Spiering, M.M.; Zhuang, Z.; Benkovic, S.J.; Croquette, V. Coupling DNA unwinding activity with primer synthesis in the bacteriophage T4 primosome. *Nat. Chem. Biol.* **2009**, *5*, 904–912. [CrossRef] [PubMed]

75. Lee, J.-B.; Hite, R.K.; Hamdan, S.M.; Xie, X.S.; Richardson, C.C.; Van Oijen, A.M. DNA T4 primase acts as a molecular brake in DNA replication. *Nature* **2006**, *439*, 621–624. [CrossRef] [PubMed]

76. Pandey, M.; Syed, S.; Donmez, I.; Patel, G.; Ha, T.; Patel, S.S. Coordinating DNA replication by means of priming loop and differential synthesis rate. *Nature* **2009**, *462*, 940–943. [CrossRef] [PubMed]

77. Kadyrov, F.A.; Drake, J.W. Conditional coupling of leading-strand and lagging-strand DNA synthesis at bacteriophage T4 replication forks. *J. Biol. Chem.* **2001**, *276*, 29559–29566. [CrossRef] [PubMed]

78. Trakselis, M.A.; Roccasecca, R.M.; Yang, J.; Valentine, A.M.; Benkovic, S.J. Dissociative properties of the proteins within the bacteriophage T4 replisome. *J. Biol. Chem.* **2003**, *278*, 49839–49849. [CrossRef] [PubMed]

79. Trakselis, M.A.; Alley, S.C.; Abel-Santos, E.; Benkovic, S.J. Creating a dynamic picture of the sliding clamp during T4 DNA polymerase holoenzyme assembly by using fluorescence resonance energy transfer. *Proc. Natl. Acad. Sci. USA* **2001**, *98*, 8368–8375. [CrossRef] [PubMed]

80. Nelson, S.W.; Kumar, R.; Benkovic, S.J. Rna primer handoff in bacteriophage T4 DNA replication the role of single-stranded DNA-binding protein and polymerase accessory proteins. *J. Biol. Chem.* **2008**, *283*, 22838–22846. [CrossRef] [PubMed]

81. Yang, J.; Nelson, S.W.; Benkovic, S.J. The control mechanism for lagging strand polymerase recycling during bacteriophage T4 DNA replication. *Mol. Cell* **2006**, *21*, 153–164. [CrossRef] [PubMed]

82. Carver, T.E.; Sexton, D.J.; Benkovic, S.J. Dissociation of bacteriophage t4 DNA polymerase and its processivity clamp after completion of okazaki fragment synthesis. *Biochemistry* **1997**, *36*, 14409–14417. [CrossRef] [PubMed]

83. Tougu, K.; Marians, K.J. The interaction between helicase and primase sets the replication fork clock. *J. Biol. Chem.* **1996**, *271*, 21398–21405. [PubMed]

84. Kurth, I.; Georgescu, R.E.; O'Donnell, M.E. A solution to release twisted DNA during chromosome replication by coupled DNA polymerases. *Nature* **2013**, *496*, 119–122. [CrossRef] [PubMed]

85. Yuan, Q.; McHenry, C.S. Cycling of the *E. coli* lagging strand polymerase is triggered exclusively by the availability of a new primer at the replication fork. *Nucl. Acids Res.* **2014**, *42*, 1747–1756. [CrossRef] [PubMed]

86. Geertsema, H.J.; van Oijen, A.M. A single-molecule view of DNA replication: The dynamic nature of multi-protein complexes revealed. *Curr. Opin. Struct. Biol.* **2013**, *23*, 788–793. [CrossRef] [PubMed]

87. Chen, D.; Yue, H.; Spiering, M.M.; Benkovic, S.J. Insights into okazaki fragment synthesis by the T4 replisome the fate of lagging-strand holoenzyme components and their influence on Okazaki fragment size. *J. Biol. Chem.* **2013**, *288*, 20807–20816. [CrossRef] [PubMed]

3

Bacteriophage-Derived Vectors for Targeted Cancer Gene Therapy

Md Zahidul Islam Pranjol [1] **and Amin Hajitou** [2,*]

[1] Institute of Clinical and Biomedical Science, University of Exeter Medical School, Exeter, Devon EX1 2LU, UK; E-Mail: z.pranjol@exeter.ac.uk

[2] Phage Therapy Group, Department of Medicine, Burlington Danes Building, Imperial College London, Hammersmith Hospital, Du Cane Road, London W12 0NN, UK

*   Author to whom correspondence should be addressed; E-Mail: a.hajitou@imperial.ac.uk

Academic Editor: Eric O. Freed

Abstract: Cancer gene therapy expanded and reached its pinnacle in research in the last decade. Both viral and non-viral vectors have entered clinical trials, and significant successes have been achieved. However, a systemic administration of a vector, illustrating safe, efficient, and targeted gene delivery to solid tumors has proven to be a major challenge. In this review, we summarize the current progress and challenges in the targeted gene therapy of cancer. Moreover, we highlight the recent developments of bacteriophage-derived vectors and their contributions in targeting cancer with therapeutic genes following systemic administration.

Keywords: bacteriophage; CMV; Grp78; AAVP; *HSVtk*; RGD4C-AAVP/*CMV-HSVtk*; RGD4C-AAVP/*Grp78-HSVtk*; glioblastoma

1. Introduction

1.1. Overview of Gene Therapy and Its Historical Perspective

Gene therapy describes the delivery of a functional therapeutic gene to target cells, which may be used to knockdown expression of a particular macromolecule, over-express a desired protein, directly induce cell death, or replace a defective or mutant gene to allow expression of a normal protein product. The concept of gene therapy arose in the early 1960s, as Joshua Lederberg conceived of the idea of a direct control of nucleotide sequences in human chromosomes, coupled with recognition, selection, and integration of the desired gene [1]. He was the first to mention the grafting of polynucleotide sequences onto a virus by chemical procedures. In the mid- to late-1960s, scientists showed the transformation of normal cells to a neoplastic phenotype by covalently, stably, and heritably integrating genetic information of the papovavirus SV40 into the genomes of target cells [2]. Subsequently came the recombinant DNA era, which provided a promising platform for developing efficient methods of gene transfer and specific genes in cloned forms. However, it was not until 1990, when the first clinical study using gene transfer was reported, whereby a retroviral vector (RV) was used to transfer the neomycin resistance marker gene into tumor-infiltrating lymphocytes obtained from five patients with metastatic melanoma [3].

Although originally conceived to treat congenital diseases, today, over 66% of clinical trials in gene therapy are designed to treat cancer [4]. Since the developmental scheme of gene therapy began in the 1970s, the practice has expanded into numerous cancer research areas with the ambition of developing a universal anti-cancer vector that can be safely administered, tolerated within normal human physiology, and selectively target tumorigenic cells. Cancer, a highly heterogeneous disease, remains a leading cause of death worldwide, accounting for 7.6 million deaths in 2008 (World Health Organization, WHO). Chemotherapy and radiotherapy are the current treatments against the disease, with chemotherapy remaining the most potent offence. However, poor drug uptake by tumor cells due to therapeutic resistance, high interstitial pressure, and the irregular tumor vasculature, are ever-present challenges which allow cancer cells to find a way to evade even the most effective anti-cancer therapies presently in place. Additionally, systemic administrations of conventional chemotherapeutic treatments against cancer present the risk of contributing to the appearance of secondary tumors.

Recently, several cancer gene therapy vectors have undergone preclinical and clinical trials, and several other vectors have the potential to be tested in the near future. In this review, we aim to summarize the targeting strategies used in cancer gene therapy and the development of eukaryotic delivery vectors with their advantages and disadvantages. Next, we will emphasize the chimeric bacteriophage (phage) vector, named adeno-associated virus/phage (AAVP), as a promising candidate in targeted systemic gene therapy of cancer.

2. Cancer Gene Therapy and Specificity

Although numerous gene therapy strategies have been developed to treat cancer, a major challenge has been to generate a systemic gene delivery vector that can both selectively and efficiently target tumor tissue. To date, most clinical trials involved intratumoral injection, of either viral or nonviral vector systems, to avoid transgene expression in normal healthy tissue. A local delivery of the transgene is necessary as proof-of-principle, but systemic administration is clinically beneficial as it permits the vehicle delivery to metastases and some of the primary tumors, which necessitate invasive procedures for local access. Indeed, development of efficient systemic vectors would ensure safety, with only the desired tissues are targeted, while sparing the healthy neighboring cells and organs. The major challenges of clinical gene therapy trials are vector targeting and the high cost of vector production. Therefore, the development of targeted vectors, which mediate efficient and long-term expression of the therapeutic gene in tumor tissues after systemic administration, should provide a major advancement in cancer gene therapy. To date, two main approaches have been used for tumor targeting: (i) transcriptional targeting, which uses promoters that are only active in the target tumors, and (ii) ligand-targeting of vectors to specific receptors expressed within the tumor tissue [5]. Each of these approaches has been attempted individually, by numerous studies, and showed promising results for vector delivery and transgene expression in preclinical tumor models.

2.1. Transcriptional Targeting

Selective targeting and killing of tumor cells is a major goal of current cancer gene therapy. Placing a transgene under the influence of a cell-type-specific promoter raises the chance of its expression in that particular cell [6]. Usually these promoters are highly expressed in certain tissues (malignant tumor) and remain at a low basal expression level in normal tissues, which can reduce or even eliminate potential toxic side effects of the therapeutic gene in these normal tissues. The use of promoters, such as *cytomegalovirus* (CMV) promoter in adenoviral (Ad) vectors based gene therapy, has been well characterized for its constitutive activity, both *in vitro* and *in vivo*. However, the promoter itself does not significantly discriminate between cell types, rather, is expressed at high levels in a range of mammalian tissues [7], which is problematic in cancer gene therapy. On the other hand, promoters, which are active in tumor cells, have only been used for targeted gene delivery to tumors, in order to selectively deliver transgene expression to the tumor tissue.

Transcriptional targeting was first reported by two studies in 1997. Rodriguez *et al.* restricted transgene expression in prostate-specific antigen (PSA) producing prostate cells by applying the prostate-specific antigen promoter into adenovirus type 5 DNA to drive transgene expression, which they referred to as "attenuated replication-competent adenovirus" [8]. Moreover, using the albumin promoter, replication of the herpes simplex virus type 1 (HSV-1) vector was restricted to albumin-expressing liver cells [9]. Since then, numerous promoters have been characterized and used in targeted cancer gene therapy (Table 1).

2.1.1. Tissue-Specific Promoter

This group of promoters is active and mediates transgene expression in only specific tissues. Several tissue-specific promoters that target tumors of a single origin were characterized and used in cancer gene therapy [10]. Examples include the ovarian-specific promoter to target ovarian cancer [11], the albumin promoter to target hepatocellular carcinoma [12], and the thyroglobulin promoter for thyroid carcinomas [13]. The tyrosine kinase promoter has been used, both *in vitro* and *in vivo*, to target melanomas [14].

Table 1. Examples of ligand and transcriptional targeting used in eukaryotic viral vectors in targeted cancer therapy.

Vector	Tumor-Specific Promoters	Ligand Targeting	References
Ad	PSA, GH, TRE, rTG, AFP, VEGFR-2, flt-1, hTERT	RGD, NGR, SIGYPLP, CGKRK, SIKVAV	[13,15–19]
AAV	PRC1, RRM2, BIRC5	NGR, GFE	[20,21]
HSV-1	Albumin, ANGPTL-3, E2F-1	ND *	[9,22]
LV	PSA/E	αCD20, hSCF	[17,23]
RV	CEA, GRP78, kdr, E-selectin	ND *	[17,24,25]
MV	ND *	HSNS, HAA	[26]

AAV: adeno-associated virus; ANGPTL-3: human angiopoietin-like 3; BIRC5: Baculoviral IAP repeat-containing 5; E2F-1: transcription factor 1; GH: Growth hormone; PSA/E: Prostate-specific antigen/Enhancer; hSCF: human stem cell factor; HSNS/HAA: modified attachment protein H on MV; LV: lentivirus; MV: measles virus vector; PRC1: protein regulator of cytokinesis 1; RRM2: ribonucleotide reductase subunit 2; rTG: rat thyroglobulin; SIGYPLP, CGKRK, SIKVAV: homing peptides; RV: retrovirus; TRE: Tetracyclin response element; VEGFR2: Vascular endothelial growth factor receptor 2; * ND: Not determined.

The prostate-specific antigen (PSA) promoter was used in targeting prostate cancer. PSA is predominantly expressed in prostate cells due to transcriptional activation. Both in cell culture and *in vivo*, PSA promoter has been previously shown to express the herpes simplex virus thymidine kinase (*HSVtk*) suicide gene in PSA-positive prostate cancer cells and prostate tumors, respectively; however, no transgene expression was observed in cells that do not express PSA [27].

Although these promoters proved efficient to deliver transgene expression in tumor cells, their activity, in both normal, as well as tumor cells, is deemed to be a major drawback.

2.1.2. Tumor-Specific Promoter

Tumor-specific promoters constitute an ideal choice for targeted cancer gene therapy in order to direct the expression of therapeutic genes, as they have been shown to be highly active in tumor cells while having little or no activity in normal cells. Based on their characteristics, tumor-specific promoters have been subdivided into four groups [10]: (i) cancer-specific promoters; (ii) tumor-type-specific promoters; (iii) tumor vasculature-related promoters; and (iv) tumor microenvironment-related promoters.

Cancer-Specific Promoters

Cancer-specific promoters, such as the promoter of the telomerase gene, are active specifically in malignant cells and have the great potential in cancer gene therapy to target a wide variety of tumors. Telomerase is highly active in around 85%–90% of human cancer cells, while remaining either low or undetectable in normal tissues [28]. Telomerase is composed of two active subunits, telomerase RNA (hTR) and catalytic component human telomerase reverse transcriptase (hTERT). The promoters of these two subunits are highly active in telomerase-positive cells, such as tumor and fetal cells. Thus, these two promoters have been individually used in many targeted cancer gene therapy studies to drive therapeutic gene expression, demonstrating enhanced killing of telomerase-positive cells. Although the *hTR* or *hTERT* promoters have been broadly utilized for transcriptional regulation of therapeutic genes, they still have some limitations in clinical use as they possess low activity and some potential toxicity to certain normal cells has also been reported [15].

Tumor-Type-Specific Promoter

Tumor-type-specific promoters are the promoters of oncofetal genes that are often overexpressed in certain types of tumors and are silent in normal tissues. The most well-characterized promoters of this group include α-fetoprotein (AFP) promoter that is active in fetal liver and hepatocellular carcinomas [29], and carcinoembryonic antigen (CEA) promoter, which is active in a proportion of breast, lung, colorectal and pancreatic cancers. This promoter was extensively used in different vector systems to selectively deliver various therapeutic genes, such as cytosine deaminase or *HSVtk* expression in CEA-positive cells, and the results demonstrated significant tumor growth suppression or regression, with no toxicity to liver and other normal organs, following prodrug 5-fluorocytosine or ganciclovir (GCV) administration, respectively [30]. Although these promoters mediate transgene expression in tumor tissues, and may, therefore, be good candidates for transcriptional targeting in cancer gene therapy, their application still remains limited since they cannot be administered for a variety of tumors.

Tumor Vasculature-Related Promoters

The genes encoding this group of promoters are overexpressed in proliferating endothelia tumor microvasculature. Examples of such promoters that have been used in targeted gene therapy are E-selectin and endothelial-specific kinase insert domain receptor (KDR/flk-1) which are upregulated in tumor endothelium. The promoters of these genes efficiently expressed tumor necrosis factor-α (TNF-α) by 10-fold increase in endothelioma cells compared to non-endothelioma cells [24]. In addition, the promoters of genes encoding Flt-1, vascular endothelial growth factor receptor 1, and human preproendothelin-1 have also been used in gene therapy vector systems and proved efficient to drive transgene expression in the vasculature of tumors and metastases [16]. While the usage of these promoters in targeted cancer gene therapy proved efficient to deliver transgene expression in the tumor vasculature, they still have limitations as some of these promoters were shown to be active in small vessels and upregulated in injured vessels as well [31].

Tumor Microenvironment-Related Promoters

Tumor microenvironment-related promoters belong to the genes that are upregulated in response to the tumor microenvironment and physiology. Compared to normal cells, tumor cells demonstrate high growth rate and an increased glucose metabolism. Moreover, neovascularization and angiogenesis may fail to keep pace with tumor growth which creates a "cancer microenvironment"-hypoxia, acidosis and glucose deprivation-characterizing poorly vascularized solid tumors. As a part of the cancer cell's response to adapt to these conditions, the promoters of some genes such as hypoxia response elements (HREs) and some of the heat shock family genes become induced.

Recently, the heat shock genes, such as the gene of the *Glucose regulated protein 78* (Grp78), have gained increasing interest in targeted cancer gene therapy because of their activation in a wide variety of tumors. The activity of its promoter and its ability to drive transgene expression within areas of tumor hypoxia, which are highly resistant to current forms of treatment, makes it even a more attractive promoter to use in targeted cancer gene therapy. Indeed, therapeutic transgene expression driven by this promoter is induced in response to insufficient blood supply and tumor necrosis and reached high levels leading to complete tumor eradication in preclinical models [32].

For cancer targeting gene therapy, the Grp78 promoter seems to be an ideal promoter to restrict expression of the therapeutic gene within the tumor tissue, and is, therefore, worthwhile a critical evaluation.

3. Grp78 as an Endogenous Macromolecule in Cancer

Grp78 gene encodes a 78-kDa protein (Grp78) that acts as an endoplasmic reticulum (ER) stress response chaperone. It has 60% amino acid homology to the 70-kDa heat shock protein (HSP70), and, hence, Grp78 has been categorized within the HSP70 family. Despite this homology, Grp78 is not induced by heat stress and it also primarily localizes in the endoplasmic reticulum (ER) lumen to function as a molecular chaperone in an ATP-dependent manner. Recently, however, this protein was shown to be present in the cytoplasm and expressed on the cell surface membranes; thus, it can be utilized as a biomarker for stressed cells, such as in tumors [33].

Also known as immunoglobulin heavy chain binding protein (BiP), Grp78 orchestrates unfolded protein response (UPR) by binding to unfolded, misfolded, and incorrectly glycosylated proteins in the ER lumen. In eukaryotic cells, when the protein production exceeds the folding capacity of the ER, the misfolded proteins elicit UPR. Under normal conditions, Grp78 remains bound to three transmembrane sensor proteins: protein kinase-like ER kinase (PERK), inositol requiring enzyme 1 (IRE1), and activating transcription factor 6 (ATF6). As the threshold of the incorrectly folded protein load exceeds a certain level, Grp78 dissociates from the sensors and binds these proteins. This dissociation process leads to activation of a signaling cascade, UPR. The outcome is a decrease in biosynthetic burden of the ER by desensitizing the cells to ER stress and ultimately upregulating pro-survival genes via transcriptional activation in the nucleus, such as Grp78 promoter to elevate Grp78 expression [33]. If, however, the ER homeostasis cannot be re-established, the UPR induces programed cell death [34].

Tumor microenvironment and lack of pace of neovascularization and angiogenesis result in an uncontrolled production of mutant and misfolded ER proteins that lead to the accumulation of unfolded or misfolded ER proteins, which subsequently trigger the UPR. Thus, while Grp78 expression remains low in major adult organs, such as brain, heart, and lung, it is highly upregulated in transformed cells and in several tumors, such as glioblastoma (GBM), breast cancer, and prostate cancer [35], and in the endothelia of tumor vasculature [36]. This overexpression of Grp78 has been shown to correlate with an enhanced tumor recurrence risk, tumor grade, and decreased survival rate in cancer patients [34,37].

Grp78 plays a major role in cancer cell survival by activating the pro-survival pathway. *In vivo* studies with heterozygous Grp78 mice ($Grp7^{+/-}$) demonstrated marked impeded tumor progression compared to wild-type Grp78 mice, as tumor size was reduced and apoptosis was promoted. In addition to primary tumors, Grp78 levels are highly induced in metastasis and assist secondary tumor survival by maintaining neovascularization [36].

Pyrko and colleagues demonstrated a positive correlation between Grp78 overexpression and glioblastoma (GBM) cell proliferation rate by knockdown of Grp78 in GBM cells that led to a reduced proliferation [38]. In addition to tumor cell survival and proliferation, cytoplasmic Grp78 plays an important role in blocking the apoptosis of stressed cells by binding and inhibiting activation of caspases-7 and -12. Grp78 expressed on several tumor cell surfaces promoted cell proliferation, survival, and metastasis by binding to plasma proteinase inhibitor α2-macroglobulin and activating its downstream signaling pathway [33].

4. Grp78 Promoter in Cancer Gene Therapy

The use of Grp78 as a promoter in cancer gene therapy was first proposed by the group of Lee [25]. A retroviral vector construct carrying the *HSVtk* transgene was used to infect tumor cells and subsequently implanted into mice. Administration of GCV, along with *HSVtk* expression, driven by Grp78 promoter, was shown to suppress and eradicate murine and human breast tumor xenografts in mice [32]. Moreover, another group assessed the efficacy of *HSVtk* under the control of Grp78 promoter in gastroesophageal junction and gastric adenocarcinomas cells and reported significant cell death *in vitro* and tumor regression *in vivo* following GCV treatment [39].

In transgenic mice, the *LacZ* transgene, driven by the rat Grp78 promoter, showed high transgene expression in cancer cells, while remaining inactive in major adult organs [32]. In addition, Grp78 gene transcription can increase over time [40] because, unlike viral promoters, such as CMV, mammalian promoters are not silenced in eukaryotic cells, thus, resulting in stronger and long-term transgene expression from the vector.

5. Ligand-Directed Targeting Ensures Specific and Efficient Transgene Delivery

Progress in tumor vascular targeting has provided a platform to target agents safely, efficiently and selectively in tumorigenic tissues. The use of *in vivo* phage display screenings has significantly contributed to the identification of such target receptors in the affected tumor endothelium of animal models [41]. As angiogenesis is vital for tumor progression, targeting these tumor-specific and tumor associated endothelial cell-specific receptor molecules holds the potential for ligand-directed targeting in cancer gene therapy (Table 2) [42]. Viral vectors can be engineered to display homing peptides on their surface in order to target specific receptor-bearing cell types within a host. This method of targeting ensures that the vector only infects the cells bearing the receptor, while leaving non-receptor-bearing cells untouched.

It was hypothesized, long ago, that selectively interfering with the tumor blood supply would lead to strong antitumor effects [42]. In addition, vascular cells are readily accessible through the systemic circulation of the vector; hence, targeting and killing tumor vasculature has been of great interest [42]. Neovasculatures that support tumor growth express different markers on their endothelium as compared to normal quiescent ones, such as $\alpha_v\beta_3$-integrin (adhesion molecule), and an estimated 100 tumor cells are supplied by one endothelial cell [43]. Additionally, several tumorigenic tissues express these integrins, such as glioblastoma (GBM), melanoma, *etc*. This cumulative understanding of the tumor physiology and biochemistry has led to the development of ligand-directed vectors targeting the tumor vasculature via α_v-integrin receptors by using α_v-integrin binding ligands that ensure specificity and optimize effectiveness.

Table 2. List of receptors used in targeted cancer therapy [42].

Receptor	Function/Class	Localization
Grp78	Stress Response	Tumor cells
αv Integrins	Cell adhesion	EC, tumor cells
CD13	Aminopeptidase N	EC, pericytes
APA	Aminopeptidase A	Pericytes, stromal cell
NG2/HMWMAA	Proteoglycan	Pericytes, tumor cells
MMP-2/MMP-9	Metalloproteinases	EC, tumor cell
HSP90	Heat shock	EC, tumor cells

EC, endothelial cells.

5.1. Ligand Targeting of Viral Gene Therapy Vectors

In the last decade several different techniques have been developed with the aim of facilitating vector targeting to specific cells after systemic administration. However, for systemic targeted cancer gene therapy, the choice of an efficient vector system, able to deliver an appropriate expression level of the therapeutic gene in cancer cells, is the most crucial step. In this process, non-viral vectors, which involve physical methods such as a gene gun, magnetofection and chemical methods, such as lipoplexes, inorganic nanoparticles, as well as injection of naked DNA, hold some potential due to their mass production, safety, and low host immunogenicity. However, their low level of transfection and transgene expression still present great challenges [44]. In contrast to non-viral vectors, arguably, viral vectors have proven to be the vectors of choice as they naturally evolved to infect mammalian cells and can mediate high level of transduction (Table 1). Although, as of 2007, several types of vectors were used in 70% of gene therapy clinical trials, many groups are aiming at improving viral vector features for the purpose of gene therapy [45]. Most progress in viral gene therapy has involved the lentivirus (LV), adenovirus (Ad), and adeno-associated virus (AAV) [46]. In particular, Ads have been utilized as an attractive approach for cancer gene therapy and have been reported as the most frequently used vectors in clinical trials [47]. Ads contain 36 kb double-stranded DNA, which provides space for inserting large sequence fragments. In addition to the ease of vector manipulation and the capability to produce high titers, Ad vectors mediate efficient transgene expression in both dividing and non-dividing cells. However, despite all the advantages of Ad vectors, they can induce severe toxicity due to raising host immune response after systemic delivery of the required dose of vector. Thus, for selective killing of tumor cells, many studies have focused on modifying the adenoviral genome in such a way to achieve cancer specificity at different stages through the adenoviral life cycle [48]. One main approach has involved the specific deletion in Ad genome and generation of replication-incompetent Ad vectors which attenuate the viral amplification and, thus, inhibit the infection. These engineered vectors can be further modified to selectively replicate in tumor cells to mediate their lysis. ONYX-015 vector is such an example of a replication-incompetent Ad vector. This vector cannot defend itself against the p53 tumor suppressor gene products and, thus, is destroyed in healthy cells. In contrast, as the p53 gene is defective in most tumor cells, ONYX-015 can proliferate within these tumor cells and mediate their lysis [49]. Moreover, if the tumor lysis mediated by these vectors is not sufficient, therapeutic genes can be inserted into the Ad vector genome in order to enhance the antitumor efficacy [50].

AAVs, member of the parvovirus family, are also promising vectors that have been used in numerous gene therapy applications as they confer low immunogenicity within the host. Moreover, AAVs have shown to mediate stable and long-term gene expression *in vivo* [51]. Additionally, AAV vectors are purified to high titers and, like Ads, can transduce both dividing and non-dividing cells. However, one of the severe limitations of AAV vectors for gene therapy is their limited gene cloning capacity (just over 4 kb) [47]. AAVs contain 4.7 kb single-stranded DNA encoding for two genes: *rep* which is responsible for viral DNA replication and *cap*, which is in charge of packaging the viral genome. *Rep* and *cap* open reading frames (ORFs) are flanked by two inverted terminal repeats (ITRs) [52]. On entry into the host cell, the AAV single stranded DNA is converted to transcriptionally active double-stranded DNA and remains episomal [51]. To generate recombinant AAV vectors, *rep* and *cap* are replaced by transgene cassettes, leaving the ITR cis-elements which serve as origin of replication [52].

Although eukaryotic viral vectors, such as AAV, boast efficient transgene delivery and extremely stable long-term expression of the transgene, their innate, broad tropism for mammalian cells pose the risk of inducing an immune response in the host, which escalates risks of re-administration of the vector. In addition, AAV vectors have limited cloning capacity, as they cannot accommodate large-size cDNA. Moreover, due to the wide tropism for mammalian cells, eukaryotic viral vectors are frequently taken up by the liver, reticulo-endothelial system, and other unwanted tissues after intravenous administration [5].

6. Bacteriophage-Guided Gene Therapy of Cancer

Bacteriophages, the viruses that only infect bacteria, present an alternative and safer strategy for targeted systemic delivery of transgenes, as they have no intrinsic tropism for mammalian cells, but can mediate modest gene expression after genetic manipulation. Production of bacteriophages is cost-effective and can be completed at high titers. Moreover, bacteriophages are safe and can be engineered to deliver genes to mammalian cells. Additionally, bacteriophages have been used for antibiotic therapy during the pre-antibiotic era and were safely administered even in children after systemic administration. In 2006, the US Food and Drug Administration (FDA) approved the use of some bacteriophage preparations for ready-to-eat (RTE) meat and poultry products as antibacterial food additives [53]. Most importantly, unlike eukaryotic viruses, bacteriophages do not require further context modification of their capsid as targeting peptides are, in actual fact, selected and isolated directly for targeting specific cell surface receptors after screening of a phage display peptide library. However, they have been, in the past, considered to be poor gene delivery vehicles as they have evolved to infect only bacteria and have no optimized strategies to efficiently express transgenes upon entry into eukaryotic cells [5]. In order to overcome these limitations, Hajitou and colleagues [54] reported a new generation of hybrid prokaryotic-eukaryotic viral vector as a chimera between eukaryotic AAV and the filamentous M13 bacteriophage, then named AAVP (AAV/phage), both contain single stranded DNA genome (Figure 1). This vector expresses three to five copies of the cyclic RGD4C (CDCRGDCFC) ligand on the phage pIII minor coat protein allowing systemic and specific targeting to the $\alpha_v\beta_3$-integrin receptor, which are expressed primarily on tumor vasculature and tumor cells, and are absent or expressed at barely detectable levels in normal endothelium and tissues [55].

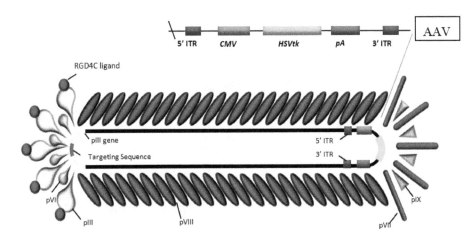

Figure 1. Structure of the hybrid vector AAV/phage (AAVP) developed by Hajitou *et al.* [54]. The particle contains a chimeric genome of a CMV-transgene cassette flanked by inverted terminal repeats, 3' ITR and 5' ITR, of AAV-2 and the genome of M13 filamentous bacteriophage. The outer capsid belongs to the M13 phage and hence lacks tropism for mammalian cells. The capsid contains a major coat protein pVIII and four minor coat proteins pIII, pVI on one side and pVII pIX on the other. The α_v-integrin binding ligand, RGD4C, is expressed on the pIII minor coat protein of AAVP in order to allow ligand-directed targeting of the tumor vasculature and tumor cells.

6.1. Novel Hybrid Gene Therapy Vector: AAVP

The hybrid vector genome was developed by inserting an engineered AAV (recombinant AAV/rAAV) transgene cassette into an intergenomic region of the phage genome, under the regulation of the CMV promoter and flanked by full-length inverted terminal repeats (ITR) from AAV serotype 2 (Figure 1). The use of AAV ITRs in the targeted AAVP (RGD4C-AAVP) improves transduction efficiency and enhances transgene expression by maintaining and forming concatemers of the eukaryotic transgene cassette [54]. *HSVtk* transgene expression under the CMV promoter (RGD4C-AAVP/*CMV-HSVtk*) and subsequent GCV treatment produces drastic suppression of established tumors in mice and rats.

Since its development, the vector has been under investigation in pre-clinical models. The National Cancer Institute of the USA (NCI) has used the ligand-targeting properties of the RGD4C-AAVP to deliver tumor necrosis factor alpha (TNF-α) to the angiogenic vasculature of human melanoma xenografts in nude mice [56]. In this systemic administration of the phage particle, the TNF-α expression was shown to be specifically localized in tumors, leading to apoptosis in tumor blood vessels and significant inhibition of tumor growth, while remaining virtually undetectable in all other tissues, notably the liver and spleen. However, the RGD4C-AAVP particles, which were found in the latter two vital organs, did not have their transgenes expressed in them. The efficacy of targeted RGD4C-AAVP expressing the TNF-α was assessed in domesticated dogs with soft tissue sarcoma [57]. Intravenous single and multidoses of the vascular-targeted RGD4C-AAVP vector was shown to be tumor specific. Repeated vector administrations resulted in complete eradication of cancer in a few dogs and stability in others, despite the presence of a high immune response against the phage viral particles. Trepel *et al.* showed the presence of a bystander effect after GCV treatment between transduced endothelial cells and non-transduced tumor cells themselves, implying that RGD4C-AAVP/*CMV-HSVtk* plus GCV therapy will not be completely limited by transduction efficiency [58]. Although the phage-based particles are known to be immunogenic, repeated administrations in domesticated dogs and immune-competent mice resulted in efficient antitumor therapy [54,57]. The selectivity and safety properties of the RGD4C-AAVP have made this novel, hybrid vector a promising tool that holds great potentials in systemic cancer gene therapy.

6.2. Development of the AAVP Viral Particles

6.2.1. Limitations Faced by the CMV Promoter in the Phage: Silencing of Gene Expression

Although the use of the CMV promoter in adenoviral vector-based gene therapy has been well characterized for its constitutive activity both *in vitro* and *in vivo*, the promoter rather remains active at high levels in a range of mammalian tissues. In addition, the silencing of gene expression from the CMV promoter in mammalian cells occurs through several mechanisms, including DNA methylation and histone deacetylation [59]. The CMV promoter activity was reported to be suppressed completely by methylating cytosine resides at 5'CpG dinucleotides within a DNA sequence using Spiroplasma methyltransferase SssI. Hypermethylation is a phenomenon that is observed in many cancers and affects genes that regulate cell cycle (p16^{INK4a}, p15^{INK4a}), DNA repair (BRCA1, MGMT), apoptosis (DAPK, TMS1), drug resistance, angiogenesis and metastasis [60]. The propensity of cancer cells towards extensive methylation may, in part, play a role in CMV methylation.

Histone deacetylation is another mechanism of CMV silencing as it ultimately leads to condensed, transcriptionally-inactive regions of chromatin. Chromatin repression and transcriptional remodeling of the *LacZ* gene in HeLa cells transduced with rAAV-*CMV*-LacZ construct occurred upon removal of the histone deacetylase inhibitor, trichostatin-A [61]. These, and other, data suggest that CMV promoter silencing over time can be an obstacle to viral vector-based gene therapy by reducing desired transgene expression in mammalian cells. This necessitates the development of a vector that demonstrates long-term transgene expression and increased systemic anti-tumor efficacy.

6.2.2. A Double-Targeted Phage Vector with the Grp78 Promoter

Combining the two tumor targeting strategies by using tumor homing ligands and tumor specific promoters, in one vector system is challenging, but would provide a major advance in targeted gene therapy of cancer. In 2012, our group has developed a double-targeted AAVP phage particle, whereby both ligand-directed, with RGD4C, and transcriptional targeting strategies were integrated into a single phage vector platform [40]. The transcriptional targeting feature of this novel phage has been achieved by substituting the CMV viral promoter by the Grp78 tumor specific promoter in the AAV transgene cassette. Grp78 is an endogenous macromolecule, which is over-expressed in several tumor cells and, therefore, the ligand-targeted AAVP carrying this promoter should only be expressed in the targeted tumor vasculature and tumor cells.

Kia and colleagues showed long-term transgene expression from AAVP under the regulation of the rat Grp78 promoter in glioblastoma cell lines, compared to CMV promoter which underwent gene silencing over time. Flow cytometry analysis of the *green fluorescent protein* (GFP) expression under

Grp78 regulation, in RGD4C-AAVP/*Grp78-GFP* transduced 9L cells, showed a drastic increase from 57% (39 days post-vector transduction) to 85% (97 days post-vector transduction). In contrast, only 37% of cells transduced by the RGD4C-AAVP/*CMV-GFP* expressed GFP at day 39, followed by a substantial drop to 11% on day 97. Significantly higher tumor cell killing over time was also observed by *HSVtk*/GCV therapy under Grp78 promoter compared to CMV, both *in vitro* and *in vivo*. *In vivo* studies revealed that when tumors grew back after therapy, repeated GCV treatment resulted in tumor growth inhibition in mice that received the RGD4C-AAVP/*Grp78-HSVtk*. However, little or no effect on tumors was observed in RGD4C-AAVP/*CMV-HSVtk* administered mice. Therapeutically RGD4C-AAVP/*Grp78-HSVtk* was shown to be advantageous over RGD4C-AAVP/*CMV-HSVtk* by producing a marked regression of the large tumors. *HSVtk*/GCV therapy was also shown to activate both promoters of endogenous Grp78 and of the RGD4C-AAVP/*Grp78-HSVtk* vector in Western blot analyses.

6.2.3. Current Development: Inhibition of Histone Deacetylation and DNA Methylation Restore Gene Expression under the CMV Promoter and Enhance Grp78-Regulated Gene Expression

Inhibition of histone deacetylation restored gene expression from the CMV promoter in human (U87) and rat (9L) GBM cell lines transduced with RGD4C-AAVP/*CMV* vector [62]. It was previously reported that histone deacetylases (HDACs) are upregulated in cancer. HDAC class I and II inhibitors trichostatin-A and suberoylanilide hydroxamic acid combined with RGD4C-AAVP carrying CMV or Grp78 promoter reactivated RGD4C-AAVP/*CMV* efficacy and enhanced RGD4C-AAVP/*Grp78* in cancer cells specifically, respectively.

Extensive methylation of the CMV promoter sequences has previously been reported [63], which consequently reduces gene expression. The DNA methylation inhibitor 5-Azacytidine reinstated this phenomenon in rat 9L GBM cell line carrying RGD4C-AAVP/*CMV*, although no significant difference was observed in cells transduced with RGD4C-AAVP/*Grp78*. These are important findings in progressive gene expression under the two promoters in the long-term, and which should be carefully considered for any future clinical applications.

7. Conclusions

Cancer gene therapy is a promising approach in treating cancer. However, a vector that has the potential to overcome the decade-long challenges posed by the lack of specificity and the risk of cytotoxicity for healthy cells after systemic administration is crucial. RGD4C-AAVP, with its transcriptional- and ligand-directed-targeting features and stable long-term gene expression under the eukaryotic Grp78 promoter, is a promising tool in targeted systemic cancer gene therapy, and should be brought forth in future clinical trials in cancer patients.

Acknowledgments

Our studies cited in this review were supported by a grant G0701159 of the UK Medical Research Council (MRC) and The Brain Tumour Research Campaign (BTRC).

Author Contributions

M.Z.I.P and A.H. wrote the paper.

References

1. Gillet, J.P.; Macadangdang, B.; Fathke, R.L.; Gottesman, M.M.; Kimchi-Sarfaty, C. The development of gene therapy: From monogenic recessive disorders to complex diseases such as cancer. *Methods Mol. Biol.* **2009**, *542*, 5–54.

2. Friedmann, T. A brief history of gene therapy. *Nat. Genet.* **1992**, *2*, 93–98.

3. Rosenberg, S.A.; Aebersold, P.; Cornetta, K.; Kasid, A.; Morgan, R.A.; Moen, R.; Karson, E.M.; Lotze, M.T.; Yang, J.C.; Topalian, S.L.; *et al.* Gene transfer into humans—Immunotherapy of patients with advanced melanoma, using tumor-infiltrating lymphocytes modified by retroviral gene transduction. *N. Engl. J. Med.* **1990**, *323*, 570–578.

4. Edelstein, M.L.; Abedi, M.R.; Wixon, J. Gene therapy clinical trials worldwide to 2007—An update. *J. Gene Med.* **2007**, *9*, 833–842.

5. Hajitou, A. Targeted systemic gene therapy and molecular imaging of cancer contribution of the vascular-targeted aavp vector. *Adv. Genet.* **2010**, *69*, 65–82.

6. Sadeghi, H.; Hitt, M.M. Transcriptionally targeted adenovirus vectors. *Curr. Gene Ther.* **2005**, *5*, 411–427.

7. Cheng, L.; Ziegelhoffer, P.R.; Yang, N.S. *In vivo* promoter activity and transgene expression in mammalian somatic tissues evaluated by using particle bombardment. *Proc. Natl. Acad. Sci. USA* **1993**, *90*, 4455–4459.

8. Rodriguez, R.; Schuur, E.R.; Lim, H.Y.; Henderson, G.A.; Simons, J.W.; Henderson, D.R. Prostate attenuated replication competent adenovirus (arca) cn706: A selective cytotoxic for prostate-specific antigen-positive prostate cancer cells. *Cancer Res.* **1997**, *57*, 2559–2563.

9. Miyatake, S.; Iyer, A.; Martuza, R.L.; Rabkin, S.D. Transcriptional targeting of herpes simplex virus for cell-specific replication. *J. Virol.* **1997**, *71*, 5124–5132.

10. Robson, T.; Hirst, D.G. Transcriptional targeting in cancer gene therapy. *J. Biomed. Biotechnol.* **2003**, *2003*, 110–137.

11. Bao, R.; Selvakumaran, M.; Hamilton, T.C. Targeted gene therapy of ovarian cancer using an ovarian-specific promoter. *Gynecol. Oncol.* **2002**, *84*, 228–234.

12. Su, H.; Lu, R.; Chang, J.C.; Kan, Y.W. Tissue-specific expression of herpes simplex virus thymidine kinase gene delivered by adeno-associated virus inhibits the growth of human hepatocellular carcinoma in athymic mice. *Proc. Natl. Acad. Sci. USA* **1997**, *94*, 13891–13896.

13. Zhang, R.; Straus, F.H.; DeGroot, L.J. Adenoviral-mediated gene therapy for thyroid carcinoma using thymidine kinase controlled by thyroglobulin promoter demonstrates high specificity and low toxicity. *Thyroid* **2001**, *11*, 115–123.

14. Fecker, L.F.; Ruckert, S.; Kurbanov, B.M.; Schmude, M.; Stockfleth, E.; Fechner, H.; Eberle, J. Efficient melanoma cell killing and reduced melanoma growth in mice by a selective replicating

adenovirus armed with tumor necrosis factor-related apoptosis-inducing ligand. *Hum. Gene Ther.* **2011**, *22*, 405–417.

15. Xiong, J.; Sun, W.J.; Wang, W.F.; Liao, Z.K.; Zhou, F.X.; Kong, H.Y.; Xu, Y.; Xie, C.H.; Zhou, Y.F. Novel, chimeric, cancer-specific, and radiation-inducible gene promoters for suicide gene therapy of cancer. *Cancer* **2012**, *118*, 536–548.

16. Bauerschmitz, G.J.; Nettelbeck, D.M.; Kanerva, A.; Baker, A.H.; Hemminki, A.; Reynolds, P.N.; Curiel, D.T. The flt-1 promoter for transcriptional targeting of teratocarcinoma. *Cancer Res.* **2002**, *62*, 1271–1274.

17. Glinka, E.M. Eukaryotic expression vectors bearing genes encoding cytotoxic proteins for cancer gene therapy. *Plasmid* **2012**, *68*, 69–85.

18. Mizuguchi, H.; Koizumi, N.; Hosono, T.; Utoguchi, N.; Watanabe, Y.; Kay, M.A.; Hayakawa, T. A simplified system for constructing recombinant adenoviral vectors containing heterologous peptides in the hi loop of their fiber knob. *Gene Ther.* **2001**, *8*, 730–735.

19. Kim, J.; Kim, P.H.; Kim, S.W.; Yun, C.O. Enhancing the therapeutic efficacy of adenovirus in combination with biomaterials. *Biomaterials* **2012**, *33*, 1838–1850.

20. Yun, H.J.; Cho, Y.H.; Moon, Y.; Park, Y.W.; Yoon, H.K.; Kim, Y.J.; Cho, S.H.; Lee, Y.I.; Kang, B.S.; Kim, W.J.; *et al.* Transcriptional targeting of gene expression in breast cancer by the promoters of protein regulator of cytokinesis 1 and ribonuclease reductase 2. *Exp. Mol. Med.* **2008**, *40*, 345–353.

21. Grifman, M.; Trepel, M.; Speece, P.; Gilbert, L.B.; Arap, W.; Pasqualini, R.; Weitzman, M.D. Incorporation of tumor-targeting peptides into recombinant adeno-associated virus capsids. *Mol. Ther.* **2001**, *3*, 964–975.

22. Foka, P.; Pourchet, A.; Hernandez-Alcoceba, R.; Doumba, P.P.; Pissas, G.; Kouvatsis, V.; Dalagiorgou, G.; Kazazi, D.; Marconi, P.; Foschini, M.; *et al.* Novel tumour-specific promoters for transcriptional targeting of hepatocellular carcinoma by herpes simplex virus vectors. *J. Gene Med.* **2010**, *12*, 956–967.

23. Froelich, S.; Ziegler, L.; Stroup, K.; Wang, P. Targeted gene delivery to cd117-expressing cells *in vivo* with lentiviral vectors co-displaying stem cell factor and a fusogenic molecule. *Biotechnol. Bioeng.* **2009**, *104*, 206–215.

24. Jaggar, R.T.; Chan, H.Y.; Harris, A.L.; Bicknell, R. Endothelial cell-specific expression of tumor necrosis factor-alpha from the kdr or e-selectin promoters following retroviral delivery. *Hum. Gene Ther.* **1997**, *8*, 2239–2247.

25. Little, E.; Ramakrishnan, M.; Roy, B.; Gazit, G.; Lee, A.S. The glucose-regulated proteins (grp78 and grp94): Functions, gene regulation, and applications. *Crit. Rev. Eukaryot. Gene Expr.* **1994**, *4*, 1–18.

26. Allen, C.; Vongpunsawad, S.; Nakamura, T.; James, C.D.; Schroeder, M.; Cattaneo, R.; Giannini, C.; Krempski, J.; Peng, K.W.; Goble, J.M.; *et al.* Retargeted oncolytic measles strains entering via the egfrviii receptor maintain significant antitumor activity against gliomas with increased tumor specificity. *Cancer Res.* **2006**, *66*, 11840–11850.

27. Gotoh, A.; Ko, S.C.; Shirakawa, T.; Cheon, J.; Kao, C.; Miyamoto, T.; Gardner, T.A.; Ho, L.J.; Cleutjens, C.B.; Trapman, J.; *et al.* Development of prostate-specific antigen promoter-based gene therapy for androgen-independent human prostate cancer. *J. Urol.* **1998**, *160*, 220–229.

28. Broccoli, D.; Young, J.W.; de Lange, T. Telomerase activity in normal and malignant hematopoietic cells. *Proc. Natl. Acad. Sci. USA* **1995**, *92*, 9082–9086.

29. Ishikawa, H.; Nakata, K.; Mawatari, F.; Ueki, T.; Tsuruta, S.; Ido, A.; Nakao, K.; Kato, Y.; Ishii, N.; Eguchi, K.; *et al.* Utilization of variant-type of human alpha-fetoprotein promoter in gene therapy targeting for hepatocellular carcinoma. *Gene Ther.* **1999**, *6*, 465–470.

30. Zhang, G.; Liu, T.; Chen, Y.H.; Chen, Y.; Xu, M.; Peng, J.; Yu, S.; Yuan, J.; Zhang, X. Tissue specific cytotoxicity of colon cancer cells mediated by nanoparticle-delivered suicide gene *in vitro* and *in vivo*. *Clin. Cancer Res.* **2009**, *15*, 201–207.

31. Herz, K.; Heinemann, J.C.; Hesse, M.; Ottersbach, A.; Geisen, C.; Fuegemann, C.J.; Roll, W.; Fleischmann, B.K.; Wenzel, D. Live monitoring of small vessels during development and disease using the flt-1 promoter element. *Basic Res. Cardiol.* **2012**, *107*, 1–14.

32. Dong, D.; Dubeau, L.; Bading, J.; Nguyen, K.; Luna, M.; Yu, H.; Gazit-Bornstein, G.; Gordon, E.M.; Gomer, C.; Hall, F.L.; *et al.* Spontaneous and controllable activation of suicide gene expression driven by the stress-inducible grp78 promoter resulting in eradication of sizable human tumors. *Hum. Gene Ther.* **2004**, *15*, 553–561.

33. Zhang, L.H.; Zhang, X. Roles of grp78 in physiology and cancer. *J. Cell. Biochem.* **2010**, *110*, 1299–1305.

34. Pfaffenbach, K.T.; Lee, A.S. The critical role of grp78 in physiologic and pathologic stress. *Curr. Opin. Cell Biol.* **2011**, *23*, 150–156.

35. Lee, A.S. Grp78 induction in cancer: Therapeutic and prognostic implications. *Cancer Res.* **2007**, *67*, 3496–3499.

36. Dong, D.; Stapleton, C.; Luo, B.; Xiong, S.; Ye, W.; Zhang, Y.; Jhaveri, N.; Zhu, G.; Ye, R.; Liu, Z.; *et al.* A critical role for grp78/bip in the tumor microenvironment for neovascularization during tumor growth and metastasis. *Cancer Res.* **2011**, *71*, 2848–2857.

37. Wang, G.; Yang, Z.Q.; Zhang, K. Endoplasmic reticulum stress response in cancer: Molecular mechanism and therapeutic potential. *Am. J. Transl. Res.* **2010**, *2*, 65–74.

38. Pyrko, P.; Schonthal, A.H.; Hofman, F.M.; Chen, T.C.; Lee, A.S. The unfolded protein response regulator grp78/bip as a novel target for increasing chemosensitivity in malignant gliomas. *Cancer Res.* **2007**, *67*, 9809–9816.

39. Azatian, A.; Yu, H.; Dai, W.; Schneiders, F.I.; Botelho, N.K.; Lord, R.V. Effectiveness of hsv-tk suicide gene therapy driven by the grp78 stress-inducible promoter in esophagogastric junction and gastric adenocarcinomas. *J. Gastrointest. Surg.* **2009**, *13*, 1044–1051.

40. Kia, A.; Przystal, J.M.; Nianiaris, N.; Mazarakis, N.D.; Mintz, P.J.; Hajitou, A. Dual systemic tumor targeting with ligand-directed phage and grp78 promoter induces tumor regression. *Mol. Cancer Ther.* **2012**, *11*, 2566–2577.

41. Pasqualini, R.; Ruoslahti, E. Organ targeting *in vivo* using phage display peptide libraries. *Nature* **1996**, *380*, 364–366.

42. Hajitou, A.; Pasqualini, R.; Arap, W. Vascular targeting: Recent advances and therapeutic perspectives. *Trends Cardiovasc. Med.* **2006**, *16*, 80–88.

43. Folkman, J. Addressing tumor blood vessels. *Nat. Biotechnol.* **1997**, *15*, 510, doi:10.1038/nbt0697-510.

44. Li, S.D.; Huang, L. Gene therapy progress and prospects: Non-viral gene therapy by systemic delivery. *Gene Ther.* **2006**, *13*, 1313–1319.

45. Waehler, R.; Russell, S.J.; Curiel, D.T. Engineering targeted viral vectors for gene therapy. *Nat. Rev. Genet.* **2007**, *8*, 573–587.

46. Bouard, D.; Alazard-Dany, D.; Cosset, F.L. Viral vectors: From virology to transgene expression. *Br. J. Pharmacol.* **2009**, *157*, 153–165.

47. Kay, M.A. State-of-the-art gene-based therapies: The road ahead. *Nat. Rev. Genet.* **2011**, *12*, 316–328.

48. Green, N.K.; Seymour, L.W. Adenoviral vectors: Systemic delivery and tumor targeting. *Cancer Gene Ther.* **2002**, *9*, 1036–1042.

49. McCormick, F. Onyx-015 selectivity and the p14arf pathway. *Oncogene* **2000**, *19*, 6670–6672.

50. Oosterhoff, D.; Pinedo, H.M.; Witlox, M.A.; Carette, J.E.; Gerritsen, W.R.; van Beusechem, V.W. Gene-directed enzyme prodrug therapy with carboxylesterase enhances the anticancer efficacy of the conditionally replicating adenovirus addelta24. *Gene Ther.* **2005**, *12*, 1011–1018.

51. Duan, D.; Sharma, P.; Yang, J.; Yue, Y.; Dudus, L.; Zhang, Y.; Fisher, K.J.; Engelhardt, J.F. Circular intermediates of recombinant adeno-associated virus have defined structural characteristics responsible for long-term episomal persistence in muscle tissue. *J. Virol.* **1998**, *72*, 8568–8577.

52. Buning, H.; Perabo, L.; Coutelle, O.; Quadt-Humme, S.; Hallek, M. Recent developments in adeno-associated virus vector technology. *J. Gene Med.* **2008**, *10*, 717–733.

53. Lang, L.H. Fda approves use of bacteriophages to be added to meat and poultry products. *Gastroenterology* **2006**, *131*, 1370, doi:10.1053/j.gastro.2006.10.012.

54. Hajitou, A.; Trepel, M.; Lilley, C.E.; Soghomonyan, S.; Alauddin, M.M.; Marini, F.C., 3rd; Restel, B.H.; Ozawa, M.G.; Moya, C.A.; Rangel, R.; *et al.* A hybrid vector for ligand-directed tumor targeting and molecular imaging. *Cell* **2006**, *125*, 385–398.

55. Hajitou, A.; Rangel, R.; Trepel, M.; Soghomonyan, S.; Gelovani, J.G.; Alauddin, M.M.; Pasqualini, R.; Arap, W. Design and construction of targeted aavp vectors for mammalian cell transduction. *Nat. Protoc.* **2007**, *2*, 523–531.

56. Tandle, A.; Hanna, E.; Lorang, D.; Hajitou, A.; Moya, C.A.; Pasqualini, R.; Arap, W.; Adem, A.; Starker, E.; Hewitt, S.; *et al.* Tumor vasculature-targeted delivery of tumor necrosis factor-alpha. *Cancer* **2009**, *115*, 128–139.

57. Paoloni, M.C.; Tandle, A.; Mazcko, C.; Hanna, E.; Kachala, S.; Leblanc, A.; Newman, S.; Vail, D.; Henry, C.; Thamm, D.; *et al.* Launching a novel preclinical infrastructure: Comparative oncology trials consortium directed therapeutic targeting of tnfalpha to cancer vasculature. *PLoS One* **2009**, *4*, e4972.

58. Trepel, M.; Stoneham, C.A.; Eleftherohorinou, H.; Mazarakis, N.D.; Pasqualini, R.; Arap, W.; Hajitou, A. A heterotypic bystander effect for tumor cell killing after adeno-associated virus/phage-mediated, vascular-targeted suicide gene transfer. *Mol. Cancer Ther.* **2009**, *8*, 2383–2391.

59. Grassi, G.; Maccaroni, P.; Meyer, R.; Kaiser, H.; D'Ambrosio, E.; Pascale, E.; Grassi, M.; Kuhn, A.; di Nardo, P.; Kandolf, R.; *et al.* Inhibitors of DNA methylation and histone deacetylation activate cytomegalovirus promoter-controlled reporter gene expression in human glioblastoma cell line u87. *Carcinogenesis* **2003**, *24*, 1625–1635.

60. Das, P.M.; Singal, R. DNA methylation and cancer. *J. Clin. Oncol.* **2004**, *22*, 4632–4642.

61. Chen, W.Y.; Townes, T.M. Molecular mechanism for silencing virally transduced genes involves histone deacetylation and chromatin condensation. *Proc. Natl. Acad. Sci. USA* **2000**, *97*, 377–382.

62. Kia, A.; Yata, T.; Hajji, N.; Hajitou, A. Inhibition of histone deacetylation and DNA methylation improves gene expression mediated by the adeno-associated virus/phage in cancer cells. *Viruses* **2013**, *5*, 2561–2572.

63. Hsu, C.C.; Li, H.P.; Hung, Y.H.; Leu, Y.W.; Wu, W.H.; Wang, F.S.; Lee, K.D.; Chang, P.J.; Wu, C.S.; Lu, Y.J.; *et al.* Targeted methylation of cmv and e1a viral promoters. *Biochem. Biophys. Res. Commun.* **2010**, *402*, 228–234.

Effect of Bacteriophage Infection in Combination with Tobramycin on the Emergence of Resistance in *Escherichia coli* and *Pseudomonas aeruginosa* Biofilms

Lindsey B. Coulter [1], **Robert J. C. McLean** [2], **Rodney E. Rohde** [1] and **Gary M. Aron** [2],*

[1] Clinical Laboratory Science Program, Texas State University, 601 University Drive, San Marcos, TX 78666, USA; E-Mails: lcoulter87@gmail.com (L.B.C.); rrohde@txstate.edu (R.E.R.)

[2] Department of Biology, Texas State University, San Marcos, 601 University Drive, TX 78666, USA; E-Mail: McLean@txstate.edu (R.J.C.M.)

* Author to whom correspondence should be addressed; E-Mail: garyaron@txstate.edu

External Editor: Rob Lavigne

Abstract: Bacteriophage infection and antibiotics used individually to reduce biofilm mass often result in the emergence of significant levels of phage and antibiotic resistant cells. In contrast, combination therapy in *Escherichia coli* biofilms employing T4 phage and tobramycin resulted in greater than 99% and 39% reduction in antibiotic and phage resistant cells, respectively. In *P. aeruginosa* biofilms, combination therapy resulted in a 60% and 99% reduction in antibiotic and PB-1 phage resistant cells, respectively. Although the combined treatment resulted in greater reduction of *E. coli* CFUs compared to the use of antibiotic alone, infection of *P. aeruginosa* biofilms with PB-1 in the presence of tobramycin was only as effective in the reduction of CFUs as the use of antibiotic alone. The study demonstrated phage infection in combination with tobramycin can significantly reduce the emergence of antibiotic and phage resistant cells in both *E. coli* and *P. aeruginosa* biofilms, however, a reduction in biomass was dependent on the phage-host system.

Keywords: biofilms; bacteriophage; antibiotic; resistance; mixed therapy

1. Introduction

Bacterial biofilms are populations of cells adherent to an abiotic or biotic surface that can grow to be several millimeters thick [1]. Biofilms, such as those formed by *Escherichia coli* and *Pseudomonas aeruginosa*, have been found on medical implants, such as catheters and artificial hips [2,3], as well as pulmonary infections within lungs of cystic fibrosis patients [2,4,5]. The community of cells are coated in a sticky matrix made of extracellular DNA, secreted proteins, and polysaccharides [4,6,7] collectively called extracellular polymeric substances (EPS) that allow for the adherence to surfaces as well as provide protection from antimicrobial agents [8]. Biofilms have been shown to be highly tolerant to antimicrobials [4,9], a feature that was first ascribed to the ability of the biofilm matrix to restrict antimicrobial penetration [10]. Other mechanisms that have been described include the establishment of slow-growing antibiotic-resistant subpopulations within biofilms (persister cells) [11], and biofilm-specific gene expression [12].

Infections due to biofilms are difficult to treat because of their high tolerance to antimicrobials [13]. Bacteriophage therapy has been suggested for the treatment of biofilms [14]. Although effective in decreasing biofilm mass, phage therapy alone has not been shown to eradicate biofilms [14,15]. Phage therapy was a common method of treating infections before replaced by the discovery of antibiotics, and is still being used in parts of Eastern Europe for the treatment of a variety of afflictions ranging from acne and urinary tract infections to methicillin-resistant *Staphylococcus aureus* [14,15]. A major concern of phage therapy is the development of phage resistance. To reduce the chance of resistant cells from emerging, combination therapy is often used which is more effective than the use of a single treatment [16,17]. Tré-Hardy *et al.* [16] demonstrated the use of tobramycin in combination with clarithromycin to be more effective in decreasing mature *P. aeruginosa* biofilms and Parra-Ruiz *et al.*, [18] found combinations of daptomycin or moxifloxacin with clarithromycin to be effective at decreasing *Staphylococcus aureus* biofilms. In phage therapy, Fu *et al.* [19] found pre-treating catheters with a cocktail of bacteriophage can decrease the formation of *P. aeruginosa* biofilms compared to the use of a single phage. There have also been reports on phage and antibiotic combinational therapy on biofilms. Bedi *et al.* [20] found amoxicillin and lytic phage on *Klebsiella pneumoniae* B5055 biofilms produced a greater reduction in biofilm (~1–2 log) compare to either treatment alone. Verma *et al.* [21] continued the work on *K. pneumoniae* B5055 using ciprofloxacin and depolymerase-producing phage KPO1K2 and found the combination to be significantly more effective. *S. aureus* biofilms treated with phage SAP-26 and rifampicin reduced the biofilm by ~5 logs, while phage SAP-26 and rifampicin only reduced the biofilm by ~3 log and ~4 log, respectively [22]. There are no reports on the emergence of resistance in biofilms treated with a combination of phage and antibiotics. This study compares the effect of bacteriophage infection with tobramycin on the emergence of phage and antibiotic resistant cells and biofilm survival. The results demonstrated in both *E. coli* and *P. aeruginosa* biofilms phage infection in combination with tobramycin reduced the emergence of antibiotic and phage resistant cells, however, biomass reduction was dependent on the phage-host system.

2. Materials and Methods

2.1. Bacteria and Bacteriophage

Escherichia coli B (ATCC 11303) and *Pseudomonas aeruginosa* PAO1 (obtained from V. Deretic, University of New Mexico) were grown in Luria-Bertani (LB) broth (Accumedia Manufacturers, Inc., Lansing, Michigan, MI, USA) at 37 °C in an orbital rotating shaker water bath (Lab-Line Instruments, Inc. model 3540 Orbital Shaker Bath, Melrose Park, IL, USA). Bacteriophage T4 (ATCC 11303-B4) was used to infect *E. coli* and bacteriophage PB-1 [23] (ATCC 15692-B3) was used to infect *P. aeruginosa*. Bacteriophage T4 and PB-1 stocks were prepared by infecting early log phase of *E. coli* or *P. aeruginosa*, respectively, at a multiplicity of infection (MOI) of approximately 1000. Infected cultures were placed into a 37 °C reciprocal shaking water bath (Blue M Electric Company Magni Whirl MSB-1122A-1 Shaker Bath, Blue Island, IL, USA) until the cultures cleared, or 2 h. Infected cultures were then placed into a glass centrifuge tube with 0.5 mL of chloroform at 4 °C. After 5 min at 4 °C the infected cultures were shaken for 1 min and placed at 4 °C for 5 min. The cultures were centrifuged (Eppendorf Centrifuge model 5810 R, Hamburg, Germany) at 3000 × *g* for 20 min at 4 °C. The supernatant was filtered (0.45 µm) and phage titers were determined by a soft-agar overlay plaque assay [24].

2.2. Antibiotic Preparation

Tobramycin (T4014, Sigma-Aldrich Co., St. Louis, MO, USA) stock solution of 10 mg.mL^{-1} were prepared by diluting tobramycin in deionzed water and filter sterilizing (0.22 µm; Fisher 25 mm syringe filter; ThermoFisher Scientific Inc., Waltham, MA, USA).

2.3. Biofilm Growth

Silicone rubber disks, 7 × 1 mm (Dapro Rubber Inc., Tulsa, OK, USA), were placed into 50 mL of LB broth and inoculated with overnight cultures of *E. coli* or *P. aeruginosa* to a cell density 10^6 CFU.mL^{-1} (OD$_{600nm}$ = 0.1). Monocultures were incubated in an orbital shaking water bath at 100 rpm for 48 h at 37 °C. Biofilm growth was measured by removing colonized disks from the culture vessel, dipping in sterile phosphate buffered saline (PBS) (Sigma-Aldrich, Co., St Louis, MO, USA) to remove unattached bacteria, then placing in a scintillation vial containing 5 mL H$_2$O and using the bath sonication, dilution plating protocol described by Corbin *et al.* [25].

2.4. Determination of MOI and Antibiotic Concentration on Cell Survival

E. coli and *P. aeruginosa* biofilms were grown as described previously. After 48 h, disks containing biofilms were rinsed in phosphate buffered saline (PBS) (Sigma-Aldrich, Co., St Louis, MO, USA) and placed in individual 20 mL scintillation vials containing LB broth and phage at various MOI (0.0001–10) or tobramycin (0.25–4 µg.mL^{-1}) [26]. Treated biofilms were incubated in a reciprocal shaking water bath at 100 rpm for 24 h at 37 °C. Biofilm enumeration following treatments were performed as described in Section 2.3. Cell density was determined by dilution plating on LB agar plates.

2.5. Antibiotic Treatment

Forty-eight-hour biofilms were gently rinsed with 5 mL of PBS in sterile test tubes to remove planktonic cells. Biofilms were placed in individual 20 mL scintillation vials containing 10 mL of LB broth with 2 $\mu g.mL^{-1}$ of tobramycin for *E. coli* and 0.5 $\mu g.mL^{-1}$ of tobramycin for *P. aeruginosa*. Biofilms were incubated at 37 °C at 70 cycles per minute in a reciprocal shaking water bath. After 24-hour exposure to tobramycin, the biofilm-coated disks were removed from the culture vial, dipped in sterile PBS to remove antibiotics and unattached cells, then placed into a scintillation vial containing 5 mL sterile PBS, and enumerated using the sonication and dilution plating protocol described in Section 2.3. Following dilution plating, cells were cultured on LB to determine the total number of cells and LB plus tobramycin (2 $\mu g.mL^{-1}$ for *E. coli* and 0.5 $\mu g.mL^{-1}$ for *P. aeruginosa*) to determine resistance.

2.6. Antibiotic and Bacteriophage Treatment

Forty-eight-hour biofilms were gently rinsed with 5 mL of PBS in sterile test tubes to remove planktonic cells. Biofilms were placed in individual 20 mL scintillation vials containing 10 mL of LB broth with tobramycin (2 $\mu g.mL^{-1}$) and T4 (MOI of 0.01) for *E. coli* and tobramycin (0.5 $\mu g.mL^{-1}$) and PB-1 (MOI of 0.01) for *P. aeruginosa*. Biofilms were incubated 70 cycles per minute in a reciprocal shaking water bath at 37 °C. After 24 h biofilm growth was measured using the sonication and dilution plating protocol described in Section 2.3. Tobramycin resistant cells were determined as described in Section 2.5. Phage resistant cells were determined by plating bacterial dilutions onto LB agar followed by an agar overlay containing 0.1 mL of 10^8 $PFU.mL^{-1}$ and measuring bacterial colonies following 24-hour incubation.

2.7. Data Analysis

A minimum of three biological replicates were performed for each experiment. CFU data was log transformed to ensure a normal distribution. When bacterial cell counts were less than the detection limit (<20 $CFU.mL^{-1}$), the experiment was repeated, and, if the result was confirmed, the CFU data was recorded as 1 ($\log_{10} (1) = 0$). Data was analyzed by ANOVA using Sigma Plot v 12.5 [27] and Holm-Sidak *post hoc* tests were used to compare the effects of antibiotic and phage treatment results against untreated controls.

3. Results and Discussion

3.1. Effect of Antibiotic and Phage Concentrations on Cell Survival

E. coli and *P. aeruginosa* 48 h biofilms were treated with varying concentrations of tobramycin (0.25–4 $\mu g.mL^{-1}$) and phage concentrations (from 10^{-4} to 10^1 multiplicity of infection (MOI)) for 24 h followed by viable cell assays to determine cell survival. During preliminary experiments [28], antibiotic and phage concentrations were chosen so that individual treatments resulted in a ten-fold reduction (*i.e.*, 90% kill) in CFU, and potential phage-antibiotic interactions could be assessed. As a result of preliminary tests, antibiotic concentrations of 2 $\mu g.mL^{-1}$ and 0.5 $\mu g.mL^{-1}$ were used to treat

E. coli and *P. aeruginosa* biofilms respectively. Similarly, a MOI of 0.01 was used for phage treatment of *E. coli* biofilms (T4 phage) and *P. aeruginosa* biofilms (PB-1 phage).

3.2. Effect of Tobramycin and T4 on E. Coli *Biofilm Cell Survival*

E. coli biofilms were exposed to tobramycin or a combination of tobramycin and T4 for 24 h followed by viable cell counts to determine the number of surviving cells (Table 1 and Figure 1). Tobramycin or the combination resulted in 2.1 ± 0.66 and −1.8 ± 0.43 CFU.mm^{-2} (log mean ± SE) of the biofilm remaining, respectively. The combination of phage and antibiotic led to ~99.99% decrease on the survival of *E. coli* biofilms compared to the use of tobramycin alone. Similar results have been found in studies on *Klebsiella pneumoniae* and *Staphylococcus aureus* biofilms using an antibiotic combined with phage [20–22].

Table 1. Survival of *E. coli* and *P. aeruginosa* biofilm cells treated with bacteriophage and tobramycin. Untreated (control) values are also listed.

	E. coli			**P. aeruginosa**		
Time (h)			Log CFU.mm^{-2} ± SE			
6 (Control)		4.39 ± 0.16			2.96 ± 0.15	
24 (Control)		4.48 ± 0.15			3.63 ± 0.15	
Treatment	Tobramycin [a]	T4 [a]	Tob + T4 [a]	Tobramycin [b]	PB-1 [b]	Tob + PB-1 [b]
6	−0.27 ± 0.55	−0.10 ± 0.21	−1.6 ± 0.32	2.7 ± 0.30	2.4 ± 0.21	2.0 ± 0.35
24	2.1 ± 0.66	−0.79 ± 0.20	−1.8 ± 0.43	1.8 ± 0.36	3.2 ± 0.17	1.6 ± 0.33

[a]: log Mean ± SE (*n* = 12); [b]: log Mean ± SE (*n* = 15).

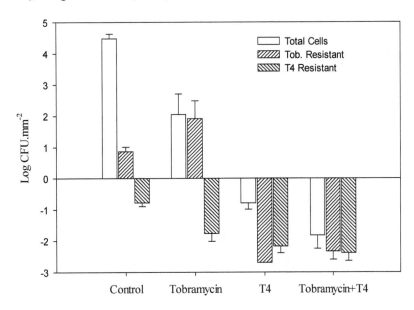

Figure 1. Effect of 24-hour exposure of preformed 48-hour-old *E. coli* biofilms to tobramycin, T4, and the combination of tobramycin and T4 phage on total cell concentrations and concentrations of tobramycin and phage (T4) resistant populations. The differences of bacterial populations (total cells), as well as tobramycin-resistant and phage-resistant populations, were all significantly different (*p* < 0.05) from their respective untreated (control) values. Bars indicate standard error of the mean.

Biology and Applications of Bacteriophages

3.3. Determination of E. coli Biofilm Cell Resistance

Following 24 h exposure to tobramycin, T4, or their combination, biofilms were assayed to determine tobramycin resistant cells and T4 resistant cells (Table 2 and Figure 1). There was a >99.99% decrease observed in tobramycin resistant cells and a 39% decrease in the number of T4 resistant cells in the combination treatment.

Table 2. Resistance to tobramycin and bacteriophage in *E. coli* and *P. aeruginosa* biofilms.

	E. coli		*P. aeruginosa*	
% decrease	Tobramycin [a]	T4 [b]	Tobramycin [c]	PB-1 [d]
in resistance	>99.99%	39%	60%	99%

[a]: % decrease in resistance following Tob + T4 treatment compared to tobramycin alone; [b]: % decrease in resistance following Tob + T4 treatment compared to T4 alone; [c]: % decrease in resistance following Tob + PB-1 treatment compared to tobramycin alone; [d]: % decrease in resistance following Tob + PB-1 treatment compare to PB-1 alone.

3.4. Effect of Tobramycin and PB-1 on P. aeruginosa *Biofilm Cell Survival*

P. aeruginosa biofilms were exposed to tobramycin or a combination of tobramycin and PB-1 for 24 h and the number of surviving cells were determined (Table 1 and Figure 2). Treatment with tobramycin or the combination resulted in 1.8 ± 0.36 and 1.6 ± 0.33 CFU.mm^{-2} (log Mean ± SE) remaining, respectively. The combination of tobramycin and PB-1 on *P. aeruginosa* biofilms was just as effective as tobramycin alone in decreasing biofilm mass. The high level of resistance observed to phage PB-1 in *P. aeruginosa* biofilms may be due to EPS restricting the ability of the phage to access receptors [29].

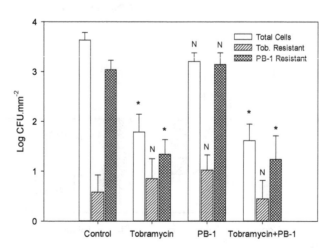

Figure 2. Effect of 24-hour exposure of preformed 48 h old *P. aeruginosa* biofilms to tobramycin, PB-1, and the combination of tobramycin and PB-1 on total cell concentrations and concentrations of tobramycin (Tob.) and phage (PB-1) resistant populations. Differences between untreated control values Figureand treated values are denoted by * (statistically significant $p < 0.05$) or N (not statistically significant, $p > 0.05$). Bars indicate standard error of the mean.

3.5. Determination of P. aeruginosa Biofilm Cell Resistance

Following 24-hour exposure to tobramycin, PB-1, and their combination, biofilms were assayed to determine tobramycin resistant cells and PB-1 resistant cells (Table 2 and Figure 2). To determine the percent decrease in resistant cells, the number of resistant cells in the combination treatment was compared to the number of resistant cells in the tobramycin and PB-1 treatment. There was a 60% decrease in the number of tobramycin resistant cells and a 99% decrease in the number of PB-1 resistant cells in the combination treatment.

The current study found the combination of tobramycin and T4 on *E. coli* biofilms led to a >99.99% decrease in tobramycin resistant cells and a 39% decrease in T4 resistant cells compared to use the use of tobramycin or phage alone, respectively. Treating *P. aeruginosa* biofilms with the combination of tobramycin and PB-1 resulted in a 60% decrease in tobramycin resistant cells and a 99% decrease in phage resistant cells compared to the use of tobramycin or phage alone, respectively. These results agree with other studies which showed that combinations of either antibiotics alone or phage alone can result in a decrease in the emergence of resistant cells [16,17,19].

As stated earlier, there is considerable work showing how phage can move through bacterial EPS components of biofilm matrices, and attach to cell surface targets by virtue of their production of depolymerase enzymes [21,29,30]. However, most of these studies are associated with pure bacterial cultures. One issue that needs to be addressed in future work is the efficacy of combination antibiotic-phage therapy in treating polymicrobial biofilm infections. Polymicrobial infections including those associated with wounds [31,32] are often much more recalcitrant to antibiotic treatment than are monoculture infections. In a recent study, we observed that a phage-susceptible bacterial species can be protected within a mixed culture biofilm by virtue of the phage-host bacterium association with a different species that would produce a chemically-different EPS, not susceptible to the highly specific phage depolymerase [33]. In this context, future studies on combination therapy involving bacteriophage should explore the effectiveness against mixed cultures.

Acknowledgments

This work was funded by the Homer E. Prince Microbiology Endowment to GMA and a Texas State University research enhancement grant to RJCM. We thank F. Weckerley for help with statistical analysis.

Author Contributions

Lindsey B. Coulter, Robert J. C. McLean, Rodney E. Rohde, and Gary M. Aron conceived and designed the experiments; LC conducted most of the experiments; Lindsey B. Coulter, Coulter, Robert J. C. McLean and Gary M. Aron did the experimental analysis; and Lindsey B. Coulter, Robert J. C. McLean, Rodney E. Rohde, and Gary M. Aron wrote the manuscript.

References

1. Costerton, J.W.; Cheng, K.J.; Geesey, G.G.; Ladd, T.I.; Nickel, J.C.; Dasgupta, M.; Marrie, T.J. Bacterial biofilms in nature and disease. *Annu. Rev. Microbiol.* **1987**, *41*, 435–464.
2. Donlan, R.M. Biofilms and device-associated infections. *Emerg. Infect. Dis.* **2001**, *7*, 277–281.
3. Costerton, J.W.; Stewart, P.S.; Greenberg, E.P. Bacterial biofilms: A common cause of persistent infections. *Science* **1999**, *284*, 1318–1322.
4. De la Fuente-Núñez, C.; Reffuveille, F.; Fernandez, L.; Hancock, R.E.W. Bacterial biofilm development as a multicellular adaptation: Antibiotic resistance and new therapeutic strategies. *Curr. Opin. Microbiol.* **2013**, *16*, 580–589.
5. Singh, P.K.; Schaefer, A.L.; Parsek, M.R.; Moninger, T.O.; Welsh, M.J.; Greenberg, E.P. Quorum-sensing signals indicate that cystic fibrosis lungs are infected with bacterial biofilms. *Nature* **2000**, *407*, 762–764.
6. Steinberger, R.E.; Holden, P.A. Extracellular DNA in single- and multiple-species unsaturated biofilms. *Appl. Environ. Microbiol.* **2005**, *71*, 5404–5410.
7. Absalon, C.; Ymele-Leki, P.; Watnick, P.I. The bacterial biofilm matrix as a platform for protein delivery. *MBio* **2012**, *3*, e00127–12, doi:10.1128/mBio.00127-12.
8. Stewart, P.S.; Costerton, J.W. Antibiotic resistance of bacteria in biofilms. *Lancet* **2001**, *358*, 135–138.
9. Nickel, J.C.; Ruseska, I.; Wright, J.B.; Costerton, J.W. Tobramycin resistance of *Pseudomonas aeruginosa* cells growing as a biofilm on urinary catheter material. *Antimicrob. Agents Chemother.* **1985**, *27*, 619–624.
10. Hoyle, B.D.; Alcantara, J.; Costerton, J.W. *Pseudomonas aeruginosa* biofilm as a diffusion barrier to piperacillin. *Antimicrob. Agents Chemother.* **1992**, *36*, 2054–2056.
11. Lewis, K. Persister cells, dormancy and infectious disease. *Nat. Rev. Microbiol.* **2007**, *5*, 48–56.
12. Gupta, K.; Marques, C.N.H.; Petrova, O.E.; Sauer, K. Antimicrobial tolerance of *Pseudomonas aeruginosa* biofilms is activated during an early developmental stage and requires the two-component hybrid SagS. *J. Bacteriol.* **2013**, *195*, 4975–4981.
13. Hoiby, N. A personal history of research on microbial biofilms and biofilm infections. *Pathog. Dis.* **2014**, *70*, 205–211.
14. Krylov, V.N. Bacteriophages of *Pseudomonas aeruginosa*: Long-term prospects for use in phage therapy. *Adv. Virus Res.* **2014**, *88*, 227–278.
15. Sulakvelidze, A.; Alavidze, Z.; Morris, J.G., Jr. Bacteriophage therapy. *Antimicrob. Agents Chemother.* **2001**, *45*, 649–659.
16. Tré-Hardy, M.; Nagant, C.; El Manssouri, N.; Vanderbist, F.; Traore, H.; Vaneechoutte, M.; Dehaye, J.P. Efficacy of the combination of tobramycin and a macrolide in an *in vitro Pseudomonas aeruginosa* mature biofilm model. *Antimicrob. Agents Chemother.* **2010**, *54*, 4409–4415.
17. Bonhoeffer, S.; Lipsitch, M.; Levin, B.R. Evaluating treatment protocols to prevent antibiotic resistance. *Proc. Natl. Acad. Sci. USA* **1997**, *94*, 12106–12111.
18. Parra-Ruiz, J.; Vidaillac, C.; Rose, W.E.; Rybak, M.J. Activities of high-dose daptomycin, vancomycin, and moxifloxacin alone or in combination with clarithromycin or rifampin in a novel *in vitro* model of *Staphylococcus aureus* biofilm. *Antimicrob. Agents Chemother.* **2010**, *54*, 4329–4334.

19. Fu, W.; Forster, T.; Mayer, O.; Curtin, J.J.; Lehman, S.M.; Donlan, R.M. Bacteriophage cocktail for the prevention of biofilm formation by *Pseudomonas aeruginosa* on catheters in an *in vitro* model system. *Antimicrob. Agents Chemother.* **2010**, *54*, 397–404.

20. Bedi, M.S.; Verma, V.; Chhibber, S. Amoxicillin and specific bacteriophage can be used together for eradication of biofilm of *Klebsiella pneumoniae* B5055. *World J. Microbiol. Biotechnol.* **2009**, *25*, 1145–1151.

21. Verma, V.; Harjai, K.; Chhibber, S. Structural changes induced by a lytic bacteriophage make ciprofloxacin effective against older biofilm of *Klebsiella pneumoniae*. *Biofouling* **2010**, *26*, 729–737.

22. Rahman, M.; Kim, S.; Kim, S.M.; Seol, S.Y.; Kim, J. Characterization of induced *Staphylococcus aureus* bacteriophage SAP-26 and its anti-biofilm activity with rifampicin. *Biofouling* **2011**, *27*, 1087–1093.

23. Bradley, D.E.; Robertson, D. The structure and infective process of a contractile *Pseudomonas aeruginosa* bacteriophage. *J. Gen Virol* **1968**, *3*, 247–254.

24. Adams, M.H. *Bacteriophages*; Interscience Publishers: New York, NY, USA, 1959.

25. Corbin, B.D.; McLean, R.J.C.; Aron, G.M. Bacteriophage T4 multiplication in a glucose-limited *Escherichia coli* biofilm. *Can J. Microbiol.* **2001**, *47*, 680–684.

26. Ceri, H.; Olson, M.E.; Stremick, C.; Read, R.R.; Morck, D.W.; Buret, A. The Calgary Biofilm Device: New technology for rapid determination of antibiotic susceptibilities of bacterial biofilms. *J. Clin. Microbiol.* **1999**, *37*, 1771–1776.

27. *SigmaPlot* version 12.5. Systat Software Inc., San José, CA, USA, **2013**.

28. Coulter, L.B. Synergistic effect of antibiotics and bacteriophage infection on *Escherichia coli* and *Pseudomonas aeruginosa* mixed biofilm communities. MS Thesis, Texas State University, San Marcos, TX, USA, **2012**.

29. Hanlon, G.W.; Denyer, S.P.; Olliff, C.J.; Ibrahim, L.J. Reduction in exopolysaccharide viscosity as an aid to bacteriophage penetration through *Pseudomonas aeruginosa* biofilms. *Appl. Environ. Microbiol.* **2001**, *67*, 2746–2753.

30. Sutherland, I.W.; Hughes, K.A.; Skillman, L.C.; Tait, K. The interaction of phage and biofilms. *FEMS Microbiol. Lett.* **2004**, *232*, 1–6.

31. Korgaonkar, A.; Trivedi, U.; Rumbaugh, K.; Whiteley, M. Community surveillance enhances *Pseudomonas aeruginosa* virulence during polymicrobial infection. *Proc. Natl. Acad. Sci. USA* **2012**, *110*, 1059–1064.

32. Watters, C.; Everett, J.A.; Haley, C.; Clinton, A.; Rumbaugh, K. Insulin treatment modulates the host immune system to enhance *Pseudomonas aeruginosa* wound biofilms. *Infect. Immun.* **2014**, *82*, 92–100.

33. Kay, M.K.; Erwin, T.C.; McLean, R.J.C.; Aron, G.M. Bacteriophage ecology in Escherichia coli and Pseudomonas aeruginosa mixed biofilm communities. *Appl. Environ. Microbiol.* **2011**, *77*, 821–829.

5

The Role of the Coat Protein A-Domain in P22 Bacteriophage Maturation

David S. Morris and Peter E. Prevelige, Jr. *

Department of Microbiology, University of Alabama at Birmingham, 845 19th Street S, BBRB 414, Birmingham, AL 35294, USA; E-Mail: davidsm@uab.edu

* Author to whom correspondence should be addressed; E-Mail: prevelig@uab.edu

Abstract: Bacteriophage P22 has long been considered a hallmark model for virus assembly and maturation. Repurposing of P22 and other similar virus structures for nanotechnology and nanomedicine has reinvigorated the need to further understand the protein-protein interactions that allow for the assembly, as well as the conformational shifts required for maturation. In this work, gp5, the major coat structural protein of P22, has been manipulated in order to examine the mutational effects on procapsid stability and maturation. Insertions to the P22 coat protein A-domain, while widely permissive of procapsid assembly, destabilize the interactions necessary for virus maturation and potentially allow for the tunable adjustment of procapsid stability. Future manipulation of this region of the coat protein subunit can potentially be used to alter the stability of the capsid for controllable disassembly.

Keywords: procapsid; bacteriophage; maturation; recombineering

1. Introduction

Bacteriophage P22 is a dsDNA virion of the *Podoviridae* family that infects *Salmonella enterica* serovar *typhimurium*. In the native host, co-expression of the major structural proteins, coat, portal and scaffolding (gp5, gp1 and gp8, respectively), among other minor proteins, results in the assembly of a T = 7 L icosahedral quasi-equivalent procapsid structure 50 nm in diameter [1–3]. Packaging of the phage DNA into the procapsid results in the exit of scaffolding protein and the subsequent expansion of the capsid shell (Figure 1A). This maturation process requires significant changes in the intersubunit contacts of the phage coat protein, including the closing off of a "pore" in the center of each coat protein capsomere (Figure 1B). Recently, phage procapsids have been utilized for a number of alternative purposes, including biomedical applications, such as nanoparticle targeted drug delivery [4–6]. To make P22, an effective vector for nanomedicine, we were interested in developing a phage display system on the surface of the P22 capsid by manipulating sequences of the P22 coat protein *in vivo*. The A-domain of the coat protein was chosen for these manipulations, because previous work has described this loop as flexible and solvent exposed, and manipulations of this region at residue T183 *in vitro* were highly permissive of procapsid assembly [7,8]. Residue T183 was chosen in these studies, because models of the P22 procapsid structure place this residue at the most axial position in each capsomere. While several A-domain peptide insertion sequences had been examined *in vitro*, the focus initially was on the tri-peptide sequence: RGD. The RGD sequence (Arg-Gly-Asp) interacts preferentially with the $\alpha_V\beta_3$ integrin, a common cell surface receptor [9,10]. While $\alpha_V\beta_3$ integrin is expressed on many cell surfaces, it is highly upregulated in cancerous cells and has been examined as a potential target for the drug delivery of toxic chemotherapies. As a proof of concept for developing the phage display library, the RGD sequence flanked by two tri-glycine linkers (GGGRGDGGG) was inserted at residue T183 into the A-domain of P22 coat protein, and the resulting procapsids were thoroughly characterized, both *in vitro* through a BL21 expression system that produces non-infectious P22 procapsids, as well as *in vivo* utilizing a stable lysogen of P22 and lambda-red recombineering. Upon finding that the RGD sequence allowed for assembly, but inhibited virus maturation, insertions of decreasing complexity were made in the A-domain in order to determine the sequence permissiveness of the region. Our results show how the manipulation of the coat protein *in vivo* can be used to study the biophysical properties of virus maturation. The insight gained from this work indicates that structural changes in the A-domain of the P22 coat protein can controllably manipulate the ability of the phage to mature, while allowing for procapsid assembly.

2. Materials and Methods

2.1. Producing P22 Procapsids

BL21 cells containing the fluorescent P22 assembler plasmids (GFP/131-303gp8-5 or mCHERRY/131-303gp8-5 contained within a pET11b vector) were grown to 0.6 OD under antibiotic selection [11]. Protein expression was induced by adding 1 mM IPTG to the culture and allowing for an additional 3.5–4 h incubation at 37 °C while shaking. Following expression, procapsids were purified as previously described. Pelleted cells were subjected to three freeze/thaw cycles and then lysed by sonication. Cell debris was cleared through centrifugation at 13,000 RPM for 1 h in a JA-20 rotor. Procapsids in solution were then pelleted through 20% sucrose at high speed (40 K 2 h 70Ti rotor) and re-suspended in Buffer B (50 mM Tris, 25 mM NaCl, 2 mM EDTA) prior to being subjected to an isopynic 5%–20% sucrose gradient designed to separate out procapsids from aggregated material (35 K, 35 min, SW55 rotor). The procapsid band was then isolated and dialyzed against Buffer B for short-term storage. The P22 assembler plasmids were mutated using the Quickchange site-directed mutagenesis method. The primers used are described in Table S1.

Figure 1. (**A**) Depiction of an immature (**left**) and mature (**right**) coat protein shell. Pentameric capsomeres are highlighted for reference. (**B**) Hexameric capsomeres of the immature (**left**) and the mature structure with the surface res1and the A-domain residue T183 highlighted. Upon maturation, the expansion of the virus results in a conformational shift that restricts the size of the pore in the middle of each capsomere (see Movie S1). The figures were generated from Protein Data Bank (PDB) 2XYZ and 2XYY using the UCSF Chimera package [12,13].

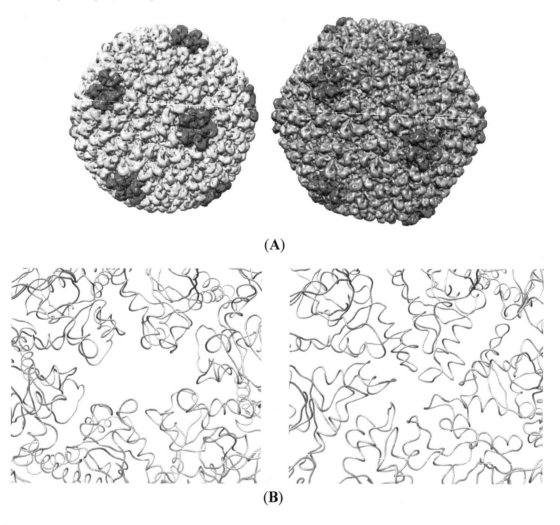

(**A**)

(**B**)

2.2. Heat Expansion Assay

Isolated procapsids were buffer exchanged to 50 mM NaPO$_4$, 1 mM MgCl$_2$, pH 7.4, and the concentration of gp5 was standardized to 20 mM by arithmetically removing the contribution to absorbance generated by the fluorescent scaffolding fusion. Fifty-microliter aliquots of the sample were heated at the times and temperatures described using thin-walled PCR tubes in a Bio-Rad iCycler themocycler held at a constant temperature. Post-heated samples were cooled to 4 °C and then run on a 1% agarose gel at 50 V for 1 h. Gels were stained overnight with Coomassie stain followed by destaining with a 10% acetic acid, 30% methanol solution until thoroughly destained. The quantification of bands was performed using ImageJ software by plotting intensities for each lane and integrating under each curve after baseline subtraction [14].

2.3. Recombineering

S. enterica serovar *typhimurium* LT2 strains containing stable P22 lysogens were obtained from Sherwood Casjens. UB-1757 (leuA–414, ΔFels2, r-, sup° containing P22 15–ΔSC302::KanR 13–amH101) and UB-2158 (leuA-414, ΔFels2, r-, sup°, ΔGalK:TetRA containing P22 c1-7, 13-amH101, ΔsieA-1, orf25::CamR-EG1) were used. Lamda-red recombineering [15,16] was performed on P22 gene 5 using the primers described in the supplementary figures. Either the TetRA tetracycline resistance cassette (Tn10dTc) or the galK galactose kinase cassette (from pgalK) was PCR amplified using primers with 3' end homology to the cassette and 40 bp 5' homology to the gp5 sequences specific to the A-domain (TetRP22for/rev or galKP22for/rev). The resulting amplicon was then PCR purified and electroporated into UB-1757 or UB-2158 electrocompetent cells, containing an arabinose induced PKD46 plasmid. Cells were then selected for tetracycline resistance or galK activity, and resulting colonies were tested by PCR and sequencing. Mutant sequences with flanking homology to the target gene were PCR amplified using the oligo extension technique, and subsequent lambda-red recombineering into the TetRA or galK interrupted strain allows for the restoration of the gene coding sequence with the mutation of interest. Selection for tetracycline sensitive strains was done twice on Bochner-Maloy media and reconfirmed by stippling onto tetracycline and kanamycin agar plates. Selection for ΔgalK strains was performed on M63 minimal media agar in the presence of glucose and 2-deoxy-galactose (2-DOG) as previously described [16].

2.4. Phage Induction and Titering

Induction of the lysogen containing *S. typhimurium* strains was performed by growing cells to 0.6 OD in LB followed by at least two hours of induction with carbodox (1 μg/mL) or mitomycin C (0.5 μg/mL) at 37 °C, unless mentioned otherwise. Cells were pelleted and incubated with a saturating volume of chloroform to complete lysis and release of infectious phage. Cell debris was pelleted by centrifugation, and the phage containing supernatant was isolated. Tailspike protein is poorly produced in the UB-1757 P22 lysogen, which required the addition of saturating amounts of the exogenously produced tailspike to get accurate titers. Phages were titered on the MS-1363 suppressor strain (leuA–am414 supE) on soft agar containing 10 mM citrate in order to assist with cell lysis. Supplementation of exogenous tailspike and plating in the presence of 10 mM citrate were not necessary for phage produced from the UB-2158 strain. Phage product was concentrated by centrifugation and sedimented through a 5%–20% sucrose gradient using the same methods as shown above for procapsids. Imaging of the fractions was performed by negative stain TEM (2% uranyl acetate) after dialysis into Buffer B.

3. Results

3.1. In Vitro Maturation by Heat Expansion

Heat expansion of P22 procapsids has been used previously to mimic the process of maturation *in vitro* [17]. Incubation of procapsids for 12 min at 65 °C produces expanded shells. Native gel electrophoresis can readily resolve unexpanded, expanded and wiffle ball forms. To characterize the ability of mutant RGD procapsids to mature *in vitro*, heat expansion followed by native agarose gel electrophoresis was performed. Shells standardized to 20 µM of P22 coat protein were heated in thin-walled PCR tubes at constant temperature and analyzed by migration on a 1% native agarose gel. Wild-type procapsids containing scaffolding protein behaved in a similar manner to previously published results [17] (Figure 2A).

Figure 2. (**A**) Heat expansion of wild-type procapsids *in vitro*. Three forms can be distinguished based on migration on native agarose gel. Immature procapsids migrate the fastest, followed by expanded shells and, finally, by the "wiffle ball" form. (**B**) RGD procapsids (T183GGGRGDGGG) were heat expanded at constant temperature and electrophoresed on native agarose. Bands associated with the migration of expanded capsids and unexpanded procapsids were quantified by densitometry using ImageJ, and the relative values were plotted. (**C**) Heating RGD procapsids at 72 °C results in a reduction of overall staining intensity, suggesting shell dissociation. The "wiffle ball" morphology is not observed.

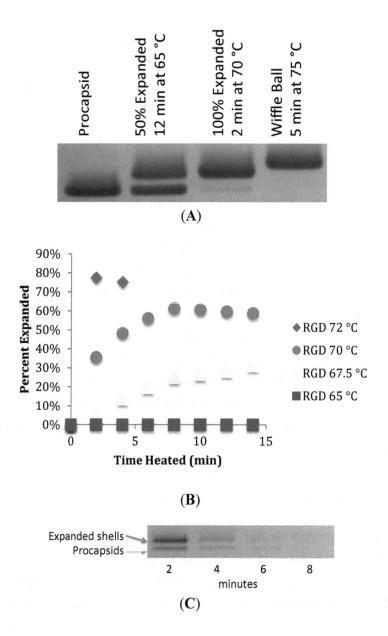

Heat expansion at 65 °C for 12 min results in approximately 50% expansion, while at 70 °C, full expansion occurs within 2 min. Heating at 75 °C results in the formation of the wiffle ball form, resulting in the loss of coat protein capsomeres at each of the five-fold vertices of the icosahedral T = 7 shell. In contrast, we did not observe expansion of the RGD procapsids when heated at 65 °C, even when heating was continued for 15 min (Figure 2B). To further define the kinetics of expansion and to determine the relative stability of the capsid, heat expansion was performed over a range of temperatures. The transition to the expanded form became apparent when RGD procapsids were heated at 67.5 °C and 70 °C. However, even at 70 °C, the transition was slower than was observed for the wild-type indicating, that the RGD insertion to the A-domain loop increases the activation energy required to transition to the expanded form. When heating at 72 °C or higher, the RGD procapsids were destabilized and dissociated completely (Figure 2C). There were no conditions at which the "wiffle ball" form of the RGD procapsid was observed. Taken together, these observations demonstrate that the RGD insertion at the A-domain loop both increases the energy required to transition to the mature-like morphology, while also destabilizing this form of the capsid.

3.2. Testing the A-Loop Insertion in Vivo

With the goal of utilizing the A-domain loop as a site for random insertions to develop a P22 phage display system, we inserted a tetracycline resistance cassette (tetRA) at residue T183 of the A-domain of the P22 lysogen by recombineering. Verification of the correct positioning of the cassette was performed by PCR using strategically-placed primers and determining the amplicon size (Figure S1). As expected, the induction of this lysogen did not result in the production of infectious phage. Amplicons encoding both the RGD sequence, the wild-type sequence and an alternatively-coded wild-type sequence (termed same-sense) were generated (Figure 3A) and recombineered back into the tetRA disrupted lysogen. The population of cells was induced, and the resultant phage titered (Figure 3B).

Both the wild-type and same-sense reactions produced phage with an efficiency of approximately three orders of magnitude less than the undisrupted lysogen. This presumably reflects the efficiency of the recombineering reaction. Sequencing of the isolated same-sense phage confirmed that the infectious phenotype was due to the insertion of the PCR product into the coding sequence. In contrast, no viable phage were detected in the RGD recombinant. To insure that recombination had occurred, counter-selection for tetracycline sensitivity using Bochner-Maloy plates was performed to isolate clonal populations of *S. typhimurium* [15,18,19]. Sequencing confirmed that a clonal population of tetracycline-sensitive cells harbored the RGD lysogen. A recombineering-based back-cross in which the wild-type sequence was reintroduced into the RGD lysogen produced infectious phage, confirming that the only mutation blocking infectivity was the RGD insertion. Thus, while the RGD insertion does not appear to block procapsid-like particle assembly in an expression system and those particles are capable of undergoing heat-induced expansion, the RGD lysogen is incompatible with the production of infectious phage.

Figure 3. (**A**) The custom-designed same-sense nucleotide sequence as compared with the wild-type. Red nucleotides are different from wild-type without changing the amino acid coding. (**B**) Phage titer results post-recombineering indicated that recombineering was successful due to the presence of titer for both the wild-type and same-sense reactions. RGD does not produce infectious phage. Titers are representative of multiple recombineering reactions.

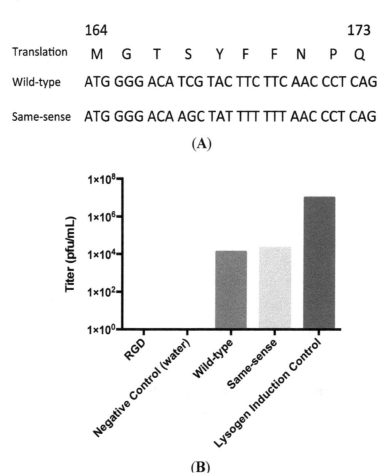

	164							173		
Translation	M	G	T	S	Y	F	F	N	P	Q
Wild-type	ATG	GGG	ACA	TCG	TAC	TTC	TTC	AAC	CCT	CAG
Same-sense	ATG	GGG	ACA	AGC	TAT	TTT	TTT	AAC	CCT	CAG

(A)

(B)

3.3. Biochemical Characterization of the RGD Lysogen

The absence of infectious phage particles could be caused by any number of failures in the assembly and maturation pathway of P22. Procapsid assembly was assayed to determine if the RGD insertion was preventing folding or assembly of the coat protein subunits. Strains of *S. typhimurium* containing both the wild-type and RGD lysogens were induced and the products analyzed by sedimentation of the lysates through a sucrose gradient. Western blots of the resulting fractions demonstrated peaks of coat and scaffolding protein co-sedimenting to a position typical of procapsids in both samples and a peak of coat protein in the phage position only for the wild-type sample (Figure 4A). Examination of the corresponding procapsid and virion fractions by TEM demonstrate that both the RGD and wild-type procapsids have a similar morphology (Figure 4B). Mature phage particles were not observed in the pellet fraction of RGD, whereas they were observed frequently in the corresponding wild-type fraction. This data suggests that RGD containing coat protein can co-assemble with scaffolding protein into procapsids, but not mature into phage. Incorporation of the portal dodecamer complex into the assembled procapsid is required for the procapsid to be capable of packaging DNA [20]. To determine whether or not portal was incorporated into the RGD procapsids, the sucrose gradient fractions were probed for the presence of portal (Figure 4C). As expected, in the case of the wild-type gradient, portal protein was detected in both the virion and the procapsid forms. Portal protein was also detected in the RGD procapsid, effectively proving that portal can be successfully assembled into RGD versions of the coat protein. The escape of scaffolding protein from the procapsid is required for DNA packaging. To gauge if the scaffolding exit was playing a role in the blocking maturation, scaffold escape was mimicked *in vitro* using guanidine hydrochloride. Treatment with guanidine hydrochloride indicated that the scaffolding protein contained within the RGD procapsid-like particle is retained more robustly than in the wild-type (Figure S2).

Figure 4. (**A**) Sucrose gradient fractionation of procapsid and mature morphologies indicates that the RGD strain only produces procapsids. (**B**) TEM of particles isolated from the procapsid fraction of the sucrose gradient confirm the similar morphology between wild-type and RGD morphologies; 2% uranyl acetate. (**C**) Western blot probing for P22 portal protein shows that RGD procapsids have portal protein present.

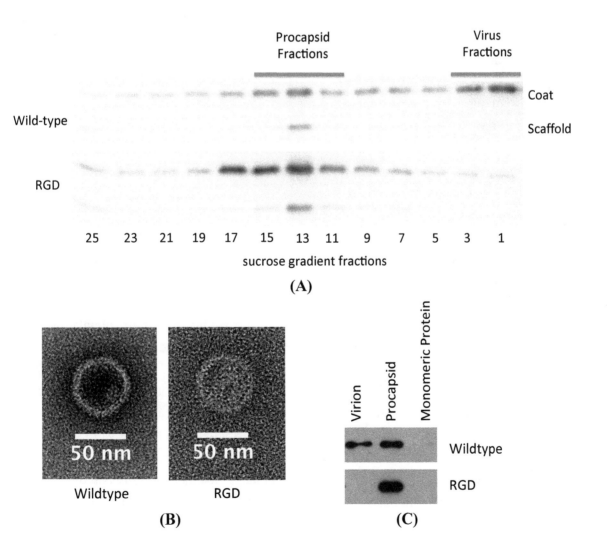

3.4. Additional Manipulations of the A-Domain

A series of additional insertions designed to discriminate between sequence specific and steric effects were recombineered into the tetRA interrupted lysogen (Table 1).

Table 1. Mutations of P22 coat generated by lambda-red recombineering.

Mutations Generated in 1757/TetRA	Mutations Generated in 2158/galK
T183GGGRGDGGG	T183AAAA
T183GGGGGGGGG	T183AAA
T183RGD	T183AA
T183GGG	T183A

In all cases, clonal populations of cells with the interrupted lysogen were isolated and confirmed by sequencing. Replacing the glycine flanked RGD sequence (T183GGGRGDGGG) with nine glycine residues did not produce infectious particles, suggesting that steric factors were paramount. Insertion of unflanked RGD (T183RGD) or T183GGG similarly abolished infectious phage production (Figure S3). This data strongly suggests that the blocking of phage production was most likely a steric hindrance effect on maturation. To further define the extent to which steric hindrance plays a role in the failed maturation of A-domain manipulated phage, we generated alanine insertions of decreasing length (T183AAAA, T183AAA, T183AA and T183A) at the A-domain loop. Clonal strains of *S. typhimurium* containing each of the four alanine mutant lysogenic P22 genomes were isolated and induced to produce phage (Figure 5). While both the single- and double-alanine insertions were capable of producing near wild-type phage titers, the triple alanine insertion resulted in a significant drop in titer, and the quadruple insertion was generally unable to produce plaques or did so at near detection limit levels (<10 pfu/mL). Plaques obtained from the alanine mutants were sequenced to determine the genotype of the resulting phage. The wild-type, single- and double-alanine mutants all contained the respective nucleotide sequences. Sequencing of plaques obtained from the triple and quadruple insertion mutants indicated that the genotype of these rare plaques were actually revertant double-alanine insertions, suggesting a strong selection pressure against insertions of three residues or greater. In addition to normal plaque morphology, small pinprick plaques were occasionally observed on the triple-alanine mutant plates, which suggested the possibility that the triple-alanine insertion generated a temperature sensitive (ts) phenotype. Plating of several induction preparations of triple alanine phage and incubating at either 30 °C or 37 °C resulted in a roughly 3–4 log increase in titer when plated at the lower temperature, confirming the hypothesis that these phages were ts (Figure 6). The wild-type phage did not exhibit this trait, since titers remained consistent at both 30 °C and 37 °C. To further classify the ts triple-alanine mutant, we asked if the triple-alanine phages were stable at 37 °C. Phage stocks were split and incubated for 2 h at either room temperature or at 37 °C, then plated and titered at 30 °C. While the wild-type phage preparations maintained their stability and did not decrease the titer, the triple-alanine mutant decreased the titer greater than 1.2 logs when incubated at 37 °C (Figure 7).

Figure 5. Induction and titer of phages produced from clonal mutants. 2158 is the original strain containing the wild-type lysogen. 2158 galK is the background strain for recombineering that contains a galK cassette within the coat gene. Ala1-4 are the single-, double-, triple- and quadruple-alanine insertions generated by lambda-red recombineering. Results are representative of multiple induction preparations.

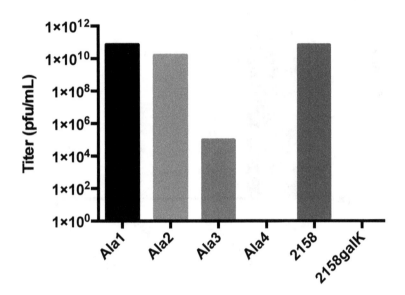

Figure 6. Temperature sensitivity test. Strains containing either wild-type or triple-alanine lysogens were induced at 30 °C for three hours and titered on DB-1636 at 37 °C or 30 °C. No significant differences were observed at either temperature for the wild-type phage. Triple-alanine insertion mutants exhibited a large increase in titer, confirming the temperature-sensitive phenotype. Relative titer results are representative of multiple inductions.

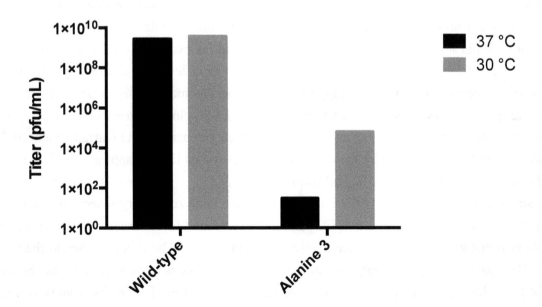

Figure 7. Heat stability test. Induced phage from 30 °C induction were isolated and incubated at room temperature or heated to 37 °C for 2 h. Titer incubation was performed at the permissive temperature, 30 °C. Heating the mutant phage results in a greater than 1.2 log reduction in the overall phage titer. Relative titer results are representative of multiple inductions.

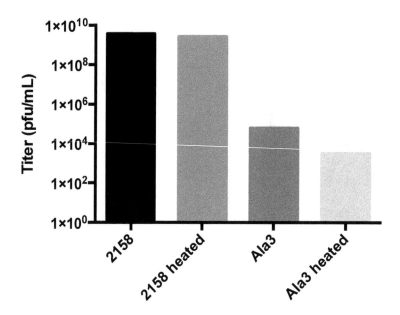

4. Discussion

4.1. Stabilizing Effect of the A-Domain and Restriction of Scaffolding Protein Escape

The A-domain of the P22 coat protein was originally identified as a protease-sensitive site that undergoes substantial conformational rearrangement during expansion and maturation. Protease sensitivity experiments demonstrated that the region becomes protease inaccessible upon maturation [21]; hydrogen/deuterium exchange studies demonstrate that the A-domain becomes highly exchange protected upon expansion [8], and the entire lattice becomes more stable.

Thermodynamic studies indicate that while the expansion is exothermic [17], the activation energy required for expansion is quite high [22]. Strikingly, cleavage or tethering of the loop appears to lower the activation energy for expansion, suggesting that rearrangement of the loop poses an energetic barrier [8,21]. In this study, we found that the insertion of a glycine-flanked RGD sequence increased the activation energy required for expansion, adding to the evidence that the rearrangement of this loop is a barrier to maturation. However, the expanded RGD particles formed appear to be less stable, as it proved impossible to obtain the wiffle ball form.

The structure of the wild-type wiffle ball form has been analyzed by cryo-electron microscopy and three-dimensional image reconstruction [23] and has been found to be largely similar to the structure of the *in vivo* mature phage lattice, except absent pentamers [12]. The failure to obtain the wiffle ball for the RGD particles is presumably due to the fact that the interactions stabilizing the hexamer are perturbed by the insertions with the consequence that the entire lattice dissociates concurrently. In the mature lattice, the hexamers and the A-domain loop have become symmetrical, which closes the capsomere pores. In contrast to the wiffle ball form, no three-dimensional structure of the expanded form has been solved, despite having a similar diameter in negatively-stained micrographs [17]. Biochemical experiments in which procapsids were loaded with GFP and were heat expanded demonstrated that the expanded form retains the pentamers [11]. It is well documented that stress in the capsid is transmitted to the pentamers, which is presumed to play a role in penton dissociation during wiffle ball formation [24,25].

Some insights into the expansion pathway can be gained from studies with HK97 [26–31]. Despite sharing no sequence similarity, HK97 and P22 share a similar coat protein fold and arrangement of coat protein subunits within the lattice. HK97 undergoes a multi-step maturation process. In the transition between Prohead II to the first expansion intermediate, EI-1, the skew character of the hexons has been resolved, and presumably, all of the A-domain interactions have become equivalent though the spine helix and the P-domain remain in a strained conformation. This strain gives rise to an exothermic transition through the second expansion intermediate, EI-2, that drives the subsequent crosslinking to stabilize the capsid and completes the HK97 maturation process. The expanded form of P22 also occurs after an exothermic transition. This favors the notion that the expanded structure correlates with the HK97 expansion intermediate, EI-2, and therefore, the lack of stability in the RGD particles is a result of the inability to form stabilizing A-domain loop interactions. Comparing the alanine mutants, it is clear that insertions of up to two alanine residues maintains these stabilizing A-loop interactions, as is suggested by the near wild-type titers for these mutants. The drop in titer and temperature sensitivity for inductions of the triple-alanine mutant (T183AAA), as well as the absence of titer for the unflanked RGD demonstrate the structural limit to these stabilizing A-domain interactions.

In vivo, the sucrose gradient experiments suggest that the RGD insertions allow procapsid formation, but block DNA packaging. The GuHCl extraction studies suggest that release of the scaffolding protein is made more difficult by the insertions, perhaps because of steric effects. Mutational studies have shown that preventing scaffolding exit inhibits DNA packaging [32]. The steric occlusion model is consistent with the observation that the T183A and T183AA insertions are capable of producing phage, while the T183AAA is temperature sensitive for stability.

4.2. A-Domain Has Strong Selection Pressure toward Near-Wild-Type Sequences in Vivo

There are many phage with capsid proteins that adopt the HK-97-like fold, yet the sequence identity and overall structures of these capsids are heterogeneous [33]. This common fold implies that these phage have undergone strong selection pressure over time to adopt these conformations. Such strong selection pressure to maintain the HK-97-like fold suggests that the A-domain would only be permissive of one residue at the 183 codon and that any further additions to this region would ultimately result in reversion to a wild-type-like sequence length. This study demonstrates that both T183A and T183AA mutations are capable of producing phage, but any additional insertions result in large drops in titer. Since producing infectious phage is inherently a strong selection pressure, it was interesting to see that revertant phages from the induction of T183AAA and T183AAAA lysogens always resulted in double-alanine genotypes.

4.3. A-Domain Manipulations Can Produce Conditionally-Stable Procapsids

One of the main requirements of utilizing viruses as nanomedicine delivery devices is that the capsid must controllably dissociate to release drug [6]. The generation of a temperature-sensitive phenotype for the T183AAA mutant phage indicated that, in addition to interrogating maturation, the A-domain could be used to alter structural stability. Previous work has generated bone binding P22 procapsids that increase the charge localization on the surface of the immature structure by adding Glu residues at the A-domain loop [5]. The robust nature of assembly-permissive mutations at the A-domain loop coupled with the role this region plays in maturation allows for the design of capsids that are controllably and conditionally stable at certain temperatures, pH and other factors.

Acknowledgments

The authors would like to thank Sherwood Casjens and Eddie Gilcrease for their helpful advice with recombineering and the generosity in providing the UB-1757 and UB-2158 strains, as well as the PKD46 plasmid; and William Britt for providing the pgalK plasmid. Funding for this work was provided through a sub-contract on NIH R01 EB01202703.

Author Contributions

David S. Morris performed all experiments and wrote the manuscript. Peter E. Prevelige, Jr. discussed the analysis and edited the manuscript.

References and Notes

1.   Prevelige, P.E., Jr.; Thomas, D.; King, J.; Towse, S.A.; Thomas, G.J., Jr. Conformational states of the bacteriophage P22 capsid subunit in relation to self-assembly. *Biochemistry (Mosc).* **1990**, *29*, 5626–5633.

2.   Caspar, D.L.; Klug, A. Physical principles in the construction of regular viruses. *Cold Spring Harb. Symp. Quant. Biol.* **1962**, *27*, 1–24.

3.   King, J.; Chiu, W. *The procapsid-to-capsid transition in double-stranded DNA bacteriophages*; Oxford University Press: New York, NY, USA, 1997.

4.   Kang, S.; Uchida, M. Implementations of P22 Viral Capsids as Nanoplatforms. *Biomacromolecules* **2010**, *11*, 2804–2809.

5.   Culpepper, B.K.; Morris, D.S.; Prevelige, P.E.; Bellis, S.L. Engineering nanocages with polyglutamate domains for coupling to hydroxyapatite biomaterials and allograft bone. *Biomaterials* **2013**, *34*, 2455–2462.

6.   Maham, A.; Tang, Z.; Wu, H.; Wang, J.; Lin, Y. Protein-based nanomedicine platforms for drug delivery. *Small* **2009**, *5*, 1706–1721.

7.   Kang, S.; Prevelige, P.E., Jr. Domain study of bacteriophage p22 coat protein and characterization of the capsid lattice transformation by hydrogen/deuterium exchange. *J. Mol. Biol.* **2005**, *347*, 935–948.

8. Kang, S.; Hawkridge, A.M.; Johnson, K.L.; Muddiman, D.C.; Prevelige, P.E., Jr. Identification of subunit-subunit interactions in bacteriophage P22 procapsids by chemical cross-linking and mass spectrometry. *J. Proteome Res.* **2006**, *5*, 370–377.

9. Svensen, N.; Walton, J.G.; Bradley, M. Peptides for cell-selective drug delivery. *Trends Pharmacol. Sci.* **2012**, *33*, 186–192.

10. Humphries, J.D.; Byron, A.; Humphries, M.J. Integrin ligands at a glance. *J. Cell Sci.* **2006**, *119*, 3901–3903.

11. O'Neil, A.; Reichhardt, C.; Johnson, B.; Prevelige, P.E.; Douglas, T. Genetically programmed *in vivo* packaging of protein cargo and its controlled release from bacteriophage P22. *Angew Chem. Int. Ed. Engl.* **2011**, *50*, 7425–7428.

12. Chen, D.H.; Baker, M.L.; Hryc, C.F.; DiMaio, F.; Jakana, J.; Wu, W.; Dougherty, M.; Haase-Pettingell, C.; Schmid, M.F.; Jiang, W.; *et al.* Structural basis for scaffolding-mediated assembly and maturation of a dsDNA virus. *Proc. Natl. Acad. Sci. USA* **2011**, *108*, 1355–1360.

13. Pettersen, E.F.; Goddard, T.D.; Huang, C.C.; Couch, G.S.; Greenblatt, D.M.; Meng, E.C.; Ferrin, T.E. UCSF Chimera—A visualization system for exploratory research and analysis. *J. Comput. Chem.* **2004**, *13*, 1605–1612.

14. Schneider, C.A.; Rasband, W.S.; Eliceiri, K.W. NIH Image to ImageJ: 25 years of image analysis. *Nat. method.* **2012**, *9*, 671–675.

15. Karlinsey, J. λ-Red Genetic Engineering in Salmonella enterica serovar Typhimurium. *Method. Enzymol.* **2007**, *421*, 199–209.

16. Warming, S.; Costantino, N.; Court, D.L.; Jenkins, N.A.; Copeland, N.G. Simple and highly efficient BAC recombineering using galK selection. *Nucl. Acids Res.* **2005**, *33*, e36.

17. Galisteo, M.L.; King, J. Conformational transformations in the protein lattice of phage P22 procapsids. *Biophys. J.* **1993**, *65*, 227–235.

18. Bochner, B.R.; Huang, H.C.; Schieven, G.L.; Ames, B.N. Positive selection for loss of tetracycline resistance. *J. Bacteriol.* **1980**, *143*, 926–933.

19. Maloy, S.R.; Nunn, W.D. Selection for loss of tetracycline resistance by Escherichia coli. *J. Bacteriol.* **1981**, *145*, 1110–1111.

20. King, J.; Lenk, E.V.; Botstein, D. Mechanism of head assembly and DNA encapsulation in Salmonella phage P22. II. Morphogenetic pathway. *J. Mol. Biol.* **1973**, *80*, 697–731.

21. Lanman, J.; Tuma, R.; Prevelige, P.E., Jr. Identification and characterization of the domain structure of bacteriophage P22 coat protein. *Biochemistry (Mosc.)* **1999**, *38*, 14614–14623.

22. Tuma, R.; Prevelige, P.E., Jr.; Thomas, G.J., Jr. Mechanism of capsid maturation in a double-stranded DNA virus. *Proc. Natl. Acad. Sci. USA* **1998**, *95*, 9885–9890.

23. Parent, K.N.; Khayat, R.; Tu, L.H.; Suhanovsky, M.M.; Cortines, J.R.; Teschke, C.M.; Johnson, J.E.; Baker, T.S. P22 coat protein structures reveal a novel mechanism for capsid maturation: Stability without auxiliary proteins or chemical crosslinks. *Structure* **2010**, *18*, 390–401.

24. Zandi, R.; Reguera, D. Mechanical properties of viral capsids. *Phys. Rev. Stat. Nonlinear Soft Matter Phys.* **2005**, *72*, 021917.

25. Teschke, C.M.; McGough, A.; Thuman-Commike, P.A. Penton release from P22 heat-expanded capsids suggests importance of stabilizing penton-hexon interactions during capsid maturation. *Biophys. J.* **2003**, *84*, 2585–2592.

26. Veesler, D.; Quispe, J.; Grigorieff, N.; Potter, C.S.; Carragher, B.; Johnson, J.E. Maturation in Action: CryoEM Study of a Viral Capsid Caught during Expansion. *Structure* **2012**, *20*, 1384–1390.

27. Veesler, D.; Khayat, R.; Krishnamurthy, S.; Snijder, J.; Huang, R.K.; Heck, A.J.; Anand, G.S.; Johnson, J.E. Architecture of a dsDNA Viral Capsid in Complex with Its Maturation Protease. *Structure* **2014**, *22*, 230–237.

28. Gertsman, I.; Fu, C.Y.; Huang, R.; Komives, E.A.; Johnson, J.E. Critical salt bridges guide capsid assembly, stability, and maturation behavior in bacteriophage HK97. *Mol. Cell. Proteomics* **2010**, *9*, 1752–1763.

29. Huang, R.K.; Khayat, R.; Lee, K.K.; Gertsman, I.; Duda, R.L.; Hendrix, R.W.; Johnson, J.E. The Prohead-I structure of bacteriophage HK97: Implications for scaffold-mediated control of particle assembly and maturation. *J. Mol. Biol.* **2011**, *408*, 541–554.

30. Gertsman, I.; Gan, L.; Guttman, M.; Lee, K.; Speir, J.A.; Duda, R.L.; Hendrix, R.W.; Komives, E.A.; Johnson, J.E. An unexpected twist in viral capsid maturation. *Nature* **2009**, *458*, 646–650.

31. Gertsman, I.; Komives, E.A.; Johnson, J.E. HK97 maturation studied by crystallography and H/2H exchange reveals the structural basis for exothermic particle transitions. *J. Mol. Biol.* **2010**, *397*, 560–574.

32. Greene, B.; King, J. Binding of scaffolding subunits within the P22 procapsid lattice. *Virology* **1994**, *205*, 188–197.

33. Hendrix, R.W. Bacteriophages: Evolution of the majority. *Theor. Popul. Biol.* **2002**, *61*, 471–480.

6

Photodynamic Inactivation of Mammalian Viruses and Bacteriophages

Liliana Costa [1], Maria Amparo F. Faustino [2], Maria Graça P. M. S. Neves [2], Ângela Cunha [1] and Adelaide Almeida [1,*]

[1] Department of Biology and CESAM, University of Aveiro, Aveiro 3810-193, Portugal;
   E-Mails: lcosta@ua.pt (L.C.); acunha@ua.pt (A.C.)
[2] Department of Chemistry and QOPNA, University of Aveiro, Aveiro 3810-193, Portugal;
   E-Mails: faustino@ua.pt (M.A.F.F.); gneves@ua.pt (M.G.P.M.S.N.)

* Author to whom correspondence should be addressed; E-Mail: aalmeida@ua.pt

Abstract: Photodynamic inactivation (PDI) has been used to inactivate microorganisms through the use of photosensitizers. The inactivation of mammalian viruses and bacteriophages by photosensitization has been applied with success since the first decades of the last century. Due to the fact that mammalian viruses are known to pose a threat to public health and that bacteriophages are frequently used as models of mammalian viruses, it is important to know and understand the mechanisms and photodynamic procedures involved in their photoinactivation. The aim of this review is to (i) summarize the main approaches developed until now for the photodynamic inactivation of bacteriophages and mammalian viruses and, (ii) discuss and compare the present state of the art of mammalian viruses PDI with phage photoinactivation, with special focus on the most relevant mechanisms, molecular targets and factors affecting the viral inactivation process.

Keywords: bacteriophages; mammalian viruses; photodynamic therapy; photosensitizer; viral photoinactivation process

Nomenclature

AlPcS$_4$	Aluminum phthalocyanine tetrasulfonate
AZT	Azidothymidine
BVDV	Bovine viral diarrhea virus
DMTU	Dimethylthiourea
EMCV	Encephalomyocarditis virus
HAV	Hepatitis A virus
HBV	Hepatitis B virus
HCV	Hepatitis C virus
HIV	Human immunodeficiency virus
HPV	Human papillomatosis virus
HSV	Herpes simplex virus
LED	Light emitting diode
MB	Methylene blue
NM	Not mentioned
NQ	Not quantified
Pc$_4$	Silicon phthalocyanine
PDI	Photodynamic inactivation
PS	Photosensitizer
ROS	Reactive oxygen species
SFV	Semliki Forest virus
SHV	Suid herpes virus
SOD	Superoxide dismutase
SSB	Singlet strand breaks
Tri-Py$^+$-Me-PF	5-(pentafluorophenyl)-10,15,20-tris(1-methylpyridinium-4-yl)porphyrin tri-iodide
VSV	Vesicular stomatitis virus
VZV	Varicella zoster virus
$^1O_2$	Singlet oxygen
$^3O_2$	Molecular oxygen
^1PS	Ground state photosensitizer
$^3PS*$	Triplet excited state photosensitizer

1. Introduction

Humans are exposed to pathogenic viruses through various routes and the development of viral-induced diseases is a common occurrence.

Although the transmission of viral diseases has been reduced by the development of good water supplies and hygienic-based procedures for a whole range of human activities [1], pathogenic viruses are still the causative agents of many diseases in humans and other species. The most usual human diseases caused by viruses include the common cold (coronaviruses), influenza (influenza viruses), chickenpox (varicella zoster virus), cold sores (herpes simplex virus), gastroenteritis and diarrhoea (caliciviruses, rotaviruses and adenoviruses) [2,3]. Pathogenic viruses are also implicated in serious diseases, such as Ebola (Ebola virus), AIDS (immunodeficiency viruses), avian influenza and sudden acute respiratory syndrome (SARS) (SARS-coronavirus), and they are also an established cause of cancer (papillomavirus, hepatitis B and C viruses, Epstein–Barr virus, Kaposi's sarcoma-associated herpes virus, human T-lymphotropic virus, and Merkel cell polyomavirus) [4].

The enhanced implication of viruses in severe infectious diseases and the increasing knowledge about the complex mechanisms of viral pathogenesis have greatly contributed to the rapid development of antiviral drugs. Consequently, the use of antivirals has largely increased in the last years and resistance to antiviral drugs is now well documented for several pathogenic viruses [5–10]. Moreover, as viruses are genetically flexible, they may mutate quickly and mutations come as no surprises, leading to the development of resistance to conventional antiviral drugs. Consequently, the emergence of antiviral drug can become a great problem, such the resistance observed for bacteria relative to antibiotics. So, alternative methods unlikely to cause resistance are required. Photodynamic inactivation (PDI) of viruses represents a promising and inexpensive potential alternative to meet that need.

The sensitivity of viruses to photodynamic procedures was reported in the 1930s [11,12] but only within the last 30 years, with the development of new active molecules, namely photosensitizers (PS), and an increment of light technologies (lasers, LED, portability, *etc.*), have photodynamic techniques for the inactivation of viruses received growing attention [13]. Most of the clinical applications of PDI for treatment of infections have so far been directed to viral lesions [14]. Clinical PDI was first applied to the treatment of herpes infection in the early 1970s [15], particularly for herpes genitalis. Since then, a great variety of viruses has been effectively inactivated by photodynamic treatment using *in vitro* conditions [16] but, considering the clinical use of viral PDI, the procedures are limited to the treatment of papillomatosis, caused by human papillomatosis virus (HPV), like laryngeal papillomatosis [17] and epidermodysplasia verruciformis [18] and, in a small scale, to the treatment of viral complications in AIDS patients [19,20]. However, considerable progress has been made in the viral photodynamic disinfection of blood products. The major threat of viral contamination in blood and blood products comes from the immunodeficiency viruses (HIV) [21], hepatitis viruses [21–23], cytomegalovirus [23], human parvovirus B19 [24] and human T-cell lymphotropic virus type I and type II [23]. HIV has been inactivated *in vitro* following a photodynamic procedure [25–39]. The photoinactivation of hepatitis viruses in blood products has also been successfully tested against the hepatitis C virus (HCV) [37,40–42], hepatitis B virus (HBV) [43] and hepatitis A virus (HAV) [44]. Inactivation of cytomegalovirus [45], human parvovirus B19 [46] and human T-cell lymphotropic virus [47] in blood products was also efficiently achieved after photodynamic treatment.

The availability of a simple and quantitative assay to follow the viral photoinactivation process is important. Traditional viral quantification techniques, such as *in vitro* viral cultures, are time-consuming and labor-intensive processes. Molecular quantitative methods such as nucleic acid amplification procedures, including real time PCR, are rapid and sensitive but detect only viral nucleic acid and do not determine infectivity. When the virucidal properties of different photosensitizing compounds are initially evaluated, bacteriophages can be useful as surrogates of mammalian viruses. The reasons for their use are: (i) the detection methods are much simpler, faster and cheaper than those of mammalian viruses, avoiding the advanced facilities and equipment needed for propagating human pathogens; (ii) they are non-pathogenic to humans; (iii) they can be grown to higher titers than most mammalian viruses and, therefore, enhancing the sensitivity of the assay; (iv) the results of bacteriophages assays are available within several hours post-inoculation, instead of the days or weeks required by mammalian viruses infectivity-based assays; (v) they are at least as resistant as the mammalian viruses to environmental factors and to water treatment [48].

It has been shown that enveloped viruses are significantly more sensitive to photodynamic destruction than non-enveloped viruses [49,50]. As most of the bacteriophages are non-enveloped, they are more difficult to suffer photoinactivation than the enveloped viruses. In general, this property makes them good indicators to evaluate the efficiency of viral PDI. A PDI protocol that is effective to inactivate a non-enveloped phage will most likely be effective against enveloped mammalian viruses.

Several bacteriophages were used in photoinactivation studies as surrogates for mammalian viruses, e.g., MS2 [44], M13 [51,52], PM2 [53], Qβ [54–56], PRD1 [57], λ [58,59], φ6 [60], R17 [60], *Serratia* phage *kappa* [61], T5 [62], T3 [63], T7 [57,64] and T4-like [65–68], and the results show that they are effectively photoinactivated.

2. Antimicrobial PDI

PDI is a simple and controllable method for the inactivation of microorganisms based on the production of reactive oxygen species (ROS) (free radicals and singlet oxygen). This technology requires the combined action of oxygen, light and a photosensitizer (PS), which absorbs and uses the energy from light to produce those ROS [69]. Therefore, the photodynamic effects depend on multiple variables including: the structural features of the PS, the concentrations of PS and molecular oxygen, and the properties of the light used (e.g., wavelength, type, dose and fluence rate) [66,67,69–72]. Changes in any of these parameters will affect the rate of microbial photoinactivation [66,67,73,74].

The majority of PS used in PDI is derived from tetrapyrrolic macrocycles known as porphyrins. These chromophores and their analogs, such as chlorins and bacteriochlorins, are involved in very important biological functions, such as respiration (heme group) and photosynthesis (chlorophyll and bacteriochlorophyll (Figure 1). Based on these macrocycles, the scientific community was able to develop a number of synthetic analogs, such as *meso*-tetraarylporphyrins, phthalocyanines, texaphyrins, porphycenes and saphyrins, which proved to have very promising features for being used as PS (Figure 2) [16]. Also, non-tetrapyrrolic derivatives, such as the naturally occurring hypericin, or synthetic dyes like toluidine blue O, rose bengal, eosin, methylene blue (MB) and fullerenes, were considered in many PDI studies (Figure 3) [71].

In order to be efficient, photosensitizing agents used for viral PDI must bind specifically to vital viral components, such as lipid envelope (when present), the protein coat or to the nucleic acids [55].

Figure 1. Structure of some tetrapyrrolic macrocycles with natural occurrence.

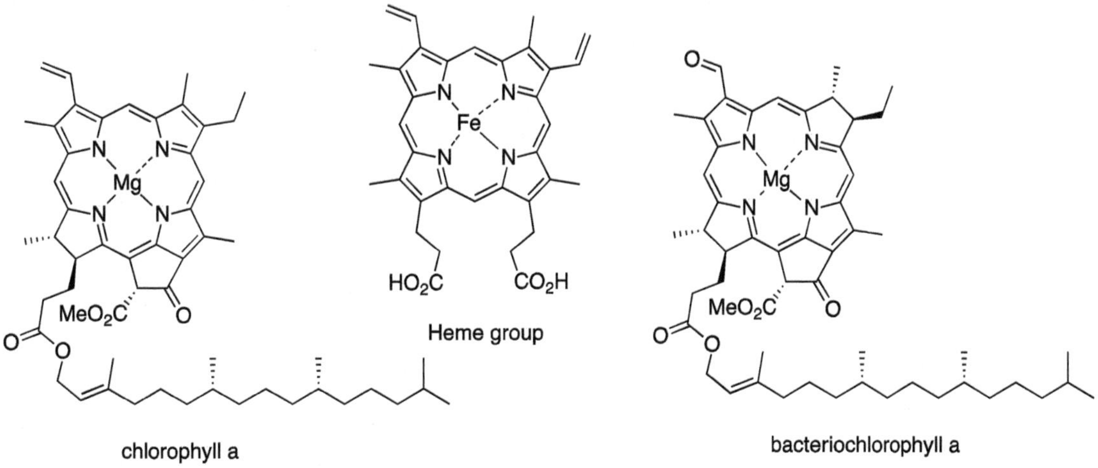

chlorophyll a

Heme group

bacteriochlorophyll a

Figure 2. Skeletons of some synthetic pyrrolic macrocycles used as photosensitizers.

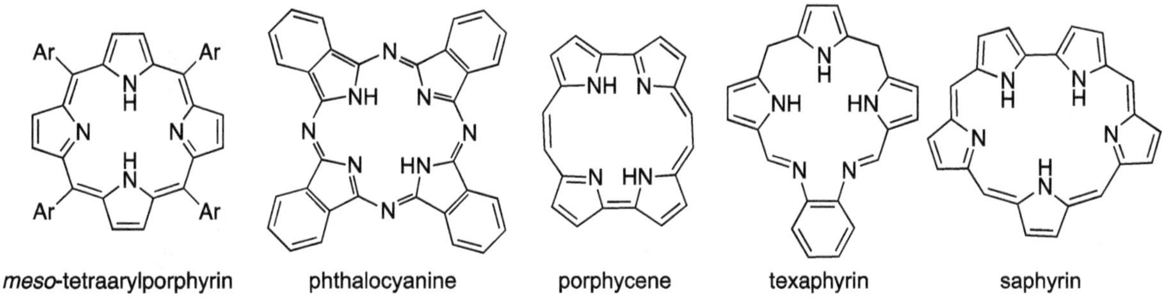

meso-tetraarylporphyrin

phthalocyanine

porphycene

texaphyrin

saphyrin

Figure 3. Structure of some non-tetrapyrrolic photosensitizers.

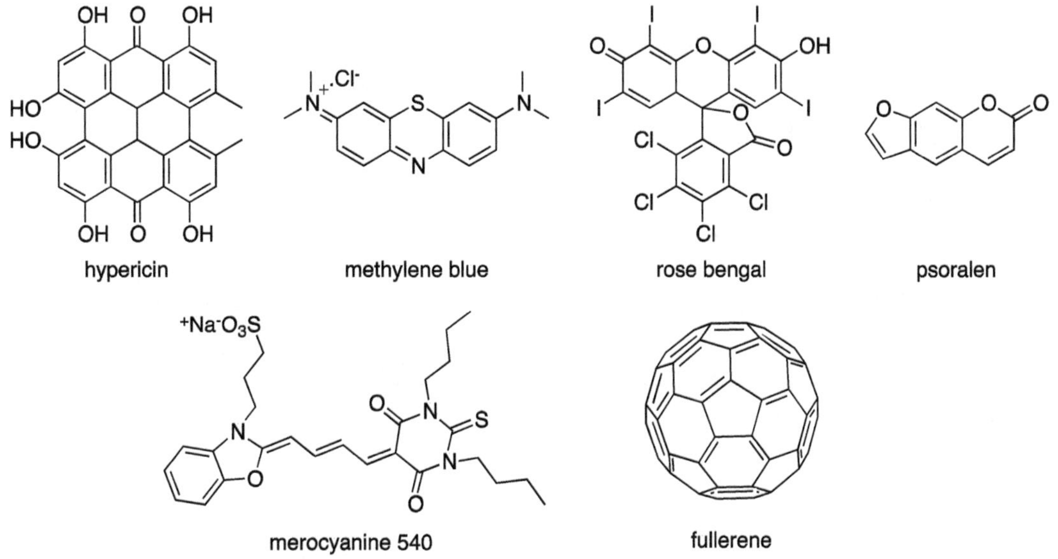

hypericin

methylene blue

rose bengal

psoralen

merocyanine 540

fullerene

The efficiency of mammalian viruses and bacteriophages PDI has been described for porphyrin derivatives, chlorin derivatives, chlorophyll derivatives, phthalocyanine derivatives, hypericin, methylene blue, rose bengal, merocyanine 540, proflavine, and fullerene derivatives (Table 1).

Table 1. Some PS used for mammalian viruses and bacteriophages PDI.

Photosensitizer	Microorganism	PDI	Reference
Mammalian viruses			
Hematoporphyrin derivative	HSV-1	7 log	[75]
	HSV-1	<0.8 log	[36]
Uroporphyrin	Adenovirus	7 log	[76]
Natural metalloporphyrin derivatives	HIV-1	<0.8 log	[36]
Chlorophyll derivatives	VSV	~6 log	[77]
7-despropionate-7-hydroxypropylmesopyropheophorbide a	BVDV	~5 log	[78]
	EMCV	~0.2 log	
Benzoporphyrin derivative monoacid ring A	HIV-1	>4 log	[33]
Glycoconjugated *meso*-tetraarylporphyrin derivatives	HSV-1	6 log	[79]
	HSV-2	6 log	
Metallo tetrasulfonated *meso*-tetraarylporphyrin derivatives	HIV-1	≤2 log	[36]
Tetrasulfonated *meso*-tetraarylporphyrin derivatives	HIV-1	≤2 log	[36]
	HAV	~4 log	[44]
meso-Tetrakis(1-methylpyridinium-4-yl)porphyrin	HAV	~4 log	[44]
meso-Tetrakis(1-butylpyridinium-4-yl)porphyrin	HAV	>3.8 log	[44]
meso-Tetrakis(1-octylpyridinium-4-yl)porphyrin	HAV	>3.9 log	[44]
Cationic β-vinyl substituted *meso*-tetraphenylporphyrin derivatives	HSV-1	<3 log	[80]
Aluminum dibenzodisulfophthalocyanine	HIV-1	3.7 log	[49]
Aluminum phthalocyanine tetrasulfonate	HIV-1	>5 log	[49]
	VSV	4.2 log	[82]
	Adenovirus	4 log	[76]
Silicon phthalocyanine derivative	VSV	4 log	[82]
Cationic phthalocyanines	HIV-1	>5 log	[49]
	HSV-1	≥5 log	[83]
Hypericin	HIV-1	NQ	[30]
	VSV	4–5 log	
	Influenza virus	NQ	
	Sendai virus	NQ	

Table 1. *Cont.*

Photosensitizer	Microorganism	PDI	Reference
Mammalian viruses			
Methylene blue	VSV	4.7 log	[81]
	HSV-1	5 log	[84]
	SHV-1	2.5 log	[84]
	HCV	<2 log	[41]
	HIV-1	<2 log	[41]
	Adenovirus	7 log	[76]
	Dengue virus	5–6.4 log	[74]
	Enterovirus 71	~8 log	[85]
	Vaccinia virus	5 log	[86]
Phenothiazine derivatives	VSV	>4.4 log	[60]
Rose bengal	Vaccinia virus	5 log	[86]
	HIV-1	NQ	[30]
	VSV	4–5 log	
	Influenza virus	NQ	
	Sendai virus	NQ	
	Adenovirus	7 log	[76]
Buckminsterfullerene	SFV	7 log	[50]
	VSV	7 log	
Merocyanine 540	HSV-1	5–6 log	[45]
Bacteriophages			
Glycoconjugated *meso*-tetraarylporphyrins	T7 phage	<3 log	[64]
	T7 phage	<3.5 log	[87]
Tetrasulfonated *meso*-tetraarylporphyrin derivatives	MS2 phage	>3.8 log	[44]
meso-Tetrakis(1-methylpyridinium-4-yl)porphyrin	λ phage	<7 log	[58]
	MS2 phage	>4.1 log	[44]
	T4 phage	7 log	[66,67]
	T7 phage	<4 log	[88]
5-(pentafluorophenyl)-10,15,20-tris(1-methylpyridinium-4-yl)porphyrin	T4 phage	7 log	[66–68]
5-(4-methoxicarbonylphenyl)-10,15,20-tris(1-methylpyridinium-4-yl)porphyrin	T4 phage	7 log	[66]
5-(4-carboxyphenyl)-10,15,20-tris(1-methylpyridinium-4-yl)porphyrin	T4 phage	3.9 log	[66]
5,10-bis(4-carboxyphenyl)-15,20-bis(1-methylpyridinium-4-yl)porphyrin	T4 phage	1.4 log	[66]
5,15-bis(4-carboxyphenyl)-10,20-bis(1-methylpyridinium-4-yl)porphyrin	T4 phage	1.2 log	[66]
5,10,15-tris(1-methylpyridinium-4-yl)-20-phenylporphyrin	T7 phage	1.7 log	[88]

Table 1. *Cont.*

Photosensitizer	Microorganism	PDI	Reference
Methylene blue	*Serratia* phage *kappa*	>4 log	[61]
	M13 phage	2.2 log	[52,81]
	f2 phage	5 log	[56]
	Qβ phage	7–8 log	[56]
	Qβ phage	7–8 log	[89]
Phenothiazine derivatives	R17 phage	4–7 log	[60]
	φ6	4–6.5 log	
Rose bengal	PRD1 phage	~3.5 log*	[57]
	T7 phage	~4.5 log*	
Riboflavin	λ phage	<4 log	[59]
Proflavine	*Serratia* phage *kappa*	4 log	[61]
	T3 phage	7–11 log	[63]
Polyhydroxylated fullerene	MS2 phage	~4 log	[90]
	PRD1 phage	~2.5 log*	[57]
	T7 phage	~3.5 log*	
	MS2 phage	~5 log*	

*log(N/N0).

Besides this, viral PDI has also been described for phthalocyanine derivatives [81], methylene blue [53,62,91,92], toluidine blue O [53,62,93], neutral red [93], proflavine [93], azure B [53] and merocyanine 540 [45,47,94].

3. Mechanisms of Photodynamic Inactivation

The mechanisms of PDI are based on the ability of the PS to absorb energy from light and transfer that energy to molecular oxygen. In the dark, the electronic configuration of a PS exists in the so-called ground state. The absorption, by the PS, of a photon at an appropriate wavelength initially leads to the production of an unstable, electronically-excited state of the PS molecule (the lifetime of this state ranges from 10^{-9} to 10^{-6} s) [95]. The excited PS molecule can then decay to the ground state by emission of light (radiative pathway - fluorescence) or by intersystem crossing, affording the excited triplet state which has a longer lifetime (10^{-3} to 10 s) [95]. At this point, the PS can reach the ground state either by spin inversion followed by phosphorescence emission, or by a non-radiative process. Due to the longer lifetime of the PS triplet state, this excited state can also react in one of two ways (Figure 2): by initiating photochemical reactions that can directly generate reactive oxygen species (ROS) (type I pathway), or indirectly by energy transfer to molecular oxygen (type II pathway), leading to the formation of singlet oxygen (Figure 4). These events afford toxic species which are responsible for the irreparable oxidative damages induced to important biological targets [1,69,95,96].

Figure 4. Schematic representation of the photosensitization process (adapted from [97]).

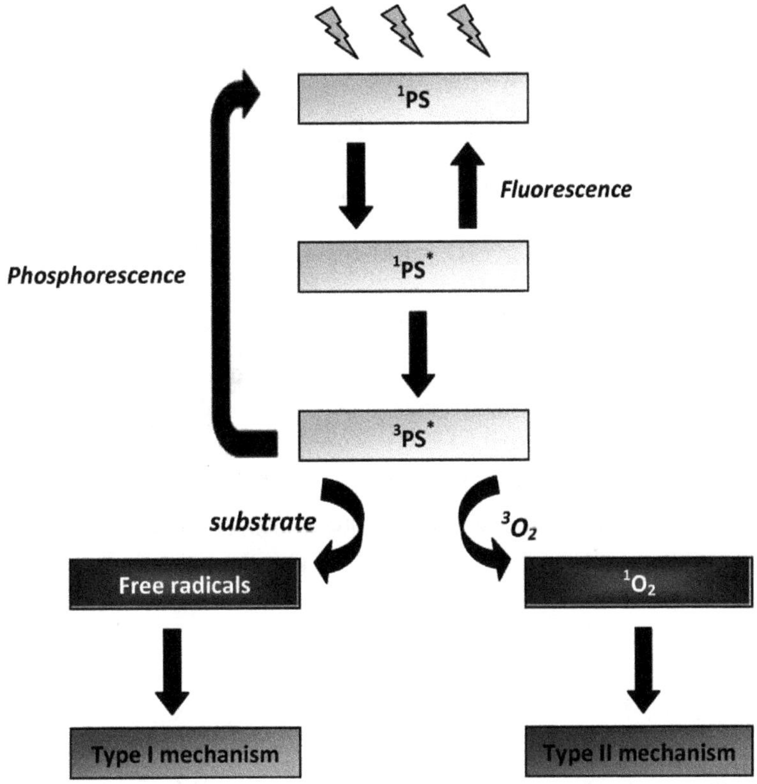

3.1. Type I and Type II Mechanisms

Type I mechanism involves hydrogen-atom abstraction or electron-transfer between the excited PS and a substrate, yielding free radicals [Equations (1) and (2)]. These radicals can react with oxygen to form active oxygen species, such as the superoxide radical anion [Equation (3)]. Superoxide is not particularly reactive in biological systems but, when protonated, can lead to the production of hydrogen peroxide and oxygen [Equations (4) and (5)] or highly reactive hydroxyl radicals [Equations (6)–(8)] [98]. Type II photooxidation is considerably less complex mechanistically than type I and in general there are far fewer products [99]. In this pathway, the excited triplet state PS ($^3PS^*$) can transfer the excess energy to molecular oxygen ($^3O_2$) and relax to its ground state (^1PS) creating an excited singlet molecular oxygen ($^1O_2$) [Equation (9)] [69]. $^1O_2$ is highly electrophilic and can interact with numerous enzymes, leading to the inhibition of protein synthesis and molecular alteration of DNA strands, which alters the transcription of the genetic material during its replication (mutagenic effect) and, in this way, leading to microbial death [Equation (10)] [98,100]. Like nucleic acids and proteins, unsaturated lipids are also prominent targets of $^1O_2$ and free radical attack. Lipid peroxidation-ensuing reactions can alter surrounding proteins, nucleic acids and other molecules, in addition to the lipids themselves [98]. Therefore, it is likely that damage of different kinds caused to the viral envelope is important in the process of microbial inactivation [13].

$$\text{SubstrateH}_2 + \text{PS} \rightarrow \text{PSH}^\bullet + \text{SubstrateH}^\bullet \tag{1}$$

$$\text{PS*} + \text{Substrate} \rightarrow \text{PS}^{\bullet-} + \text{Substrate}^{\bullet+} \text{ or } \text{PS*} + \text{Substrate} \rightarrow \text{PS}^{\bullet+} + \text{Substrate}^{\bullet-} \tag{2}$$

$$PS^{\bullet -} + {}^3O_2 \rightarrow PS + O_2^{\bullet -} \tag{3}$$

$$O_2^{\bullet -} + H^+ \rightleftharpoons HOO^{\bullet} \tag{4}$$

$$2\,HOO^{\bullet} \rightarrow H_2O_2 + O_2 \tag{5}$$

$$H_2O_2 + Fe^{2+} \rightarrow HO^{\bullet} + OH^- + Fe^{3+} \tag{6}$$

$$BiomoleculeH + HO^{\bullet} \rightarrow Biomolecule^{\bullet} + H_2O \tag{7}$$

$$Biomolecule^{\bullet} + {}^3O_2 \rightarrow Biomolecule\text{-}OO^{\bullet} \rightarrow products \tag{8}$$

$$PS^* + {}^3O_2 \rightarrow PS + {}^1O_2 \tag{9}$$

$$Biomolecules + {}^1O_2 \rightarrow oxidative\ products \tag{10}$$

Both type I and type II mechanisms can occur simultaneously or exclusively, and the ratio between these processes depends on the PS used and on the concentrations of substrate and oxygen [95]. The competition between organic substrates and molecular oxygen for the ${}^3PS^*$ determines whether the reaction pathway is type I or type II and the predominant mechanism can be changed during the course of the PDI process [101].

3.2. Evaluation of the Specific Involvement of Type I and Type II Mechanisms

An important goal in the investigation of viral PDI is to identify the type of mechanism involved (type I or type II) in the presence of a selected PS [102]. The simple detection of a reactive species does not necessarily explain the mechanism by which a specific PS induces the toxic effect. It is generally easier to draw a negative conclusion, *i.e.*, if singlet oxygen is absent, it cannot be the reactive species responsible for the photodynamic effect [103]. The simplest approach for determining whether singlet oxygen (type II mechanism) or free radicals (type I mechanism) is involved in the photodynamic process is to study the inhibitory effects of various scavengers, *i.e.*, compounds that can intercept these ROS at high rates and in a putatively selective manner [99,101,104].

3.2.1. Type I Mechanism Scavengers

A first line of defence against ROS is, of course, the protection against their formation. However, the interception of the damaging species once formed, to prevent it from further deleterious reactions, is also a deactivation strategy of defence. In general, free radical scavengers neutralize the radical species by donating one of their own electrons. The quenching agents themselves are not particularly toxic before and after the electron donation [105].

Three different types of quenching are possible, which include the transfer of the radical character with the formation of a reactive scavenger-derived radical; trapping of free radicals with the formation of a stable or inert free radical trap; and molecules which mimic quenching enzyme activities. In general, scavenger molecules either prevent free radicals from being formed or remove them before they can damage vital molecular components [105].

Several free radical scavengers have been used to evaluate the specific involvement of type I mechanism during mammalian viruses and bacteriophages PDI with different PS (Table 2).

Table 2. Free radical scavengers used in mammalian viruses and bacteriophages PDI.

PS	Scavenger	Microorganism	Scavenger protection	Reference
Mammalian viruses				
Aluminum phthalocyanine tetrasulfonate	Reduced glutathione Mannitol Glycerol SOD	VSV	Little/no effect Little/no effect Little/no effect Little/no effect	[106]
Polyhydroxylated fullerene	Glutathione (2.0 mM)	SFV VSV	no effect no effect	[50]
	Hydroquinone (2.0 mM)	SFV VSV	no effect no effect	[50]
Merocyanine 540	Glutathione (10 and 30 mmol L^{-1}) Cysteamine (10 and 30 mmol L^{-1}) SOD (1.5 to 29 U mL^{-1})	HSV-1	30-50% 60-70% no effect	[45]
Methylene blue	Mannitol (100 mM) Glycerol (10 mM) SOD (300 U mL^{-1}) Catalase (30 U mL^{-1})	HSV-1	24% 24% 24% 24%	[84]
Bacteriophages				
5,10,15-(4-β-D-glucosylphenyl)-20-phenylporphyrin	DMTU (0.1–5.0 mM)	T7 phage	44%	[64]
5,10.15,20-Tetrakis(4-β-D-glucosylphenyl) porphyrin	DMTU (0.1–5.0 mM)	T7 phage	79%	[64]
5,10,15-(4-β-D-galactosylphenyl)-20-(pentafluorophenyl)-porphyrin	DMTU (0.1–5.0 mM)	T7 phage	89%	[87]
5-(pentafluorophenyl)-10,15,20-tris(1-methylpyridinium-4-yl)porphyrin	D-mannitol (100 mM)	T4 phage Qβ	20% no effect	[107]
	L-cysteine (100 mM)	T4 phage	9%	[107]
5,10,15,20-tetrakis(1-methylpyridinium-4-yl)porphyrin	D-mannitol (100 mM)	T4 phage	no effect	[107]
Proflavine	L-cysteine (0.025 M)	T3 phage	75–80%	[63]
Polyhydroxylated fullerene	SOD	MS2 phage	no effect	[90]

3.2.1.1. Free Radicals in PDI of Mammalian Viruses

Free radical species had, in general, little or no effect on the photoinactivation of the studied mammalian viruses (Table 2). In fact, it can be observed that the rate of inactivation of

HSV [45,84,106], influenza virus [108], Semliki Forest virus (SFV) and VSV [50] in the presence of different PS and scavengers like glutathione, D-mannitol, glycerol, superoxide dismutase (SOD), catalase and hydroquinone was not significantly affected. Although this data suggest that free radicals are not major players in the viral inactivation process, the participation of type I reaction pathways cannot be ruled out, as was shown by the considerable level of protection afforded by glutathione and cysteamine when merocyanine 540 was used as PS for inactivation of HSV-1 [45].

3.2.1.2. Free Radicals in PDI of Bacteriophages

The photoinactivation rate of some bacteriophages can be reduced in the presence of free radical scavengers, suggesting a contribution of radical species in the inactivation process (Table 2). In particular, it was reported that the inhibition of T7 phage photoinactivation in the presence of glycoconjugated *meso*-tetraarylporphyrins varied according to the structure of the PS and the concentration of dimethylthiourea (DMTU) [64,87]. In fact, T7 phage PDI by *meso*-tetrakis(4-β-D-glucosylphenyl)porphyrin [64] and 5,10,15-(4-β-D-galactosylphenyl)-20-(pentafluorophenyl)porphyrin [87] seemed to be mainly mediated by free radical species, as revealed by the protection effect of free radical scavenger DMTU, contrary to T7 phage photosensitization by 5,10,15-(4-β-D-glucosylphenyl)-20-phenylporphyrin, which revealed a significantly smaller contribution from type I mechanism. The highest inhibition was reached at about 1.0 mM of DMTU; further increase in scavenger concentration did not decrease the slope of photoinduced inactivation of phages. However, in spite of inhibiting the efficacy of the PS, DMTU did not completely inhibit T7 phage PDI [64,87]. Similar results were reported for T3 phage in the presence of L-cysteine as the scavenger and proflavine as the PS. However, the photoinactivation rate of MS2 by a polydroxylated fullerene was not affected by the presence of SOD, suggesting a negligible contribution of radical species, such as the superoxide radical anion [90]. T4-like phage PDI was also little or not affected by the presence of free radical scavengers L-cysteine and D-mannitol in the presence of porphyrin derivatives, leading to the conclusion that free radical species are not major participants in phage PDI [107].

3.2.2. Type II Mechanism Quenchers

In general, the action of chemical singlet oxygen quenchers involves the reaction of singlet oxygen with the quenching agent, producing an oxidized product. Another possibility is the deactivation of singlet oxygen to ground state ($^3O_2$) by physical quenching, achieved by either energy or charge transfer, without consumption of oxygen or product formation [101,109]. Residues of histidine, tryptophan and tyrosine in proteins are considered to be major natural quenchers of singlet oxygen [110].

Several singlet oxygen quenchers have been used to evaluate the specific involvement of type II mechanism during viral PDI with different PS (Table 3).

Table 3. Singlet oxygen quenchers used on mammalian viruses and bacteriophage PDI.

PS	Quencher	Microorganism	Quencher protection	Reference
Mammalian viruses				
Aluminum phthalocyanine tetrasulfonate	Sodium azide	VSV	significant effect	[106]
	Tryptophan	VSV	Significant effect	
Rose bengal	β-carotene	Influenza virus	Significant effect	[108]
	Sodium azide			
Hypericin	Sodium azide	HIV	Significant effect	[111]
Methylene blue	Imidazole (5.0 and 10 mM)	HSV-1	55%–75%	[84]
Bacteriophages				
5,10,15-(4-β-D-galactosylphenyl)-20-(pentafluorophenyl)porphyrin	Sodium azide (0.1–5.0 mM)	T7 phage	38%	[87]
5-(pentafluorophenyl)-10,15,20-tris(1-methylpyridinium-4-yl)porphyrin	Sodium azide (100 mM)	T4 phage	80%	[107]
		Qβ	39%	
	L-histidine (50 mM)	T4 phage	74%	
meso-tetrakis(1-methylpyridinium-4-yl)porphyrin	Sodium azide (100 mM)	T4 phage	90%	[107]
	L-histidine (100 mM)	T4 phage	78%	
5,10,15,20-Tetrakis(4-β-D-glucosylphenyl)porphyrin	1,3-diphenylisobenzofuran (0.1–5.0 mM)	T7 phage	42%	[64]
5,10,15-(4-β-D-glucosylphenyl)-20-phenylporphyrin	1,3-diphenylisobenzofuran (0.1-5.0 mM)	T7 phage	74%	[64]
Polyhydroxylated fullerene	β-carotene	T7 phage	69%	[57]
		PRD1 phage	56%	
	β-carotene (26 μM)	MS2 phage	50%–60%	[90]
Rose bengal	Sodium azide (3.5–35 mM)	M13 phage	31%	[52]

3.2.2.1. Singlet Oxygen in PDI of Mammalian Viruses

Singlet oxygen seems to be the most important mediator of virucidal activity (Table 3) on mammalian viruses. The rate of viral photoinactivation is significantly inhibited by oxygen removal or by addition of singlet oxygen quenchers, such as β-carotene, imidazole, L-histidine or sodium azide [45,84,106–108]. Hypericin may induce photochemical alterations on HIV major capsid protein p24, which are inhibited by sodium azide, suggesting that the damage results from singlet oxygen [111]. When merocyanine 540 [45], phthalocyanine derivatives [106] or rose bengal [108] were used as PS, the results suggest that $^1O_2$ is the main cytotoxic species involved in VSV photoinactivation, while type I reactants such as hydroxyl radicals are less important.

3.2.2.2. Singlet Oxygen in PDI of Bacteriophages

Considering the PDI of bacteriophages in the presence of singlet oxygen quenchers, the results (Table 3) suggest that, in most of the studied cases, singlet oxygen is an important mediator of the toxic effect induced by PDI. However, the participation of free radicals cannot be ruled out. For instance, the inactivation of M13 bacteriophage by MB was inhibited from 1.72 log to 0.54 log by sodium azide in a quencher dose-dependent mode, up to a concentration of 3.5 mM. However, photoinactivation occurred even in the presence of sodium azide, suggesting that both type I and type II mechanisms may be involved in the M13 photoinactivation process. In the presence of quencher concentrations ranging from 3.5 to 35 mM, a sodium azide protective effect was not observed, as evidenced by increasing rates of M13 phage photoinactivation, reaching a plateau thereafter [52]. Also, the effect of singlet oxygen quenchers and of hydrogen peroxide indicated singlet oxygen as the main factor responsible for the loss of biological activity of bacteriophage M13 by rose bengal [51].

The efficiency of 5,10,15-(4-β-D-galactosylphenyl)-20-(pentafluorophenyl)porphyrin to photoinactivate T7 phage decreased in 38% in the presence of sodium azide [87]. This result, and the ones obtained in the presence of DMTU (Table 2), proved that for this PS, both mechanisms play a role in T7 phage photoinactivation, with type I being the predominant one. Similar results were obtained by Gábor et al. [64] in the presence of glycoconjugated meso-tetraarylporphyrin derivatives as PS and using 1,3-diphenylisobenzofuran as the singlet oxygen quencher. When T7 phage was phototreated with 5,10,15,20-tetrakis(4-β-D-glucosylphenyl)porphyrin, the rate of inactivation decreased 42% in the presence of 1,3-diphenylisobenzofuran. When 5,10,15-(4-β-D-glucosylphenyl)-20-phenylporphyrin was used, the rate of protection substantially increased (74%). It can then be concluded that the type of PDI mechanism depends on the PS structure, with the symmetric derivative exerting its toxic effect mainly via the generation of free radicals, whether the asymmetric derivative proceeds mainly by singlet production [64]. However, in the study of Egyeki et al. [87] using the same asymmetric 5,10,15-(4-β-D-galactosylphenyl)-20-(pentafluorophenyl)porphyrin as PS, and the same phage, the toxic effect occurred mainly via free radical generation. Besides this, the contribution of type I and type II processes was PS concentration-dependent and the sum of the photoinactivation rate measured in the presence of scavengers was smaller than the one measured without the scavengers. This result may imply a synergism between singlet oxygen and hydroxyl radical-mediated damages or it can also be supposed that the efficiency of neither scavenger is 100% [64,87].

A recent study showed that irradiation of polyhydroxylated fullerene suspensions (40 μM) in the presence of β-carotene reduced the photoinactivation rate of PRD1 and T7 phages, demonstrating singlet oxygen involvement [57]. Also, when the T4-like phage was irradiated in the presence of porphyrin derivatives and singlet oxygen quenchers sodium azide and L-histidine, the rate of phage inactivation was considerably reduced, suggesting that singlet oxygen may be an important mediator of the virucidal activity of these PS [107]. However, from the data obtained, other inactivation mechanisms cannot be excluded [57,107].

Although some data about the importance of the type I and II mechanisms in PDI of bacteriophages are discrepant, in general, it seems that the type II pathway is more important than the type I mechanism in phage PDI. On the other hand, there are only a few studies focusing on the simultaneous effect of singlet oxygen and free radicals scavengers under the same protocol of viral PDI [64,84,87,90,106,107].

4. Molecular Targets of Antiviral PDI

The short-lived ROS generated by photodynamic mechanisms are responsible for the damage induced to critical molecular targets [112]. Different viral targets, such as the envelope lipids and proteins, capsid and core proteins and the nucleic acid can be attacked by singlet oxygen and/or other ROS (hydrogen peroxide, superoxide and hydroxyl radicals) to achieve the loss of infectivity [84]. For a better understanding of the photoinactivation process, the knowledge of how the molecular targets are affected by PDI assumes a great importance [113]. For this reason, a detailed photophysical and photochemical study of the interactions between the toxic species generated by the PS and key biomolecules such as lipids, proteins and nucleic acids is essential for the knowledge and prediction of photosensitization process efficiency [114]. However, the studies performed show that the primary target of PDI depends on the chemical structure of the PS, the targeted virus and the mechanism of photoinactivation [64].

4.1. Nucleic Acids

Depending upon the viruses, the nucleic acid can be either DNA or RNA (single or double stranded). The size of the nucleic acid also varies depending on the viruses. Several studies have shown that both DNA and RNA mammalian viruses and phages are efficiently inactivated by PDI. There is now considerable information that PS like MB can bind to and penetrate viral membranes, whereupon they intercalate with nucleic acids. Upon activation by light, the generated ROS can cause the destruction of the nucleic acids, particularly at guanine residues, preventing viral replication [115]. However, there is a difference in target selectivity depending on the mechanism involved: sugar moieties are usually attacked by radicals (generated via type I process) and guanine residues are the targets of singlet oxygen (generated via type II process) [97].

4.1.1. DNA Damage

From the four DNA bases, guanine is the most susceptible component to suffer a type I photosensitization reaction, due to the fact that it exhibits the lowest oxidation potential among DNA bases and it is the only base that can be oxidized by singlet oxygen (type II process) [116].

The treatment of viruses with MB and other heterocyclic dyes resulted in the damage of viral DNA [53,65,75,76] either by base modification or base loss, single strand breaks (SSB), or cross-links of DNA with proteins [34,75,81,88,117]. It is known that cationic porphyrins can bind to nucleic acids via intercalation into base pairs or self-stacking, inducing lesions upon photoinactivation due to the easy oxidation of guanine residues [118–120].

The binding of cationic porphyrins to DNA is presumably due to the electrostatic interaction between the positively-charged substituents in the porphyrin macrocycle and the negatively charged phosphate oxygen atoms of DNA [120]. However, porphyrin binding to DNA is not a prerequisite for an efficient photosensitization, since free porphyrins are more effective in virus inactivation than the

DNA-bound species [88]. This observation, which is in conflict with the generally accepted idea that the porphyrin molecule must be in close vicinity with the site of photosensitized damage, may be explained by the lower quantum yield of singlet oxygen by the bound porphyrin when compared with the free one [88].

4.1.1.1. Damages in the DNA of Mammalian Viruses

Viral DNA is thought to be a critical target structure for PDI by MB and light [93]. DNA isolated from adenovirus treated with 1.3 µM MB exhibited a smear in Southern blot analysis, indicative of random DNA fragmentation [76]. MB plus light treatment of HSV-1 gives rise to DNA damage and blocks DNA replication [121].

4.1.1.2. Damage in the DNA of Bacteriophages

An internal component of T4 phage has been suggested as an important target because MB needs to cross the outer barrier made by its protein capsids in order to produce a significant effect [65]. In fact, some of the irradiated phages could still inject functional genetic material but have lost their ability to form plaques, suggesting that their DNA was damaged. Protein synthesis was also severely impaired [65]. Treatment of M13 phage with MB and aluminum phthalocyanine tetrasulfonate (AlPcS$_4$) caused strand breaks and piperidine-labile bonds in DNA, which is correlated with the loss of infectivity. This is in agreement with the proposal that lesions of the viral genome might be responsible for the lethality induced by sensitization [81]. DNA strand cleavage was found to be MB concentration and light dose dependent. Viral inactivation and DNA damage were found to be oxygen-dependent processes. However, DNA damage was not correlated with the loss of PM2 phage infectivity, as observed in transfection studies which measured the infectivity of the extracted viral DNA, indicating that DNA from MB-treated phage was just as capable of generating progeny virus as the untreated controls [53]. The observed DNA damage is not correlated with loss of phage infectivity and may not be the prime target of viral PDI, because 100% of closed circular DNA was recovered from the MB phototreated PM2 phage [53]. Concerning the effects of PDI on isolated viral DNA, treatment of M13mp2 DNA with increasing concentrations of MB, in the presence of light, yielded increasing amounts of 8-oxo-7,8-dihydro-2'-deoxyguanosine (8-oxodguo), a prevalent adduct produced by singlet oxygen and perhaps by oxygen free radicals. At 100 µM MB, 1 residue of 8-oxodguo was produced for every 40 residues of deoxyguanosine in DNA. Thus, treatment of M13mp2 DNA with MB plus light resulted in putative alterations at deoxyguanosine residues that impede the progression of DNA synthesis *in vitro* [116].

4.1.2. RNA Damage

RNA has been suggested to be a key factor in viral PDI with many PS, but direct evidence of a correlation between RNA damage and infectivity loss has not been reported yet, as is the case of VSV when treated with phthalocyanine derivatives [81]. In RNA, as for DNA [71], guanine is suggested as the major target for oxidation by photosensitizing agents and light.

4.1.2.1. Damage in the RNA of Mammalian Viruses

VSV genome was damaged by 30 µg mL^{-1} of a chlorophyll derivative and red light illumination which caused a decrease of as much as 85% in RNA polymerase activity, which can be due to damage in the viral RNA polymerase complex, and 98% inhibition of viral RNA synthesis in 6 hours [77]. According to Moor *et al.* [82], the RNA and/or the RNA polymerase complex of VSV might be a major target for its photoinactivation by AlPcS$_4$ and MB. MB and phthalocyanine derivatives inactivated VSV and inhibited fusion of the virus envelope with Vero cells. The degree of inhibition was small compared to the extent of virus inactivation, suggesting that non-membrane targets, like the viral RNA, might be involved in VSV photoinactivation. However, there is no report of a correlation between RNA damage and loss of infectivity [81]. Photoinactivation of HIV-1 by MB and light lead to destruction of its RNA [34].

4.1.2.2. Damage in the RNA of Bacteriophages

Following MB plus light exposure, the Qβ RNA genome exhibited sufficient lethal lesions to account for phage inactivation [122]. However, the protein component of the phage also exerted some effect in viral PDI [122]. In a comparison of RNA photoinactivation using MB and rose bengal as the PS, Schneider *et al.* [54] suggested a causal relationship between 8-oxodguo formation in RNA and R17 and Qβ bacteriophage inactivation. However, no direct relationship between photodynamically induced RNA damage and viral inactivation was described [54]. 8-oxodguo formation or oxidative damage of Qβ RNA alone does not directly account for the lethal event of the virus. Directly treating extracted phage RNA with MB and light caused a loss of activity in the infectious RNA assay but there was a much greater loss of activity if the phage RNA was treated with MB and light in the phage *per se*. The results demonstrated that Qβ RNA infectious activity is significantly more affected by photoinactivation in its protein-associated virion state as compared with its purified isolated polymer state [92,122]. Inactivation of purified RNA by MB and light, in the absence of proteins, most likely occurs due to oxidative damage to the RNA at the site at which MB is bound and might involve oxidized bases such as 8-oxoguanine or strand breaks [122].

In spite of the reduced number of reports focusing on the damage induced by PDI in the nucleic acids of mammalian viruses and bacteriophages, it can be concluded that both DNA and RNA are potential targets of viral PDI. However, there are no studies specifically focusing on the damages induced to DNA and RNA of both mammalian viruses and bacteriophages under the same PDI protocol.

4.2. Outer Structures

Enveloped viruses are inactivated more rapidly than non-enveloped viruses because the destruction of the envelope structure is generally accompanied by loss of virus infectivity [13,40,94,123,124]. The damages caused by photodynamic reactions on unsaturated lipids present in their envelopes and/or on major envelope proteins, which act as PS binding-sites, modify their structure and avoid cell infection and virus replication [50,84]. However, some studies showed that non-enveloped viruses can also be efficiently inactivated by the toxic action of PS [55,56,58,62,64–67,73,81,87,88,94,122].

The higher susceptibility to PDI of enveloped viruses, relatively to non-enveloped viruses, indicates that the viral envelope may be a more important target than nucleic acids for photosensitization. It also indicates that the unsaturated lipids present in the envelope, as well as the major envelope proteins, are important PDI targets. However, as far as it is known, no studies focus on the degradation of viral envelope lipids after PDI or even on other viral internal lipids. There are, however, many studies about the effects of PDI on viral envelope proteins as well as on other core proteins.

The statement that enveloped viruses are more easily inactivated than non-enveloped ones is only based in indirect studies which compare the inactivation results of enveloped and non-enveloped viruses. The enveloped viruses used in PDI protocols [30,36,45,77,81–83] were only assayed for their protein alterations and no additional experimental work was done concerning their lipids. However, the results of PDI obtained by Lytle *et al.* [94] with the enveloped ϕ6 phage, although indirectly, are in good accordance with what is reported in the literature about the major contribution from lipids for the viral photoinactivation process.

Relative to proteins degradation by PDI, the results of different studies showed that the main damage is the formation of protein cross-links, followed by other types of damage, which include loss of proteins, alterations in protein molecular conformation, mass and charge, and alterations in protein band intensity (Table 4).

When proteins are irradiated with UV or visible light in the presence of a PS, photooxidation of sensitive amino acid residues such as cysteine, L-histidine, tyrosine, methionine and tryptophan, and covalent cross-linking of peptide chains can be observed, leading to the formation of molecular aggregates [125,126], disrupting their normal folding conformation, thus forcing them into other conformations that affect their normal functioning [127]. In fact, the formation of cross-linked/aggregated material appears to be a major consequence of photosensitized-mediated protein oxidation [128], and it has been demonstrated that the formation of protein cross-links is not a primary photodynamic event, but a secondary reaction between the photooxidation products of sensitive amino acid residues and other groups in the protein [126].

The PS *per se* can induce alterations in the folding of some enzymes, leading to the exposure of some amino acid residues normally shielded in the protein, and to the shielding of others usually exposed in the molecule. These protein modifications lead to changes in properties such as solubility, proteolytic susceptibility, absorbance, and fluorescence emission of several of their amino acids. These alterations are mainly mediated by hydrogen peroxide and hydroxyl radical generation, although singlet oxygen mediated reactions could also occur [129]. The amino acids located in the surface of the protein are photooxidized at a much faster rate than the residues buried in the interior of the molecule. If a protein is completely unfolded, susceptible amino acids may also be attacked and photodegraded [103,130].

Table 4. Degradation of viral outer structures after mammalian viruses and bacteriophages PDI.

Virus	Type of damage	PS	Reference
Enveloped-mammalian viruses			
HSV-1	Viral envelope (reduced ability to adhere to and penetrate host cells)	Merocyanine 540	[45]
	Viral envelope (prevention of viral adsorption and host penetration)	Phthalocyanine derivatives	[131]
	Glycoprotein D; loss of proteins; dimerization; protein cross-links; alterations in protein molecular mass and charge	Phthalocyanine derivatives	[83]
HSV-2	Viral envelope (prevention of viral adsorption and host penetration)	Phthalocyanine derivatives	[131]
HSV	Protein cross-links	Phthalocyanine derivatives	[132]
VZV	Viral envelope (prevention of viral adsorption and host penetration)	Phthalocyanine derivatives	[131]
HIV	Major capsid protein p24	Hypericin	[111]
HIV-1	Loss of infectivity; loss of fusion function; membrane proteins cross-links	Hypericin	[30]
	Loss of infectivity; loss of fusion function; membrane proteins cross-links	Rose bengal	[30]
	p24 and gp120 proteins; protein cross-links	MB	[34]
	Inhibition of cell fusion activity of Env proteins	Natural and sulfonated tetraarylporphyrins	[36]
VSV	Loss of infectivity; loss of fusion function; cross-linking of G and M proteins	Hypericin	[30]
	Loss of infectivity; loss of fusion function; cross-linking of G and M proteins	Rose bengal	[30]
	Inhibition of fusion of the envelope to Vero cells; G protein	MB	[81]
	Inhibition of fusion of the envelope to Vero cells; G protein	Aluminum phthalocyanine tetrasulfonate	[81]
	G and M proteins; protein cross-links	Phthalocyanine derivatives	[82]
	G, M, L and N proteins; protein cross-links	Chlorophyll derivatives	[77]
Influenza virus	Loss of infectivity; loss of fusion function; cross-linking of G and M proteins	Hypericin	[30]
	Loss of infectivity; loss of fusion function; cross-linking of G and M proteins	Rose bengal	[30]
	Loss of infectivity; HA fusion protein; protein cross-links	Rose bengal	[108]

Table 4. *Cont.*

Virus	Type of damage	PS	Reference
Enveloped-mammalian viruses			
Sendai virus	Loss of infectivity; loss of fusion function; cross-linking of G and M proteins	Hypericin	[30]
	Loss of infectivity; loss of fusion function; cross-linking of G and M proteins	Rose bengal	[30]
Vaccinia virus	Histidine residues in virus proteins	Rose bengal	[86]
Human cytomegalovirus	Viral envelope (reduced ability to adhere to and penetrate host cells)	Merocyanine 540	[45]
Sindbis virus	Viral envelope (reduced ability to adhere to and penetrate host cells)	Merocyanine 540	[47]
	Viral capsid protein	Hypericin	[133]
Friend erythroleukemia virus	Viral envelope (reduced ability to adhere to and penetrate host cells)	Merocyanine 540	[134]
Non-enveloped mammalian viruses			
Adenovirus	Not damaged	Phthalocyanine derivatives	[131]
Enterovirus 71	Appearance/disappearance of protein bands; increase of the protein band intensity	Methylene blue	[85]
T7 phage	Protein capsid; loosening of the protein-DNA interaction	Glycoconjugated *meso*-tetraarylporphyrins	[64]
	Capsid and core proteins; loosening of protein-DNA interaction	Glycoconjugated *meso*-tetraarylporphyrins	[87]
	Capsid proteins; protein cross-links	*meso*-Tetrakis(1-methylpyridinium-4-yl)porphyrin	[88]
	Capsid proteins; protein cross-links	Polyhydroxylated fullerene	[57]
M13 phage	Coat protein	Methylene blue Aluminum phthalocyanine tetrasulfunate	[81]
PRD1 phage	Capsid proteins; protein cross-links; phospholipids (less affected)	Polyhydroxylated fullerene	[57]
Qβ phage	Coat and maturation (A) proteins; formation of protein carbonyls; RNA-protein cross-links	Methylene blue	[89]
	RNA-protein cross-links	Methylene blue	[92]
MS2 phage	A protein	Polyhydroxylated fullerene	[57]

4.2.1. Damage on Mammalian Viral Outer Structures

It has been shown that enveloped viruses can be inactivated due to protein damage [30,82,83,131]. However, while the same treatment is reported to be ineffective against some non-enveloped viruses [83,131], the results from Wong et al. [85] showed that even a non-enveloped virus can be efficiently inactivated due to the damage induced by PDI to its viral proteins (Table 4).

The proteins in the viral envelope of HSV-1 were considered to be major targets of merocyanine 540 photosensitization [45]. Some phthalocyanine derivatives have been shown to induce cross-links in HSV protein that might be responsible for the observed loss of infectivity [132]. Protein analysis by SDS-PAGE, after treatment with phthalocyanine derivatives, revealed irreversible changes in the HSV-1 envelope proteins, which were reflected by the loss of many proteins, the appearance of cross-linked material on the top of the gel and by alterations in the molecular mass and molecular charge of the proteins. These alterations contribute, in all likelihood to HSV-1 inactivation [83].

In VSV treated with 3.75–30 μL mL^{-1} of chlorophyll derivatives and light, the M protein band was not detected, which was accompanied by a decrease in the intensity of the G protein band [77]. Large complexes of proteins were also detected on the top of the gel, indicating that viral PDI cross-linked the proteins [77]. Using a fusion assay and protein analysis, it was shown that MB and AlPcS$_4$ caused a decrease in the intensity of the G-protein (which is known to play a crucial role in binding VSV to the host cell) band and a slight decrease in the intensity of M protein (matrix protein) band and protein cross-links. However, the observed damage in viral proteins could not account for VSV PDI [82]. VSV was inactivated by MB and phthalocyanine derivatives, which inhibited the fusion of the virus envelope to Vero cells. However, the degree of this inhibition was small compared to the extent of virus inactivation (43% inhibition vs. 4.7 log or 99.998% inactivation, for MB) [81]. Abe and Wagner [81] also found few changes in the relative abundance of VSV G protein after MB and AlPcS4 phototreatment, and they also observed additional protein bands on SDS-PAGE analysis [81]. It was found, by Western blot analysis, that HIV-1 p24 and gp120 proteins were altered in size, possibly due to cross-linking, after MB phototreatment [34]. However, using the same PS, AlpcS$_4$ and MB, no changes in protein patterns after SDS-PAGE of the viral proteins were observed, under conditions that caused complete VSV inactivation [135].

The results from Vzorov et al. [36] indicated that the porphyrins inhibited the cell fusion activity of HIV Env proteins (a biological function that is important for viral entry as well as induction of viral cytopathic effects) when expressed from recombinant vectors. These results showed that the viral Env protein is an important target of these compounds [36].

PDI of influenza virus by rose bengal altered the HA fusion protein and led to protein cross-links [108].
Photoinactivation of vaccinia virus with rose bengal significantly altered the concentration and oxidized histidine in vaccinia virus protein, suggesting that inactivation was attributed to alterations in viral proteins, as opposed to nucleic acids [86].

Treatment of of influenza and Sindbis viruses by hypericin [30], lead to an extensive cross-linking of the envelope proteins, which may have impaired the capacity of the viruses to adhere to and penetrate the host cells.

The protein profile of the non-enveloped enterovirus 71 was considerably altered after a low dose PDI and a MB concentration ≥0.5 μM, as revealed by a smearing and the disappearance of several protein bands [85]. However, enterovirus 71 PDI was also due to damages in the viral genome [85].

4.2.2. Damages on Bacteriophage Outer Structures

In spite of the limited available data for enveloped bacteriophages, substantially higher photoinactivation rates compared with other non-enveloped phages were described [94]. The photoinactivation by merocyanine 540 of four bacteriophages, two non-enveloped phages without lipids (phi X174 and T7), a non-enveloped phage with lipids (PRD1), and an enveloped phage with an external lipoprotein envelope (phi 6) was studied by Lytle *et al.* [94]. The survival curves of the different viruses clearly demonstrated different levels of sensitivity to photoinactivation by this PS, with phi 6 being the most sensitive, followed by T7 (21-fold less sensitive). While both PRD1 and phi 6 have lipid components, only phi 6 was photoinactivated by the PS. Thus, the internal lipid components of PRD1 were not sufficient to allow photoinactivation by merocyanine 540. A higher inactivation rate with a fullerene derivative was also observed by Hotze *et al.* [57] for a phage without lipids (T7 phage) than for PRD1 phage. The dissimilarities in phage composition resulted from differential resistance to singlet oxygen by the outer structures, since PRD1 has a double capsid with an internal lipid membrane, whereas T7 has a single proteinaceous capsid lacking lipids, and both phages contain double stranded DNA with similar GC content (48% for T7 and 51% for PRD1) [57]. Phage proteins were significantly affected by photosensitization (30–92%) when compared to the relatively smaller effect on nucleic acids in both PRD1 and T7, and lipids in PRD1 phage (≤13%), as assessed by FTIR spectra analysis [57]. The higher T7 phage inactivation is consistent with greater damage to its proteinaceous capsid. Besides this, SDS-PAGE analysis further evidenced that oxidative cross-linking of capsid proteins induced by exogenous singlet oxygen is the likely cause of phage inactivation [57]. The high propensity for MS2 phage inactivation by this PS (compared to PRD1 and T7 phages) possibly arises from damage to its A protein, which is necessary for infecting its host *Escherichia coli* since it contains highly reactive amino acids such as methionine, cysteine, histidine, and tyrosine and not to damages to the nucleic acid [57]. Glycosylated substituted porphyrins led to structural changes at the protein capsid and/or loosening of the protein-DNA interaction, which can be responsible for T7 phage inactivation [64]. Besides of the alteration of the DNA structure, the phototreatment pointed to significant alterations in the protein structure and/or in the DNA-protein interaction, which may be the cause of photodynamic inactivation [87,88]. The alterations in the DNA secondary structure might also be the result of photochemical damage in phage capsid proteins and consequent disruption of the phage particle. Photomodification of core proteins can also lead to phage inactivation, even if the primary structure of the DNA part is preserved, since these proteins play an important role in the early events of infection and DNA penetration [87]. The damage of T7 nucleoprotein is a complex process and clearly both phage DNA and protein capsid are affected by photoreactions [88]. Irradiation of Qβ bacteriophage in the presence of increasing concentrations of MB resulted in exponentially increasing amounts of viral RNA-protein cross-linkage products, and this is probably the most important event in viral inactivation [92]. The RNA genome of Qβ bacteriophage contained sufficiently lethal lesions following MB plus light exposure to account for the resulting phage inactivation. Nevertheless, the data also indicate that the protein component of the phage somehow contributes to the inactivation of the phage [122]. The protein component of Qβ phage is involved in the process of photoinactivation because the formation of protein carbonyls and RNA-protein cross-links were efficiently formed by MB plus light exposure [89]. The close correlation of cross-link formation with phage inactivation and the expectation that even one such cross-link in a phage genome would be lethal makes the RNA-protein cross-link lesion a strong candidate for the primary inactivating lesion of Qβ phage exposed to MB and light [122].

Little alteration of M13 phage proteins on SDS-PAGE after MB and AlPcS$_4$ photoinactivation was observed by Abe and Wagner [81]. The results of Zupán *et al.* [136], suggested that the tetracationic porphyrin *meso*-tetrakis(1-methylpyridinium-4-yl)porphyrin did not interact with capsid proteins and did not disturb protein-DNA interaction, even if it has a strong stabilization effect on the intraphage DNA.

5. Resistance to PDI and Recovery of Viability

The development of increasing numbers of antiviral agents over the past decades, in the same way as with antibiotics, has provided the clinician with therapeutic options previously unavailable. With the increasing utilization of antiviral drugs, however, has come an enhanced appreciation of the development of antiviral resistance [1,7,137,138,139,140]. Drug resistance is costly to the health service, to the patient who fails to gain maximum therapeutic benefit, and for the community in which resistant viruses may be spread [9].

There is now an urgent need for the development of novel, convenient and inexpensive measures for combating antimicrobial-untreatable infections and limiting the development of additional antimicrobial resistant microorganisms. Photodynamic technology may provide one approach to meet this need, both in terms of therapy and in terms of sterilization, by a mechanism that is markedly different from that typical of most antimicrobials [1,141,142].

As mentioned before, photosensitization involves the generation of singlet oxygen and free radical species, which cause molecular damage. Whether microorganisms could develop resistance to these active oxygen species is still questionable [143] and, consequently, the development of microbial resistance to photosensitization is still under debate. Until now, the development of microbial resistance to PDI is not known and is thought very improbable to be developed. In general, the development of resistance to PDI by microbial strains should be considered as an unlikely event since this process is typically multi-target, with ROS causing damage to many microbial components, which is at a variance with the mechanism of action of most antimicrobial drugs [139,144,145]. In contrast to most common antimicrobials, the number of molecular alterations required to ensure survival would be too great and the microorganism would require multi-site mutations to become highly resistant, an event with significantly lower probability than single-site mutations, which is often sufficient for conferring resistance to small-molecule inhibitors [42,146]. This particular property of antimicrobial PDI is important regarding the repeated treatment of chronic and/or recurrent infections [139].

Antimicrobial PDI, when compared to standard treatments which may require application for several weeks to achieve an effective killing of the microorganism, shortly after initiation of light exposure, exhibits serious and irreversible damage of microorganisms [66,68]. This damage does not allow the creation or operation of any kind of anti-drug or mutagenic mechanism. Antimicrobial PDI is therefore very effective and, up until now, no photosensitization-resistant mutants have been found [68].

5.1. Resistance of Mammalian Viruses and Recovery of Viability after Photosensitization

Data from North *et al.* [33] show that HIV azidothymidine (AZT)-resistant strains were as susceptible as the AZT-sensitive ones to photosensitization with a benzoporphyrin derivative. This finding comes as no surprise since the mechanisms of action of AZT (inhibition of reverse transcription) and light-activated benzoporphyrin derivative are different. Thus, mutations in the virus that occur at the reverse transcriptase level will not affect photodynamic destruction [33].
Studies focusing on the possible development of viral resistance are extremely scarce and little is known about the recovery of viral viability after consecutive photodynamic treatments.

5.2. Bacteriophage Resistance and Viability Recovery after Photosensitization

Concerning bacteriophages, there is only one study focusing on the possible development of viral resistance after photosensitization [68]. After 10 consecutive cycles of photodynamic treatment, a T4-like phage, in the presence of the tricationic porphyrin 5-(pentafluorophenyl)-10,15,20-tris(1-methylpyridinium-4-yl)porphyrin (Tri-Py$^+$-Me-PF) at 5.0 μM under white light irradiation, exhibited no changes in the rate of photoinactivation during the course of the experiments, meaning that no resistance was observed. If phage resistance would occur, important reductions on phage photoinactivation efficiency would be detected between experiments. Besides that, T4-like phage did not recover its viability after exposure to Tri-Py$^+$-Me-PF during 120 min of irradiation [68]. In a preliminary study by Perdrau and Todd [12], all attempts at reactivating the inactivated *Staphylococcus* phage by MB were unsuccessful.

6. Factors Affecting Viral PDI

6.1. Effect of the Number of Charges, Symmetry, Size of Meso Substituent Groups and Photosensitizer Concentration

It has been shown that the location and binding site of the PS, which is highly dependent on the structure and intramolecular charge distribution, is an important factor in microbial PDI [143,147].

In terms of molecular structure, molecular charge is important in determining antimicrobial activity. Positively charged PS are generally more efficient and can act at lower concentrations than neutral and anionic PS molecules [144]. The positive charges on the PS molecule appear to promote a tight electrostatic interaction between the positively charged PS and the negatively charged sites at the viral capsids and envelopes, orientating the PS toward sites which are critical for the stability and metabolism of a particular microorganism [44,147,148]. This kind of association increases the efficiency of the photoinactivation process.

Cationic PS photodamage can be induced in nucleic acid or viral outer structures by PS binding or by PS localized in its vicinity [136]. For instance, it is more likely that positively charged PS will be effective in causing nucleic acid damage than will neutral or anionic congeners, which mainly act against the outer side of the microorganism [149].

The symmetry and the size of the chain of *meso* substituent groups also affect the photodynamic effect. PS with opposite charged groups are more symmetrical than PS with adjacent charged groups. The adjacent positive charges in the PS macrocycle should result in a molecular distortion due to electrostatic repulsion [150]. The toxicity of a PS can be modulated by the introduction of selected substituents on the macrocycle periphery. In this way, the physicochemical properties of a synthetic PS can be manipulated in order to enhance its interactions with the structural features of the viruses, such as viral capsids, and to minimize the interactions with plasma membranes or mammalian cell membranes [44].

The amphiphilic nature of a PS is another important feature affecting PDI efficiency and can be modulated by the introduction of adequate functionalities in the macrocycle periphery, such as different numbers of positive charges, an asymmetrical charge distribution, or introduction of aromatic hydrocarbon side chains [16,151].

PS concentration is also an important parameter that must be taken into account since viral PDI was shown to be strongly influenced by PS concentration. Increasing the PS concentration reduces the time needed to achieve complete viral inactivation, thus increasing the efficiency of a particular PDI protocol [66].

6.1.1. Mammalian Viruses PDI

Complete inactivation of VSV (4.2 log) can be obtained by treating it with 1.0 µM of the anionic phthalocyanine derivative AlPcS$_4$ and 5 min illumination with red light. For the neutral phthalocyanine derivative (Pc$_4$), complete inactivation (4 log) was achieved using a much lower amount of PS (4.5 nM) in combination with 10 min illumination [82]. The inactivation of VSV in PBS showed a linear relationship with illumination time [82]. Inactivation of the fusion activity of VSV, influenza and Sendai viruses was reached with nanomolar concentrations of hypericin and rose bengal and was absolutely dependent upon light and increased with increasing time of illumination [30]. HAV in PBS or plasma was completely inactivated within 10 min (>3.7 log) by the cationic symmetric porphyrin *meso*-tetrakis(1-methylpyridinium-4-yl)porphyrin. In contrast, inactivation of HAV to 3.6 log with the anionic symmetric porphyrin *meso*-tetrakis(4-sulfonatephenyl)porphyrin required 90 min [44].

The rate and extent of inactivation appeared to vary with the nature of the *meso* substituent groups [44]. HIV and VSV lost infectivity upon illumination with hypericin and rose bengal in a concentration-dependent manner [30].

6.1.2. Bacteriophage PDI

MS2 phage inactivation has been observed with neutral porphyrin derivatives. However, this required higher irradiation periods (30 min) than for the cationic ones (1 min) [44]. Neutral glycosylated substituted porphyrins can also significantly photoinactivate the T7 phage [64,87]. The T4-like phage PDI was achieved by exposing the phage in the presence of six cationic porphyrins at different concentrations (0.5, 1.0 and 5.0 μM) to white light for 270 min. The results showed that phage photoinactivation varied according with the PS concentration, with higher concentrations being the most efficient ones [66]. The T4-like phage PDI also varied with the number of porphyrin charges, with tri- and tetracationic porphyrin derivatives being more effective in viral inactivation that the dicationic ones, which inactivated the phage below the limit of detection. Tetra- and tricationic porphyrin derivatives (*meso*-tetrakis(1-methylpyridinium-4-yl)porphyrin and 5-(pentafluorophenyl)-10,15,20-tris(1-methylpyridinium-4-yl)porphyrin, respectively) lead to complete T4-like phage inactivation (~7 log) after 270 min of irradiation with 40 W m^{-2} [66]. This tetracationic porphyrin showed similar results in another study (7 log of reduction) for lambda phage inactivation, when irradiated with light of 658 nm [58]. Increasing porphyrin concentration at a fixed light dose leads to increased viral inactivation [58]. A concentration-dependent effect was also detected with a porphyrin derivative [87], but over 2.0 μM of PS the process was saturated. A further increase in porphyrin concentration did not lead to a higher inactivation rate of T7 phage. Aggregation and/or photobleaching of PS are likely explanations [87]. Cationic *meso*-tetrakis(1-alkylpyridinium-4-yl)porphyrin derivatives with different alkyl substituent groups were tested for MS2 phage inactivation but, with the exception of 5,10,15,20-tetrakis(4-sulfonatophenyl)porphyrin, showed toxicity even in the absence of light [44].

In a study conducted by Gábor *et al.* [64], the porphyrin derivative with symmetrical glycosylated groups was found to be twice as effective as the asymmetrical one on the inactivation process of T7 phage. According to Costa and colleagues [66], the rate of T4-like phage inactivation was also dependent on the lipophilic character of the *meso*-substituent groups. The presence of a lipophilic aryl group in one of the *meso* positions of the porphyrin core appears to have an important role in phage inactivation, affecting the rate and efficiency of T4-like phage [66]. Casteel *et al.* [44] have also observed differences in the photoinactivation rate of MS2 phage when they used PS with different alkyl substituent groups and concluded that the rate and extent of inactivation appeared to vary with the nature of the *meso* substituent groups.

6.2. Effect of Different Light Sources and Fluence Rate on Antimicrobial PDT

PDT requires a source of light to activate the PS by exposing it to visible or near-visible light at a specific wavelength [152]. The light source for PDT must also exhibit suitable spectral characteristics coinciding preferentially with the maximum absorption wavelength range of the PS, applied in order to generate enough ROS to produce an efficient toxic effect [153].

In parallel with the advances in chemistry (related with the discovery and synthesis of new and more efficient PS) there has also been much activity in developing new light sources, better suited for the photosensitization process. Briefly, these include user-friendly lasers frequently based on solid state laser diodes, as well as inexpensive light emitting diodes (LED) and filtered broad-band lamps [154].

PS activation has been achieved via a variety of light sources, such as arc plasma discharge lamps, metal halogen lamps, slide projector illumination assemblies, and a variety of lasers. For treatment of larger areas, non-coherent light sources, such as tungsten filament, quartz halogen, xenon arc, metal halide, and phosphor-coated sodium lamps, are in use. Recently, non-laser light sources, such as LED, have also been applied in PDT. These light sources are much less expensive and small, lightweight and highly flexible, its lifetime can reach up to one hundred thousands hours, and can be manufactured to wavelengths that activate commercially available PS [152,155–159].

At first glance, the available literature on fluence rate effects for PDT seems contradictory. Some studies indicate less damage at low fluence rate, others indicate more killing at lower, compared to higher, fluence rates for the same total fluence and some indicate no influence of fluence rate at all [152,157,158]. A reduction in the fluence rate lowers the rate of oxygen consumption, thereby extending the radius over which singlet oxygen may be formed and consequently increasing the phototoxic effect [159]. Qin *et al.* [160] showed that an increase in the fluence rate increases microbial damage, although, it seems to have an upper limit of photons to observe this effect. Since each PS molecule can only absorb one photon at a time, when the number of light photons bypasses the number of PS molecules, the PS will no longer be able to absorb the photons "in excess" and the rate of PDI will not increase. In fact, if the number of photons is higher than this limit, the antimicrobial effect will decrease because the dye in suspension will not absorb all the excess light [160]. Schindl *et al.* [161] referred that the biological effect of light depends on the fluence, irrespective of the time over which this dose is delivered. Maclean *et al.* [162] also indicate that the inactivating light may be applied at high irradiance over a short time or at lower irradiance over a longer time. A numerical model, assuming that the rate of photodynamic damage occurring at time t is proportional to the fluence rate at that time and the local concentrations of PS and oxygen can be established. However, according to this model, relatively low fluence rates can be nearly as effective as high fluence rate sources if applied over the same period of time [163].

There is also a direct correlation between the phototoxic effect and the PS concentration and light fluence. With a lowering of the PS concentration, more light has to be applied to achieve identical effects, and *vice versa*. Lower doses of PS require higher activating light fluences, and higher fluence requires a longer duration of light application [96].

6.2.1. Effect of Light on Mammalian Viruses PDI

The effects of dengue virus inactivation were increased with the increase of MB concentration, the enhancement of power density of the light source and the extension of illumination time, as well as the decrease of illumination distance. This enabled the narrow bandwidth light system to kill or inactivate the enveloped virus at much greater distance in much shorter time [74]. VSV in the presence of MB was rapidly inactivated by red (provided by LED incident light at 272 W cm^{-2}) or green-yellow light (provided by low-pressure sodium lamps at a fluence rate of 165 W cm^{-2}) but slower by white light (provided by a bank of fluorescent tubes at a fluence rate of 42 W cm^{-2}) [46], showing that higher power densities produce a high rate of viral inactivation than low fluence rates. Wagner *et al.* [164] also showed that red light of 9 W m^{-2}, given at a total dose of 1.8×10^4 and 3.2×10^4 J m^{-2}, inactivated MB-treated VSV by 6 and ≥ 7 log, respectively. VSV inactivation was linearly dependent on the fluence rate of red light illumination [165].

6.2.2. Effect of Light on Bacteriophage PDI

In terms of what is known about phage PDI, only one study focusing on the effect of different light sources and power densities [67] exists. In this study, cationic porphyrin derivatives (*meso*-tetrakis(1-methylpyridinium-4-yl)porphyrin and 5-(pentafluorophenyl)-10,15,20-tris(1-methylpyridinium-4-yl)porphyrin), when irradiated with different sources of light (fluorescent PAR lamps, sun light and halogen lamp) with fluence rates ranging from 40 W m^{-2} to 1690 W m^{-2}, efficiently photoinactivated non-enveloped phages. All light sources tested lead to reductions of about 7 log for the somatic T4-like phage. However, the rate and the extent of inactivation were dependent on the light source, namely when low fluence rates were used (40 W m^{-2}) and on the energy dose, being considerably more effective when light was delivered at a lower fluence rate. However, depending on the light source used, different irradiation periods were required to inactivate T4-like phage to the limits of detection. The results also showed that the efficacy of T4-like phage inactivation, using the same fluence rate, was dependent on the light source used, in particular when the light is delivered at a low fluence rate. M13 phage was phototreated with 5.0 μM MB and was inactivated in an irradiation dose-dependent manner [52]. Kastury and Platz [58] showed that increasing the concentration of a PS at a fixed light dose leads to increased viral inactivation as does an increase in the total light exposure at a fixed PS concentration. The inactivation rate of T1 bacteriophage increased with increasing fluence rate, indicating that the distance of the sample from the light source is a variable which must be controlled [73]. At higher PS concentrations, the inactivation rate reaches a maximum and then decreases, because the filtering effect of the dye decreases the effective fluence rate [73]. In a simple model purposed by Lee *et al.* [56], the phage survival ratio can also be considered as a decreasing exponential fraction of the light fluence (assuming that the fluence is uniform throughout the system).

7. Conclusion

The efficiency of different types of PS in viral PDI has been proved for different types of mammalian viruses and bacteriophages, whether they are enveloped or non-enveloped, for either DNA or RNA viruses. Even though enveloped viruses are more easily inactivated than non-enveloped ones, several studies confirm that non-enveloped mammalian viruses and phages can be efficiently inactivated by PDI. The type of viral nucleic acid has not been described as an important factor affecting viral photoinactivation but, as far as it is known, no studies specifically focus on the photoinactivation behaviour of DNA and RNA viruses. However, RNA phage MS2 was highly susceptible to photoinactivation when compared with DNA phages under the same conditions of photosensitization.

The type of mechanisms involved in the process of viral photosensitization was already elucidated and singlet oxygen and free radical species were identified as important contributors for an effective viral PDI. However, the contribution of singlet oxygen seems to be more pronounced in mammalian viruses and bacteriophage PDI. There are, however, few studies simultaneously comparing the contribution of both types of mechanisms (type I and type II) involved in viral PDI. The primary targets for the photoinactivation of viruses, whether treating mammalian viruses or phages, are the outer structures. Although there are several studies about the specific effects of PDI on viral proteins, for different types of mammalian viruses and phages, there are no studies concerning the specific effects of PDI on viral lipids. However, it has been clearly shown that enveloped viruses are more easily inactivated than their non-enveloped counterparts, which imply that the lipids present on viral envelopes are important targets of viral PDI.

PS are effective in inactivating the phages to the limits of detection in a way that they do not recover viability, avoiding the development of viral resistance. Nothing is known yet for the particular case of mammalian viruses but, as the viral targets are the same for mammalian viruses and phages, it is also expected that no resistance will be developed in the case of mammalian viruses. Besides that, antiviral PDI is equally effective whether the mammalian virus is sensitive or resistant to conventional antiviral agents. Taking into account all these advantages, PDI for viral inactivation can be regarded as a promising alternative therapy to conventional antiviral treatments, namely for the disinfection of blood and blood products, preventing viral contamination and for the treatment of wound and burn infections. Viral PDI has a fast mode of action and has also the additional benefits of being more economical and an environmental friendly technology, which might be successfully used also in the environmental field for wastewater, drinking water and fish-farming water disinfection.

Different PS concentrations and different light sources and fluence rates were tested, showing that they are important PDI parameters that must inevitably be taken into account when a viral photosensitization protocol has to be elaborated. The inactivation of mammalian viruses and phages can be attained at micromolar-level PS concentrations and different light sources are equally effective, depending on the final dose at which the viruses are exposed to. Besides that, PS can also be modulated by the addition of different *meso* substituent groups and positive charges in order to facilitate their interactions with the viruses, making them more efficient for mammalian viruses and phage PDI.

The similarity of the results obtained for mammalian viruses and bacteriophages show that they exhibit a similar behaviour when submitted to viral photoinactivation techniques: (i) the PS used for viral PDI were equally effective in the photoinactivation of mammalian viruses and bacteriophages; (ii) the mechanism of mammalian viruses and bacteriophage photosensitization involves the production of singlet oxygen (type II mechanism) with a slight contribution of free radical species (type I mechanism); (iii) singlet oxygen and free radicals were shown to affect viral nucleic acids and also the proteins and lipids present in the mammalian viruses and bacteriophage outer surfaces, with the latter being considerably more affected by PDI; and (iv) the rate and extent of mammalian viruses and phage PDI is also affected by the same factors, like the PS concentration and number of positive charges, the nature and position of *meso* substituent groups, the fluence rate and energy dose. Consequently, it is important to persist in the development of more PDI phage studies to clarify some aspects of viral PDI, such as influence of viral nucleic acid type (DNA or RNA) in the photoinactivation efficiency and the possibility of viral resistance development and viability recovery after photosensitization. It will also be important to study the synergistic effect between viral PDI and antiviral classical methodologies using bacteriophages as models of mammalian virus' photoinactivation.

Acknowledgments

Thanks are due to the University of Aveiro, Fundação para a Ciência e a Tecnologia (FCT) and FEDER for funding the QOPNA unit (project PEst-C/QUI/UI0062/2011) and to Centre for Environmental and Marine Studies (CESAM) for funding the Microbiology Research Group. Liliana Costa is also grateful to FCT for her grant (SFRH/BD/39906/2007).

References

1.   Jori, G.; Brown, S.B. Photosensitized inactivation of microorganisms. *Photochem. Photobio. Sci.* **2004**, *3*, 403–405.

2.   Van Der Poel, W.H.; Vinjé, J.; van Der Heide, R.; Herrera, M.I.; Vivo, A.; Koopmans, M.P. Norwalk-like calicivirus genes in farm animals. *Emerg. Infect. Dis.* **2000**, *6*, 36–41.

3.   Blerkom, L.V.L. Role of viruses in human evolution. *Yearbk. Phys. Anthropol.* **2009**, *46*, 14–46.

4.   Pulitzer, M.P.; Amin, B.D.; Busam, K.J. Merkel cell carcinoma: Review. *Adv. Anat. Pathol.* **2009**, *16*, 135–44.

5.   Sullivan, V.; Biron, K.K.; Talarico, C.; Stanat, S.C.; Davis, M.; Pozzi, L.M.; Coen, D.M. A point mutation in the human cytomegalovirus DNA polymerase gene confers resistance to ganciclovir and phosphonylmethoxyalkyl derivatives. *Antimicrob. Agents Chemother.* **1993**, *37*, 19–25.

6.   Smee, D.F.; Barnett, B.B.; Sidwell, R.W.; Reist, E.J.; Holy, A. Antiviral activities of nucleosides and nucleotides against wild-type and drug-resistant strains of murine cytomegalovirus. *Antivir. Res.* **1995**, *26*, 1–9.

7.   Kimberlin, D.W.; Whitley, R.J. Antiviral resistance: Mechanisms, clinical significance, and future implications. *J. Antimicrob. Chemother.* **1996**, *37*, 403–421.

8. Jabs, D.A.; Enger, C.; Forman, M.; Dunn, J.P. for The cytomegalovirus retinitis and viral resistance study group. Incidence of foscarnet resistance and cidofovir resistance in patients treated for cytomegalovirus retinitis. *Antimicrob. Agents Chemother.* **1998**, *42*, 2240–2244.

9. Pillay, D.; Zambon, M. Antiviral drug resistance. *Br. Med. J.* **1998**, *317*, 660–662.

10. Smee, D.F.; Sidwell, R.W.; Kefauver, D.; Bray, M.; Huggins, J.W. Characterization of wild-type and cidofovir-resistant strains of camelpox, cowpox, monkeypox, and vaccinia viruses. *Antimicrob. Agents Chemother.* **2002**, *46*, 1329–1335

11. Schultz, E.W.; Krueger, A.P. Inactivation of *Staphylococcus* bacteriophage by methylene blue. *Proc. Soc. Exp. Biol. Med.* **1928**, *26*, 100–101.

12. Perdrau, J.R.; Todd, C. The photodynamic action of methylene blue on certain viruses. *Proc. R. Soc. Lond. B Biol. Sci.* **1933**, *112*, 288–298.

13. Käsermann, F.; Kempf, C. Buckminsterfullerene and photodynamic inactivation of viruses. *Rev. Med. Virol.* **1998**, *8*, 143–151.

14. Hamblin, M.R.; Hasan, T. Photodynamic therapy: A new antimicrobial approach to infectious disease? *Photochem. Photobiol. Sci.* **2004**, *5*, 436–450.

15. Felber, T.D.; Smith, E.B.; Knox, J.M.; Wallis, C.; Melnick, J.L. Photodynamic inactivation of herpes simplex: Report of a clinical trial. *J. Am. Med. Assoc.* **1973**, *92*, 223–289.

16. Almeida, A.; Cunha, A.; Faustino, M.A.F.; Tomé, A.C.; Neves, M.G.P.M.S. Porphyrins as Antimicrobial Photosensitizing Agents. In *Photodynamic Inactivation of Microbial Pathogens: Medical and Environmental Applications*; Royal Society of Chemistry: Cambridge, UK, 2011; pp. 83–160.

17. Mullooly, V.M.; Abramson, A.L.; Shikowitz, M.J. Dihemato-porphyrin ether-induced photosensitivity in laryngeal papilloma patients. *Lasers Surg. Med.* **1990**, *10*, 349–356.

18. Karrer, S.; Szeimies, R.M.; Abels, C.; Wlotzke, U.; Stolz, W.; Landthaler, M. Epidermodysplasia verruciformis treated using topical 5-aminolaevulinic acid photodynamic therapy. *Br. J. Dermatol.* **1999**, *140*, 935–938.

19. Lavie, G.; Mazur, Y.; Lavie, D.; Meruelo, D. The chemical and biological properties of hypericin—A compound with a broad spectrum of biological activities. *Med. Res. Rev.* **1995**, *15*, 111–119.

20. Smetana, Z.; Malik, Z.; Orenstein, A.; Mendelson, E.; Ben-Hur, E. Treatment of viral infections with 5-aminolevulinic acid and light. *Lasers Surg. Med.* **1997**, *21*, 351–358.

21. Sloand, E.M.; Pitt, E.; Klein, H.G. Safety of the blood supply. *J. Am. Med. Assoc.* **1995**, *274*, 1368–1373.

22. Mannucci, P.M. Outbreak of hepatitis A among Italian patients with haemophilia. *Lancet* **1992**, *339*, 819.

23. Klein, H.G. Oxygen carriers and transfusion medicine. *Artif. Cell. Blood Substit. Biotechnol.* **1994**, *22*, 123–135.

24. Azzi, A.; Fanci, R.; Ciappi, S.; Zakrzewska, K.; Bosi, A. Human parvovirus B19 infection in bone marrow transplantation patients. *Am. J. Hematol.* **1993**, *44*, 207–209.

25. Asanaka, M.; Kurimura, T.; Toya, H.; Ogaki, J.; Kato Y. Anti-HIV activity of protoporphyrin. *AIDS* **1989**, *3*, 403–404.

26. Dixon, D.W.; Marzilli, L.G.; Schinazi R.F. Porphyrins as agents against the human immunodeficiency virus. *Ann. N. Y. Acad. Sci.* **1990**, *616*, 511–513.

27. Lambrecht, B.; Mohr, H.; Knuver-Hopf, J.; Schmitt, H. Photoinactivation of viruses in human fresh plasma by phenothiazine dyes in combination with visible light. *Vox Sang.* **1991**, *60*, 207–213.

28. Levere, R.D.; Gong, Y.F.; Kappas, A.; Bucher, D.J.; Wormser, G.; Abraham, N.G. Heme inhibits human immunodeficiency virus 1 replication in cell cultures and enhances the antiviral effect of zidovudine. *Proc. Natl. Acad. Sci. USA* **1991**, *88*, 1756–1759.

29. Matthews, J.L.; Sogandares-Bernal, F.; Judy, M.; Gulliya, K.; Newman, J.; Chanh, T.; Marengo-Rowe, A.J. Inactivation of viruses with photoactive compounds. *Blood Cell.* **1992**, *18*, 75–88.

30. Lenard, J.; Rabson, A.; Vanderoef, R. Photodynamic inactivation of infectivity of humam immunodeficiency virus and other enveloped viruses using hypericin and rose bengal: Inhibition of fusion and syncytia formation. *Proc. Natl. Acad. Sci. USA* **1993**, *90*, 158–162.

31. Neurath, A.R.; Strick, N.; Jiang, S. Rapid prescreening for antiviral agents against HIV-1 based on their inhibitory activity in site-directed immunoassays—Approaches applicable to epidemic HIV-1 strains. *Antivir. Chem. Chemother.* **1993**, *4*, 207–214.

32. Debnath, A.K.; Jiang, S.; Strick, N.; Lin, K.; Haberfield, P.; Neurath, A.R. 3-Dimensional structure-activity analysis of a series of porphyrin derivatives withanti-HIV-1 activity targeted to the v3 loop of the gp120 envelope glycoprotein of the human-immunodeficiency-virus type 1. *J. Med. Chem.* **1994**, *37*, 1099–1108.

33. North, J.; Coombs, R.; Levy, J. Photodynamic inactivation of free and cell-associated HIV-1 using the photosensitizer, benzoporphyrin derivative. *J. Acquir. Immune Defic. Syndr.* **1994**, *7*, 891–898.

34. Bachmann, B.K.-H.J.B.; Lambrecht, B.; Mohr, H. Target structures for HIV-1 inactivation by methylene blue and light. *J. Med. Virol.* **1995**, *47*, 172–178.

35. Song, R.; Witvrouw, M.; Schols, D.; Robert, A.; Balzarini, J.; de Clercq, E.; Bernadou, J.; Meunier, B. 1997. Anti-HIV activities of anionic metalloporphyrins and related compounds. *Antivir. Chem. Chemother.* **1997**, *8*, 85–97.

36. Vzorov, A.N.; Dixon, D.W.; Trommel, J.S.; Marzilli, L.G.; Compans, R.W. Inactivation of human immunodeficiency virus type 1 by porphyrins. *Antimicrob. Agents Chemother.* **2002**, *46*, 3917–3925.

37. Vanyur, R.; Heberger, K.; Jakus, J. Prediction of anti-HIV-1 activity of a series of tetrapyrrole molecules. *J. Chem. Inform. Comput. Sci.* **2003**, *43*, 1829–1836.

38. Dairou, J.; Vever-Bizet, C.; Brault, D. Interaction of sulfonated anionic porphyrins with HIV glycoprotein gp120: photodamages revealed by inhibition of antibody binding to V3 and C5 domains. *Antivir. Res.* **2004**, *61*, 37–47.

39. Marchesan, S.; Da Ros, T.; Spalluto, G.; Balzarini, J.; Prato, M. Anti-HIV properties of cationic fullerene derivatives. *Bioorg. Med. Chem. Lett.* **2005**, *15*, 3615–3618.

40. North, J.; Freeman, S.; Overbaugh, J.; Levy, J.; Lansman, R. Photodynamic inactivation of retrovirus by benzoporphyrin derivative: A feline leukemia virus model. *Transfusion* **1992**, 32, 121–128.

41. Müller-Breitkreutz, K.; Mohr, H. Hepatitis C and human immunodeficiency virus RNA degradation by methylene blue/light treatment of human plasma. *J. Med. Virol.* **1998**, *56*, 239–245.

42. Cheng, Y.; Tsou, L.K.; Cai, J.; Aya, T.; Dutschman, G.E.; Gullen, E.A.; Grill, S.P.; Chen, A.P.-C.; Lindenbach, B.D.; Hamilton, A.D.; *et al.* A novel class of meso-tetrakis-porphyrin derivatives exhibits potent activities against hepatitis C virus genotype 1b replicons *in vitro*. *Antimicrob. Agents Chemother.* **2010**, *54*, 197–206.

43. Lin L.; Hu, J. Inhibition of hepadnavirus reverse transcriptase RNA interaction by porphyrin compounds. *J. Virol.* **2008**, *82*, 2305–2312.

44. Casteel, B.M.J.; Jayaraj, K.; Avram, G.; Bail, L.M.; Sobsey, M.D. Photoinactivation of hepatitis A virus by synthetic porphyrins. *Photochem. Photobiol.* **2004**, *80*, 294–300.

45. O'Brien, J.M.; Gaffney, D.K.; Wang, T.P.; Sieber, F. Merocyanine 540 sensitized photoinactivation of enveloped viruses in blood products: Site and mechanism of phototoxicity. *Blood* **1992**, *80*, 277–285.

46. Mohr, H.; Bachmann, B.; Klein-Struckmeier, A.; Lambrecht, B. Virus inactivation of blood products by phenothiazine dyes and light. *Photochem. Photobiol.* **1997**, *65*, 441–445.

47. Sieber, F.; O'Brien, J.M.; Krueger, G.J.; Schober, S.L.; Burns, W.H.; Sharkis, S.J.; Sensenbrenner, L.L. Antiviral activity of merocyanine 540. *Photochem. Photobiol.* **1987**, *46*, 707–711.

48. Leclerc, H.; Edberg, S.; Pierzo, V.; Delattre, J.M. Bacteriophages as indicators of enteric viruses and public health risk in groundwaters. A review. *J. Appl. Microbiol.* **2000**, *88*, 5–21.

49. Rywkin, S.; Ben-Hur, E.; Malik, Z.; Prince, A.M.; Li, Y.S.; Kenney, M.E.; Oleinick, N.L.; Horowitz, B. New phthalocynanines for photodynamic virus inactivation in red blood cell concentrates. *Photochem. Photobiol.* **1994**, *60*, 165–170.

50. Käsermann, F.; Kempf, C. Photodynamic inactivation of enveloped viruses by buckminsterfullerene. *Antivir. Res.* **1997**, *34*, 65–70.

51. DiMascio, P.; Wefers, H.; Do-Thi, H-P.; Lafleur, M.V.M.; Sies, H. Singlet molecular oxygen causes loss of biological activity in plasmid and bacteriophage DNA and induces single strand breaks. *Biochim. Biophys. Acta* **1989**, *1007*, 151–157.

52. Abe, H.; Ikebuchi,K.; Wagner, S.J.; Kuwabara, M.; Kamo, N.; Sekiguchi, S. Potential involvement of both type I and type II mechanisms in M13 virus inactivation by methylene blue photosensitization. *Photochem. Photobiol.* **1997**, *66*, 204–208.

53. Specht, K.G. The role of DNA damage in PM2 viral inactivation by methylene blue photosensitization. *Photochem. Photobiol.* **1994**, *59*, 506–514.

54. Schneider, J.E.; Philips, J.R.; Pye, Q.; Maidt, M.L.; Price, S.; Floyd, R.A. Methylene blue and rose bengal photoinactivation of RNA bacteriophages: Comparative studies of 8-oxoguanine formation in isolated RNA. *Arch. Biochem. Biophys.* **1993**, *301*, 91–97.

55. Jockush, S.; Lee, D.; Turro, N.J.; Leonard, E.F. Photoinduced inactivation of viruses: Adsorption of methylene blue, thionine and thiopyronine on Qβ bacteriophage. *Proc. Natl. Acad. Sci. USA* **1996**, *93*, 7446–7451.

56. Lee, D.; Foux, M.; Leonard, E.F. The effects of methylene blue and oxygen concentration on the photoinactivation of Qβ bacteriophage. *Photochem. Photobiol.* **1997**, *65*, 161–165.

57. Hotze, E.M.; Badireddy, A.R.; Chellam, S.; Wiesner, M.R. Mechanisms of bacteriophage inactivation via singlet oxygen generation in UV illuminated fullerol suspensions. *Environ. Sci. Technol.* **2009**, *43*, 6639–6645.

58. Kasturi, C.; Platz, M.S. Inactivation of lambda phage with 658 nm light using a DNA binding porphyrin sensitizer. *Photochem. Photobiol.* **1992**, *56*, 427–429.

59. Martin, C.B.; Wilfong, E.; Ruane, P.; Goodrich, R.; Platz, M. An action spectrum of the riboflavin-photosensitized inactivation of lambda phage. *Photochem. Photobiol.* **2005**, *81*, 474–480.

60. Wagner, S.J.; Skripchenkol, A.; Robinenel, D.; Foley, J.W.; Cincotta, L. Factors affecting virus photoinactivation by a series of phenothiazine dyes. *Photochem. Photobiol.* **1998**, *67*, 343–349.

61. Brendel, M. Different photodynamic action of proflavine and methylene blue on bacteriophage. I. Host cell reactivation of *Serratia*phage *kappa*. *Mol. Gen. Genet.* **1970**, *108*, 303–311.

62. Yamamoto, N. Photodynamic inactivation of bacteriophage and its inhibition. *J. Bacteriol.* **1957**, *6*, 510–521.

63. Witmer, H.; Fraser, D. Photodynamic action of proflavine on coliphage T3 II. Protection by L-cysteine. *J. Virol.* **1971**, *7*, 319–322.

64. Gábor, F.; Szolnoki, J.; Tóth, K.; Fekete, A.; Maillard, P.; Csík, G. Photoinduced inactivation of T7 phage sensitized by symmetrically and asymmetrically substituted tetraphenyl porphyrin: comparison of efficiency and mechanism of action. *Photochem. Photobiol.* **2001**, *73*, 304–311.

65. Kadish, L.L.; Fisher, D.B.; Pardee, A.B. Photodynamic inactivation of free and vegetative bacteriophage T4. *Biochim. Biophys. Acta* **1967**, *138*, 57–65.

66. Costa, L.; Alves, E.; Carvalho, C.M.B.; Tomé, J.P.C.; Faustino, M.A.F.; Neves, M.G.P.M.S.; Tomé, A.C.; Cavaleiro, J.A.S.; Cunha, A.; Almeida, A. Sewage bacteriophage photoinactivation by cationic porphyrins: a study of charge effect. *Photochem. Photobiol. Sci.* **2008**, *7*, 415–422.

67. Costa, L.; Carvalho, C.M.B.; Faustino, M.A.F.; Neves, M.G.P.M.S.; Tomé, J.P.C.; Tomé, A.C.; Cavaleiro, J.A.S.; Cunha, Â.; Almeida, A. Sewage bacteriophage inactivation by cationic porphyrins: influence of light parameters. *Photochem. Photobiol. Sci.* **2010**, *9*, 1126–1133.

68. Costa, L.; Tomé, J.P.C.; Neves, M.G.P.M.S.; Tomé, A.C.; Cavaleiro, J.A.S.; Faustino, M.A.F.; Cunha, Â.; Gomes, N.C.M.; Almeida, A. Evaluation of resistance development and viability recovery by a non-enveloped virus after repeated cycles of aPDT. *Antivir. Res.* **2011**, *91*, 278–282.

69. DeRosa, M.C.; Crutchley, R.J. Photosensitized singlet oxygen and its applications. *Coord. Chem. Rev.* **2002**, *233–234*, 351–371.

70. Capella, M.A.M.; Capella, L.S.A light in multidrug resistance: Photodynamic treatment of multidrug-resistant tumors. *J. Biomed. Sci.* **2003**, *10*, 361–366.

71. Castano, A.P.; Demidova, T.N.; Hamblin, M.R. Mechanisms in photodynamic therapy: Part one-photosensitizers, photochemistry and cellular localization. *Photodiagn. Photodyn.* **2004**, *1*, 279–293.

72. Prates, R.A.; da Silva, E.G.; Yomada, A.M.; Jr.; Suzuki, L.C.; Paula, C.R.; Ribeiro, M.S. Light parameters influence cell viability in antifungal photodynamic therapy in a fluence and rate fluence dependent manner. *Laser Phys.* **2009**, *19*, 1038–1044.

73. Welsh, J.N.; Adams, M.H. Photodynamic inactivation of bacteriophage. *J. Bacteriol.* **1954**, *1*, 122–127.

74. Huang, Q.; Fu, W-L.; Chen, B.; Huang, J-F.; Zhang, X.; Xue, Q. Inactivation of dengue virus by methylene blue/narrow bandwidth light system. *J. Photochem. Photobiol. B Biol.* **2004**, *77*, 39–43.

75. Schnipper, L.E.; Lewin, A.A.; Swartz, M.; Crumpacker, C.S. Mechanisms of photodynamic inactivation of herpes simplex viruses; comparison between methylene blue, light plus electricity, and hematoporphyrin plus light. *J. Clin. Investig.* **1980**, *65*, 432–438.

76. Schagen, F.H.E.; Moor, A.C.E.; Cheong, S.C.; Cramer, S.J.; van Ormondt, H.; van der Eb, A.J.; Dubbelman T.M.A.R.; Hoeben, R.C. Photodynamic treatment of adenoviral vectors with visible light: An easy and convenient method for viral inactivation. *Gene Ther.* **1999**, *6*, 873–881.

77. Lim, D.-S.; Ko, S.-H.; Kim, S.-J.; Park, Y.-J.; Park, J.-H.; Lee, W.-Y. Photoinactivation of vesicular stomatitis virus by a photodynamic agent, chlorophyll derivatives from silkworm excreta. *J. Photochem. Photobiol. B Biol.* **2002**, *67*, 149–156.

78. Sagristá, M.L.; Postigo, F.; De Madariaga, M.A.; Pinto, R.M.; Caballero, S.; Bosch, A.; Vallés, M.A.; Mora, M. Photodynamic inactivation of viruses by immobilized chlorine-containing liposomes. *J. Porphyrin Phthalocyanines* **2009**, *13*, 578–588.

79. Tomé, J.P.C.; Neves, M.G.P.M.S.; Tomé, A.C.; Cavaleiro, J.A.S.; Mendonça, A.F.; Pegado, I.N.; Duarte, R.; Valdeira, M.L. Synthesis of glycoporphyrin derivatives and their antiviral activity against herpes simplex virus types 1 and 2. *Bioorg. Med. Chem.* **2005**, *13*, 3878–3888.

80. Silva, E.M.P.; Giuntini, F.; Faustino, M.A.F.; Tomé, J.P.C.; Neves, M.G.P.M.S.; Tomé, A.C.; Silva, A.M.S.; Santana-Marques, M.G.; Ferrer-Correia, A.J.; Cavaleiro, J.A.S.; *et al.* Synthesis of cationic β-vinyl substituted *meso*-tetraphenylporphyrins and their *in vitro* activity against herpes simplex virus type 1. *Bioorg. Med. Chem. Lett.* **2005**, *15*, 3333–3337.

81. Abe, H.; Wagner, S.J. Analysis of viral DNA, protein and envelope damage after methylene blue, phthalocyanine derivative or merocyanine 540 photosensitization. *Photochem. Photobiol.* **1995**, *61*, 402–409.

82. Moor, A.C.E.; Wagenaars-van Gompel, A.E.; Brand, A.; Dubbelman, T.M.A.R.; Van Steveninck, J. Primary targets for photoinactivation of vesicular stomatitis virus by AlPcS$_4$ or Pc$_4$ and red light. *Photochem. Photobiol.* **1997**, *65*, 465–470.

83. Smetana, Z., Ben-Hur, E.; Mendelson, E.; Salzberg, S.; Wagner, P.; Malik, Z. Herpes simplex virus proteins are damaged following photodynamic inactivation with phthalocyanines. *J. Photochem. Photobiol. B Biol.* **1998**, *44*, 77–83.

84. Müller-Breitkreutz, K.; Mohr, H.; Briviba, K.; Sies, H. Inactivation of viruses by chemically and photochemically generated singlet molecular oxygen. *J. Photochem. Photobiol. B Biol.* **1995**, *30*, 63–70.

85. Wong, T.-W.; Huang, H.-J.; Wang, Y.-F.; Lee, Y.-P.; Huang, C.-C.; Yu, C.-K. Methylene blue-mediated photodynamic inactivation as a novel disinfectant of enterovirus 71. *J. Antimicrob. Chemother.* **2010**, *65*, 2176–2182.

86. Turner, G.S.; Kaplan, C. Photoinactivation of vaccinia virus with rose bengal. *J. Gen. Virol.* **1968**, *3*, 433–443.

87. Egyeki, M.; Turóczy, G.; Majer, Zs.; Tóth, K.; Fekete, A.; Maillard, Ph.; Csík, G. Photosensitized inactivation of T7 phage as surrogate of non-enveloped DNA viruses: Efficiency and mechanism of action. *Biochim. Biophys. Acta* **2003**, *1624*, 115–124.

88. Zupán, K.; Egyeki, M.; Tóth, K.; Fekete, A.; Herényi, L.; Módos, K.; Csík, G. Comparison of the efficiency and the specificity of DNA-bound and free cationic porphyrin in photodynamic virus inactivation. *J. Photochem. Photobiol. B Biol.* **2008**, *90*, 105–112.

89. Schneider, J.E.; Jr.; Tabatabale, T.; Maidt, L.; Smith, R.H.; Nguyen, X.; Pye, Q.; Floyd, R.A. Potential mechanisms of photodynamic inactivation of virus by methylene blue I. RNA-protein crosslinks and other oxidative lesions in Qβ bacteriophage. *Photochem. Photobiol.* **1998**, *67*, 350–357.

90. Badireddy, A.R.; Hotze, E.M.; Chellam, S.; Alvarez, P.J.J.; Wiesner, M.R. Inactivation of bacteriophages via photosensitization of fullerol nanoparticles. *Environ. Sci. Tech.* **2007**, *41*, 6627–6632.

91. Marotti, J.; Aranha, A.C.C.; Eduardo, C.D.P.; Ribeiro, M.S. Photodynamic therapy can be effective as a treatment for herpes simplex labialis. *Photomed. Laser Surg.* **2009**, *27*, 357–363.

92. Floyd, R.A.; Schneider, J.E.; Dittmer, D.P. Methylene blue photoinactivation of RNA viruses. *Antivir. Res.* **2004**, *61*, 141–151.

93. Wallis, C.; Melnick, J.L. Photodynamic inactivation of animal viruses: A review. *Photochem. Photobiol.* **1965**, *4*, 159–170.

94. Lytle, C.D.; Budacz, A.P.; Keville, E.; Miller, S.A.; Prodouz, K.N. Differential inactivation of surrogate viruses with merocyanine 540. *Photochem. Photobiol.* **1991**, *54*, 489–493.

95. Via, L.D.; Magno, S.M. Photochemotherapy in the treatment of cancer. *Curr. Med. Chem.* **2001**, *8*, 1405–1418.

96. Schmidt-Erfurth, U.; Hasan, T. Mechanisms of action of photodynamic therapy with verteporfin for the treatment of age-related macular degeneration. *Surv. Ophthalmol.* **2000**, *45*, 195–214.

97. Wainwright, M. Photodynamic antimicrobial chemotherapy (PACT). *J. Antimicrob. Chemother.* **1998**, *42*, 13–28.

98. Bonnett, R. *Chemical Aspects of Photodynamic Therapy*; Gordon and Breach Science Publishers: Amsterdam, The Netherlands, 2000.

99. Girotti, A.W. Photosensitized oxidation of membrane lipids: Reaction pathways, cytotoxic effects, and cytoprotective mechanisms. *J. Photochem. Photobiol. B Biol.* **2001**, *63*, 103–113.

100. Calin, M.A.; Parasca, S.V. Light sources for photodynamic inactivation of bacteria. *Laser Med. Sci.* **2009**, *24*, 453–460.

101. Min, D.B.; Boff, J.M. Chemistry and reaction of singlet oxygen in foods. *Compr. Rev. Food Sci. Food Saf.* **2002**, *1*, 58–72.

102. Maisch, T.; Bosl, C.; Szeimies, R.M.; Lehn, N.; Abels, C. Photodynamic effects of novel XF porphyrin derivativeson prokaryotic and eukaryotic cells. *Antimicrob. Agents Ch.* **2005**, *49*, 1542–1552.

103. Ochsner, M. Photophysical and photobiological processes in the photodynamic therapy of tumours. *J. Photochem. Photobiol. B Biol.* **1997**, *39*, 1–18.

104. Wondrak, G.T.; Jacobson, M.K.; Jacobson, E.L. Identification of quenchers of photoexcited states as novel agents for skin photoprotection. *J. Pharmacol. Exp. Therapeut.* **2005**, *312*, 482–491.

105. Sies, H. Oxidative stress: Oxidants and antioxidants. *Exp. Physiol.* **1997**, *82*, 291–295.

106. Rywkin, S.; Lenny, L.; Goldstein, J.; Geacintov, N.E.; Margolis-Nunno, H.; Horowitz, B. Importance of type I and type II mechanisms in the photodynamic inactivation of viruses in blood with aluminum phthalocyanine derivatives. *Photochem. Photobiol.* **1992**, *56*, 463–469.

107. Costa, L.; Tomé, J.P.C.; Faustino, M.A.F.; Neves, M.G.P.S.; Tomé, A.C.; Cavaleiro, J.A.S.; Cunha, A.; Almeida, A. Involvement of type I and type II mechanisms on the photoinactivation of non-enveloped DNA and RNA bacteriophages. *Environ. Sci. Technol.* **2012**, submitted for publication.

108. Lenard, J.; Vanderoef, R. Photoinactivation of influenza virus fusion and infectivity by rose bengal. *Photochem. Photobiol.* **1993**, *58*, 527–531.

109. Bisby, R.H.; Morgan, C.G.; Hamblett, I.; Gorman, A.A. 1999. Quenching of singlet oxygen by trolox c, ascorbate, and amino acids: effects on pH and temperature. *J. Phys. Chem. A* **1999**, *103*, 7454–7459.

110. Baker, A.; Kanofsky, J.R. Quenching of singlet oxygen bybiomolecules from Ll210 leukemia cells. *Photochem. Photobiol.* **1992**, *55*, 523–528.

111. Degar, S.; Prince, A.M.; Pascual, D.; Lavie, G.; Levin, B.; Mazur, Y.; Lavie, D.; Ehrlich, L.S.; Carter, C.; Meruelo, D. Inactivation of the human immunodeficiency virus by hypericin: Evidence for photochemical alterations of p24 and a block in uncoating. *AIDS Res. Hum. Retrovir.* **1992**, *8*, 1929–1936.

112. Wainwright, M. Local treatment of viral disease using photodynamic therapy. *Int. J. Antimicrob. Agents* **2003**, *21*, 510–520.

113. Garcia, G.; Sarrazy, V.; Sol, V.; Morvan, C.L.; Granet, R.; Alves, S.; Krausz, P. DNA photocleavage by porphyrin–polyamine conjugates. *Bioorg. Med. Chem.* **2009**, *17*, 767–776.

114. Miranda, M.A. Photosensitization by drugs. *Pure Appl. Chem.* **2001**, *73*, 481–486.

115. Wainwright, M. The use of methylene blue derivatives in blood product disinfection. *Int. J. Antimicrob. Agents* **2000**, *16*, 381–394.

116. McBride, T.J.; Schneider, J.E.; Floyd, R.E.; Loeb, L.A. Mutations induced by methylene blue plus light in single stranded M13mp2. *Proc. Natl. Acad. Sci. USA* **1992**, *89*, 6866–6870.

117. OhUigin, C.; McConnell, D.J.; Kelly, J.M.; van der Putten, W.J.M. Methylene blue photosensitised strand cleavage of DNA: Effects of dye binding and oxygen. *Nucleic Acids Res.* **1987**, *15*, 7411–7427.

118. Mettath. S.; Munson, B.R.; Pandey, R.K. DNA interaction and photocleavage properties of porphyrins containing cationic substituents at the peripheral position. *Bioconjugate Chem.* **1999**, *10*, 94–102.

119. Kubát, P.; Lang, K.; Anzenbacher, P.; Jr.; Jursíkova, K.; Král, V.; Ehrenberg, B. Interaction of novel cationic *meso*-tetraphenylporphyrins in the ground and excited states with DNA and nucleotides. *J. Chem. Soc. Perkin Trans. 1* **2000**, *1*, 933–941.

120. Caminos, D.A.; Durantini, E.N. Interaction and photodynamic activity of cationic porphyrin derivativesbearing different patterns of charge distribution with GMP and DNA. *J. Photochem. Photobiol. A: Chem.* **2008**, *198*, 274–281.

121. Müller-Breitkreutz, K.; Mohr, H. Infection cycle of herpes viruses after photodynamic treatment with methylene blue and light. *Transfusions Medizin* **1997**, *34*, 37–42.

122. Schneider, J.E.; Jr.; Pye, Q.; Floyd, R.A. Qβ bacteriophage photoinactivated by methylene blue plus light involves inactivation of its genomic RNA. *Photochem. Photobiol.* **1999**, *70*, 902–909.

123. Smetana, Z.; Mendelson, E.; Manor, J.; Van Lier, J.E.; Ben-Hur, E.; Salzberg, S.; Malik, Z. Photodynamic inactivation of herpes simplex viruses with phthalocyanine derivatives. *J. Photochem. Photobiol. B: Biol.* **1994**, *22*, 37–43.

124. Ben-Hur, E.; Horowitz, B. Virus inactivation in blood. *AIDS* **1996**, *11*, 1183–1190.

125. Girotti, A.W. Photodynamic action of protoporphyrin IX on human erythrocytes: Cross-linking of membrane proteins. *Biochem. Biophys. Res. Comm.* **1976**, *72*, 1367–1374.

126. Verweij, H.; van Steveninck, J. Model studies on photodynamic cross-linking. *Photochem. Photobiol.* **1982**, *35*, 265–267.

127. Macdonald, I.J.; Dougherty, T.J. Basic principles of photodynamic therapy. *J. Porphyrin Phthalocyanines* **2001**, *5*, 105–129.

128. Davies, M.J. Singlet oxygen-mediated damage to proteins and its consequences. *Biochem. Biophys. Res. Commun.* **2003**, *305*, 761–770.

129. Afonso, S.G; Enriquez, S.R.; Batlle, C.A.M. The photodynamic and non photodynamic actions of porphyrins. *Braz. J. Med. Biol. Res.* **1999**, *32*, 255–266.

130. Jori, G.; Galiazzo, G.; Tamburro, A.M.; Scoffone, E. Dye-sensitized photooxidation as a tool for determining the degree of exposure of amino acid residues in proteins. *J. Biol. Chem.* **1970**, *245*, 3375–3383.

131. Malik, Z.; Ladan, H.; Nitzan, Y.; Smetana, Z. Antimicrobial and antiviral activity of porphyrin photosensitization. *Proc. SPIE* **1993**, *2078*, 305–312.

132. Malik, Z.; Smetana, Z.; Mendelson, E., Wagner, P.; Salzberg, S.; Ben-Hur, E. Alteration in herpes simplex virus proteins following photodynamic treatment with phthalocyanines. *Photochem. Photobiol.* **1996**, *63*, 59S.

133. Yip, L.; Hudson, J.B.; Gruszecka-Kowalik, E.; Zalkow, L.H.; Neil Towers, G.H.N. Antiviral activity of a derivative of the photosensitive compound hypericin. *Phytomedicine* **1996**, *2*, 185–190.

134. Sieber, F.; Krueger, G.J.; O'Brien, J.M.; Schober, S.L.; Sensenbrenner, L.L.; Sharkis, S.J. Inactivation of Friend erythroleukemia virus and Friend virus-transformed cells by merocyanine 540-mediated photosensitization. *Blood* **1989**, *73*, 345–350.

135. Melki, R.; Gaudin, Y.; Blondel, D. Interaction between tubulin and the viral matrix protein of vesicular stomatitis virus: possible implications in the viral cytopathic effect. *Virology* **1994**, *202*, 339–347.

136. Zupán, K.; Herényi, L.; Tóth, K.; Majer, Z.; Csík, G. Binding of cationic porphyrin to isolated and encapsidated viral DNA analyzed by comprehensive spectroscopic methods. *Biochemistry* **2004**, *43*, 9151–9159.

137. Yoshikawa, T.T. Antimicrobial resistance and aging: beginning of the end of the antibiotic era? *J. Am. Geriatr. Soc.* **2002**, *50*, S226–S229.

138. Malik, Z.; Gozhansky, S.; Nitzan, Y. Effects of photoactivated HPD on bacteria and antibiotic resistance. *Microbios Lett.* **1982**, *21*, 103–112.

139. Maisch, T.; Szeimies, R-M.; Jori, G.; Abels, C. Antibacterial photodynamic therapy in dermatology. *Photochem. Photobiol. Sci.* **2004**, *3*, 907–917.

140. Pillay, D. Emergence and control of resistance to antiviral drugs in resistance in herpes viruses, hepatitis B virus, and HIV. *Commun. Dis. Public Health* **1998**, *1*, 5–13.

141. Reddi, E.; Ceccon, M.; Valduga, G.; Jori, G.; Bommer, J.C.; Elisei, F.; Latterini, L.; Mazzucato, U. Photophysical properties and antibacterial activity of *meso*-substituted cationic porphyrins. *Photochem. Photobiol.* **2002**, *75*, 462–470.

142. Jori, G.; Coppellotti, O. Inactivation of pathogenic microorganisms by photodynamic techniques: Mechanistic aspects and perspective applications. *Anti-Infect. Agents Med. Chem.* **2007**, *6*, 119–131.

143. Minnock, A.; Vernon, D.I.; Schofield, J.; Griffiths, J.; Parish, J.H.; Brown, S.B. Mechanism of uptake of a cationic water-soluble pyridinium zinc phthalocyanine across the outer membrane of *Escherichia coli*. *Antimicrob. Agents Chemother.* **2000**, *44*, 522–527.

144. Demidova, T.; Hamblin, M. Effects of cell-photosensitizer binding and cell density on microbial photoinactivation. *Antimicrob. Agents Chemother.* **2005**, *6*, 2329–2335.

145. Jori, G.; Fabris, C.; Soncin, M.; Ferro, S.; Coppellotti, O.; Dei, D.; Fantetti, L.; Chiti, G.; Roncucci, G. Photodynamic therapy in the treatment of microbial infections: Basic principles and perspective applications. *Lasers Surg. Med.* **2006**, *38*, 468–481.

146. Wainwright, M. Photoantimicrobials—So what's stopping us? *Photodiagn. Photodyn.* **2009**, *6*, 167–169.

147. Merchat, M.; Bertolini, G.; Giacomini, P.; Villanueva, A.; Jori, G. *Meso*-substituted cationic porphyrins as efficient photosensitizers of Gram-positive and Gram-negative bacteria. *J. Photochem. Photobiol.* **1996**, *32*, 153–157.

148. Dowd, S.E., Pillai, S.D.; Wang, S.; Corapcioglu, M.Y. Delineating the specific influence of virus isoelectric point and size on virus adsorption and transport through sandy soils. *Appl. Environ. Microbiol.* **1998**, *64*, 405–410.

149. Wainwright, W.; Photoantimicrobials—A PACT against resistance and infection. *Drugs Future* **2004**, *29*, 85–93.

150. Kessel, D.; Raymund, L.; Vicente, M.G.H. Localization and photodynamic efficacy of two cationic porphyrins varying in charge distribution. *Photochem. Photobiol.* **2003**, *78*, 431–435.

151. Banfi, S.; Caruso, E.; Buccafurni, L.; Battini, V.; Zazzaron, S.; Barbieri, P.; Orlandi, V. Antibacterial activity of tetraaryl-porphyrin photosensitizers: An *in vivo* study on Gram negative and Gram positive bacteria. *J. Photochem. Photobiol. B Biol.* **2006**, *85*, 28–38.

152. Konopka, K.; Goslinski, T. Photodynamic therapy in dentistry. *Crit. Rev. Oral Biol. Med.* **2007**, *8*, 694–707.

153. Robertson, C.A.; Evans, D.H.; Abrahamse, H. Photodynamic therapy (PDT): A short review on cellular mechanisms and cancer research applications for PDT. *J. Photochem. Photobiol. B Biol.* **2009**, *96*, 1–8.

154. Brancaleon, L.; Moseley, H. Laser and non-laser light sources for photodynamic therapy. *Laser Med. Sci.* **2002**, *17*, 173–186.

155. Veenhuizen, R.B.; Stewart, F.A. The importance of fluence rate in photodynamic therapy: Is there a parallel with ionizing radiation dose-rate effects? *Radiother. Oncol.* **1995**, *37*, 131–135.

156. Allison, R.R.; Mota, H.C.; Sibata, C.H. Clinical PD/PDT in North America: an historical review. *Photodiagn. Photodyn.* **2004**, *1*, 263–277.

157. Juzeniene, A.; Juzena, P.; Ma, L-W.; Iani, V.; Moan, J. Effectiveness of different light sources for 5-aminolevulinic acid photodynamic therapy. *Laser Med. Sci.* **2004**, *19*, 139–149.

158. Kübler, A.C. Photodynamic therapy. *Med. Laser Appl.* **2005**, *20*, 37–45.

159. Lukšiene, Z. New approach to inactivation of harmful and pathogenic microorganisms by photosensitization. *Food Tech. Biotechnol.* **2005**, *43*, 411–418.

160. Qin, Y.; Luan, X.; Bi, L.; He, G.; Bai, X.; Zhou, C.; Zhang, Z. Toluidine blue-mediated photoinactivation of periodontal pathogens from supragingival plaques. *Laser Med. Sci.* **2008**, *23*, 49–54.

161. Schindl, A.; Rosado-Sholosser,B.; Trautinger, F. Reciprocity regulation in photobiology: An overview (in German). *Hautarzt* **2001**, *52*, 779–785.

162. Maclean, M.; MacGregor, S.J.; Anderson, J.G.; Woolsey, G.A. The role of oxygen in the visible-light inactivation of *Staphylococcus aureus*. *J. Photochem. Photobiol. B Biol.* **2008**, *92*, 180–184.

163. Langmack, K.; Mehta, R.; Twyman, P.; Norris, P. Topical photodynamic therapy at low fluence rates—theory and practice. *J. Photochem. Photobiol. B Biol.* **2001**, *60*, 37–43.

164. Wagner, S.J.; Storry, J.R.; Mallory, D.A.; Stromberg, R.R.; Benade, L.E.; Friedman, L.I. Red cell alterations associated with virucidal methylene blue phototreatment. *Transfusion* **1993**, *33*, 30–36.
165. Wagner, S.J. Virus inactivation in blood components by photoactive phenothiazine dyes. *Transfus. Med. Rev.* **2002**, *16*, 61–66.

7

Structural Aspects of the Interaction of Dairy Phages with Their Host Bacteria

Jennifer Mahony [1] **and Douwe van Sinderen** [1,2,]*

[1] Department of Microbiology, University College Cork, Western Road, Cork, Ireland;
E-Mail: j.mahony@ucc.ie

[2] Alimentary Pharmabiotic Centre, Biosciences Institute, University College Cork, Western Road, Cork, Ireland

* Author to whom correspondence should be addressed: E-Mail: d.vansinderen@ucc.ie

Abstract: Knowledge of phage-host interactions at a fundamental level is central to the design of rational strategies for the development of phage-resistant strains that may be applied in industrial settings. Phages infecting lactic acid bacteria, in particular *Lactococcus lactis* and *Streptococcus thermophilus*, negatively impact on dairy fermentation processes with serious economic implications. In recent years a wealth of information on structural protein assembly and topology has become available relating to phages infecting *Escherichia coli*, *Bacillus subtilis* and *Lactococcus lactis*, which act as models for structural analyses of dairy phages. In this review, we explore the role of model tailed phages, such as T4 and SPP1, in advancing our knowledge regarding interactions between dairy phages and their hosts. Furthermore, the potential of currently investigated dairy phages to in turn serve as model systems for this particular group of phages is discussed.

Key words: bacteriophage; milk fermentation; receptor

1. Introduction

Phages are the most abundant biological entities on Earth [1] and are responsible, at least in part, for driving the evolution of their bacterial hosts [2]. The selective pressure imposed by phages on their hosts also requires an equivalent genomic plasticity and adaptability by the phages themselves in order to successfully produce progeny in a host that tries to escape its parasites. All characterized phages known to infect lactic acid bacteria (LAB) possess a tail and therefore this review will solely consider tailed phages.

One of the first physical phage-host interactions is that between the receptor on the host bacterium and the distal end of the tail of the phage. Many isolated phages have been characterised morphologically by electron microscopy analysis [3] and the information derived from such analyses may be useful to gain an insight into the position and nature of host receptor (see below). In recent years structural information has emerged related to phage proteins that are involved in binding to these receptors, usually referred to as anti-receptors, receptor binding proteins (RBPs) or adhesins. Depending on the macromolecular nature of the receptor, *i.e.*, being a protein or carbohydrate, the phage tail tip of *Siphoviridae* phages, which contains the receptor-binding protein, appears to display a particular morphology. For example, phages that interact with protein moieties generally possess either a pointed or "stubby" end, while those that interact with carbohydrate moieties more often display a larger, more obvious structure, which is called the baseplate. While there are exceptions to this generalisation, including many undefined phage-host interactions, this may perhaps be the primary indicator of the type of moiety required by a phage to interact with its host. Here we will explore the role of electron microscopy and structural analysis in defining the interactions between phages and their hosts, and examine the various types of receptor-binding proteins encoded by phages and their corresponding host-encoded receptor. The interactions and wealth of structural data for phage RBPs that now exists will be examined with a view to understand the interactions between dairy phages and their hosts. Strains of *Lactococcus lactis* and *Streptococcus thermophilus* are the most widely utilised LAB in the dairy industry, with certain species of *Lactobacillus* also being used intensively. Consequently, phages infecting these species dominate LAB phage research. Since the most frequently isolated, and hence the most problematic phages of dairy lactic acid bacteria (LAB) are those belonging to the *Siphoviridae* (isolated for *Lactococcus*, *Streptococcus* and *Lactobacillus* spp., among others) and, to a lesser extent, *Myoviridae* families (isolated for *Lactobacillus* spp.), we will focus on the former for purposes of clarity.

2. Phages that Recognize a Proteinaceous Receptor

Phages recognizing a proteinaceous receptor can achieve a strong phage-bacterium interaction through protein-protein interactions and therefore only a single phage protein or small complex is required due to the high affinity, avidity and specificity of this interaction [4]. In electron micrographic images such phages possess a pointed or stubby tail tip, which is where the RBP is located (See Table 1 for examples). A very well established example of this is the *Bacillus subtilis*-infecting siphophage SPP1, which interacts irreversibly with its host-encoded receptor protein YueB via its pointed tail tip following an initial and reversible attraction to a carbohydrate moiety on the cell surface [4]. Additionally, the lactococcal c2-type siphophages interact with the phage infection protein (PIP) through a RBP that is presumed to be located at the blunt-ended tail tip [5]. It is noteworthy that both YueB and PIP are members of the Type VII secretion system and are closely related proteins and therefore functional analogies are perhaps not surprising. The tail adsorption protein of c2 is proposed to be encoded by orf *110*, a 75 kDa protein identified through mass spectrophotometry and immune-gold electron microscopy [6].

The bacterial receptors of the *Siphoviridae* coliphages λ and T5 (both of these phages possess pointed tail tips) have been defined and isolated (LamB and FhuA, respectively), which has permitted the development of *in vitro* DNA ejection assays to positively identify these receptor proteins [7,8]. The coliphages belonging to the *Myoviridae* T4 superfamily possess six long tail fibers that interact reversibly with lipopolysaccharide (LPS) or a porin protein, such as OmpC [9,10]. Subsequently, the baseplate is lowered toward the cell surface and the six short tail fibers are released to allow reversible binding to occur, thereby allowing the pointed puncturing device to initiate DNA injection into the host after the baseplate is switched to the star conformation [11]. Of the above-mentioned phages, T4 is one of the best studied model phages with respect to the structural analysis of the tail fibers and injection devices.

The 3D structure of the long tail fiber receptor-binding domain of gp37 has been resolved, which revealed that the aromatic and positive residues on the surface of the fiber tips are most likely receptor-binding determinants with the receptor, although the authors do not specify whether the receptor material is LPS or the OmpC porin protein [9]. The proteins constituting the T4 baseplate complex have been studied by means of cryo-electron microscopy and X-ray crystallography [11–13].

Through these analyses, at least twelve proteins have been identified as baseplate components: gp5, gp6, gp7, gp8, gp9, gp10, gp11, gp12, gp25, gp27, gp53, gp54. The six gp10 trimers are a crucial component of the baseplate and structural analysis of this protein has revealed its role in extension of the short tail fibers encoded by gp12 by means of a partner shift: the pre-infection state sees an interaction between the C-terminal end of gp10 and gp12, while infection requires an interaction between the baseplate protein gp9 and gp10 so as to release the short tail fibers, thus permitting interaction with the host receptor [13]. This fascinating and complex phage serves as an excellent model for the physical interactions between a phage and its host. However, this model is much more sophisticated than many of the phages of the *Siphoviridae* family and for this purpose phage SPP1 may be a better representative for this family of phages.

Table 1. Proven and putative phage receptors.

Host	Phage	Receptor type	Host receptor	Reference
Bacillus subtilis	SPP1	Protein	YueB	[14]
Escherichia coli	T4-like	Protein	OmpC/LPS	[10]
	T5	Protein	FhuA	[8]
	Λ	Protein	LamB	[7]
Lactococcus lactis	c2	Protein	PIP	[5]
	p2	Saccharide	Unknown	[15]
	bIL170	Saccharide	Unknown	[16]
	TP901–1	Saccharide	Unknown	[17]
	Tuc2009	Saccharide	Unknown	
Lactobacillus	LL-H	Lipoteichoic acid	Poly-Glycerophosphate LTAs	[18]
Streptococcus	OBJ	Saccharide	Glucosamine/Ribose	[19]
thermophilus	CYM	Saccharide	Glucosamine/Rhamnose	[19]

3. SPP1 as a Model for Protein-Interacting Siphophages of Gram-Positive Bacteria

SPP1 possesses a long non-contractile tail, which is the trademark property of the *Siphoviridae* phages. At the base of the tail is a narrow device, the tail tip, which interacts with a membrane-anchored protein (YueB) on the host cell surface [14]. Significant structural data has emerged in recent years relating to the capsid, tail, head-tail connector and the tail tip of SPP1, thus rendering it a well understood model for other siphophages infecting Gram-positive bacteria, including those that target LAB [20–27].

One of the first structural studies of phage SPP1 focused on the tail at pre- and post-infection stages, and detailed the loss of the tail tip following DNA ejection [14]. The tail of SPP1 is composed of an estimated 40 stacked rings (gp17.1 and gp17.1*), inside which is a channel containing the tail tape measure protein (gp18) and through which the phage genome is passed upon release from the capsid [14]. The outside of the tail rings exhibit a six-fold symmetry (gp17.1 and gp17.1*, see below), while the inner tube has a continuous appearance that is composed of several gp18 subunits [14]. The tail tip, which has no channel to allow DNA passage, interacts with YueB, which prompts its release from the tail cap, thereby triggering a sequence of conformational changes that result in the ejection of the viral genome from the capsid. Following DNA ejection, the internal channel diameter appears to undergo a size reduction from 56 Å to 42 Å due to rearrangements in the major tail proteins gp17.1 and gp17.1* and not negating the fact that the tail tape measure protein fills the channel prior to DNA ejection [14]. These gp17.1 and gp17.1* proteins are present in a 3:1 ratio and protein microsequencing has demonstrated that the product of gp17.1* contains an additional C-terminal 10 kDa region, which is thought to be exposed on the outer surface of the tail [28]. This feature also seems to be encoded by prophages of *Bacillus licheniformis* and *Bacillus halodurans*, while such extensions also appear to be present in the phage receptor-binding proteins and major tail protein-associated sequences of other *Siphoviridae* phages such as the lactococcal phage Q54 and *Listeria* phage A118 [28].

Figure 1. Schematic representation of a top view of the re-organisation of the p2 (top) and SPP1 (bottom) baseplate regions before (right) and after (left) binding to the host cell. The p2 baseplate, composed of a hexamer of ORF15 (orange), a trimer of ORF16 (green) and six trimers of ORF18 (blue) which represents the RBP. In parallel, the SPP1 distal tail region is composed of two back-to-back hexamers of gp19.1 (orange) and a trimer of gp21 (green) bound to each of the Dit (gp19.1) hexamers, one trimer in the open conformation and the other in the closed conformation. A representation of the genomic regions that encode the tail structural components are presented in the lowermost section, including the genes encoding the proposed major tail protein (pink), the tail tape measure protein (purple), the distal tail protein (Dit)/hub protein (orange), the tail spike (SPP1) or tail tip (p2) (green) and the RBP (blue, p2). Those in grey encode non-structural proteins or have not been functionally assigned.

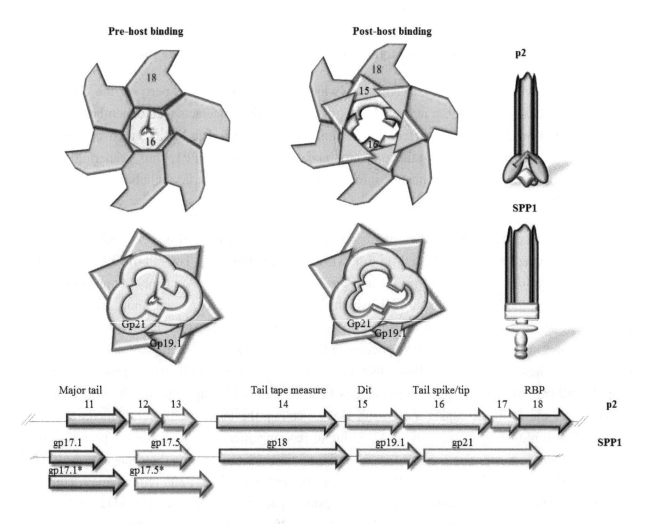

Other elements of the distal end of the tail of SPP1 have been characterised including the distal tail protein (Dit, gp19.1) and the N-terminal region of the tail spike protein (gp21) [23,25]. The SPP1 Dit appeared as a dimer of hexamers in a back-to-back orientation which is structurally similar to a component of the lactococcal phage p2 baseplate (ORF 15), however it is now clear that this apparent dimer is a crystallographic artifact in the case of SPP1 [23]. Two trimers of the SPP1 truncated tail spike protein have been shown to interact with Dit, where one trimer is presumed to be in a "closed" conformation while the other trimer is in the "open" conformation (Figure 1) [25], a system which is consistent with observations made for the lactococcal phage p2 baseplate [25,29]. The Dit-tail spike complex appears to constitute the tail cap of SPP1 and the structural conservation of these features in lactococcal phages may indicate a common lineage of such systems [25]. Calcium ions stabilise the "open" conformation of tail cap, which is also consistent with the lactococcal phage p2 [25,29]. The Dit-tail complex acts as a hub upon which the baseplate (in the case of phages that recognize a saccharidic receptor) or the tail tip/spike (for phages that bind to a proteinaceous receptor) is hinged, at the distal end of the tail. Furthermore, (multimers of) the major tail protein (MTP) and tail tape measure protein (TMP) attach to this structure to form the tail tube. It is suggested that the interaction of the RBP with the host-encoded receptor induces tail base to change to the open conformation triggering the movement of the first ring of MTP which is then sequentially continued up through the MTP rings with ensuing release of DNA from the phage capsid [14,25]. The structural conservation of the Dit-tail cap complex and conformational arrangements in SPP1 and lactococcal phages endorses SPP1 as a model for *Siphoviridae* phages, and as such is a basis for understanding the interactions between dairy phages and their hosts.

4. Structural Approaches to Understand Dairy Phage Interactions and Carbohydrate-Recognizing Phages

Lactococcal phages have, among the dairy phages, received the most extensive scientific scrutiny with regards to the structural analysis of protein/protein complexes [17,29–32]. Isolated lactococcal phages are grouped into ten distinguishable species based on DNA homology and particle morphology [33]. Of these, the 936, c2 and P335 species are most frequently encountered in dairy facilities world-wide. Members of the 936 and P335 species are believed to bind a saccharide component that is present on the surface of the cell envelope, while the c2 phages are known to bind the so-called phage infection protein (PIP) [5]. Members of the 936 and P335 species have been studied at a structural level and constitute the backbone of emerging structural data relating to dairy phages.

4.1. Lactococcal P335 and 936 Phage-Host Interactions

Over the past decade, significant efforts have been made to determine the molecular nature and mechanism of the interactions between lactococcal phages and their hosts, and members of the P335 and 936 species have been central to such studies due to their industrial relevance. It has long been suggested that these phages recognize a carbohydrate moiety on their host, a notion that is consistent with carbohydrate inhibition assays [34]. Through mutational analyses several adjacent genes within an operon, which encodes the biosynthetic machinery for a cell envelope-associated polysaccharide (also called the pellicle), have been shown to be required for adsorption of 936-type phages. The involvement of this pellicle biosynthesis operon in phage infection has been demonstrated in two separate studies for three 936-type phages, namely bIL170, 645 and sk1, although the precise role of the mutated genes as a phage receptor has not yet been established [35,36]. The 20–25 kb pellicle operon contains elements that are conserved among all sequenced lactococcal strains, while it also contains strain-specific regions that are expected to lead to the production of pellicles with a distinct saccharidic structure and composition. The so-called "adsorption genes", or receptor-encoding genes, involved in 936-type phage recognition and adsorption encode either a glycosyltransferase or a membrane-associated protein [35]. The involvement of these gene-products may imply two possibilities of the mode of interaction and DNA injection pathway of these phages. Firstly, that the glycosyltransferase inactivation may prevent the transfer and decoration of the required pellicle monosaccharide component to the surface thus negating the primary contact between the phage RBP and its saccharide receptor; or secondly, if primary contact is established and a membrane component is required for the translocation of the DNA across the membrane, the membrane-associated protein inactivation prevents internalisation of the phage DNA. Chemical mutagenesis of the lactococcal host strain 3107, which is sensitive to the P335 phages LC3 and TP901–1, resulted in the isolation of five genetically and phenotypically distinct mutants resistant to one or both phages [37]. This analysis revealed that the injection pathways employed by these phages of the same species are distinct. It is through studies such as these that more attention has been diverted towards the identification of phage RBPs on lactococcal phages in order to determine the relationship between a phage and its host(s).

The 936 phage RBPs has been identified through the isolation of a chimeric sk1/bIL170, which also was shown to exhibit a "host-range swapping" phenotype [38]. The identification of the RBPs of these phages presented the opportunity to define the structure of the corresponding proteins and their associated complexes. This study applied an approach previously used by Duplessis and Moineau (2001) in which the variable C-terminal VR2 region of ORF18 of DT1 was replaced by the corresponding fragment of MD4. The resultant chimeric DT1 infected the host for MD4, thereby elucidating the host specificity function of this protein [39]. Similarly, a chimeric derivative of phage TP901–1 (designated TP901–1C), harbouring the lower baseplate protein of phage Tuc2009 instead of its original TP901–1 equivalent, was shown to have acquired the host range of Tuc2009, thereby identifying the lower baseplate protein as the RBP of both of these P335 phages [40]. Such discoveries have been very useful in the assignment of RBPs of other P335 phages through homology and genomic positioning of the encoding gene [41]. Also noteworthy is the observation of the altered baseplate morphology in the case of the chimeric phage TP901–1C as compared to the parent phage [36]. TP901–1 has a clear double-disc baseplate while the chimeric phage TP901–1C was described to possess "hanging droplets" rather than a distinct lower baseplate and furthermore the lower disc appeared narrower than that of the parent phage (23 nm *versus* 28 nm in the parent phage) and intermediate between that of the wild-type TP901–1 and Tuc2009. The genetic plasticity of the lactococcal phages has permitted the development of recombinant phages of both the 936 and P335 species and may present a mechanism by which these phages alter their host specificity or enlarge their host spectrum as their environment dictates.

4.2. 936 Phage Baseplates

The first lactococcal phage RBP to be structurally defined was that of the 936 phage p2 [42]. The p2 RBP is a homotrimeric protein comprising three domains: the head, neck and shoulder. The actual receptor binding site was identified within the "head" domain through the application of the heavy chain neutralising llama antibody fragment $V_{HH}5$, which had previously been determined to prevent infection of the lactococcal host by phage p2 [15,42,43]. The shoulder domain possesses eight ß-strands interspersed with an α-helix between ß-strands one and two, and a coil between ß-strands two and three [27]. The neck is a ß-prism of three segments that yields a rigid structure to which analogy has been drawn to the gp12 short tail fiber structure of the coliphage T4. Finally, the head domain is a ß-barrel of seven anti-parallel ß-strands [42]. Seven charged or polar and five non-polar residues were identified in the RBP head domain interacting with the $V_{HH}5$.

The receptor binding site was further characterised through the isolation of escape mutants that could bypass the llama neutralising antibody fragments [15]. The majority of the mutations that enable the mutant phages to overcome the llama antibodies are within the interacting site. Just twelve out of fifty assessed lactococcal 936 phages were neutralised by $V_{HH}5$ antibodies, highlighting the apparent variability of RBPs found in members of this closely related phage species. Mutants overcoming the $V_{HH}5$ were readily isolated for eleven of the twelve phages neutralised and in all cases a single point mutation in the corresponding RBP-encoding gene was shown to be sufficient to prevent antibody recognition. This study also served to improve the resolution of the RBP structure of p2 from 2.3 to 1.7 Å and identified a glycerol molecule bound to the structure originating from the cryoprotectant. The binding of the glycerol molecule was observed to be tighter in the head domain than the shoulder domain where three hydrogen bonds are formed between head domain residues and the glycerol O-1 and O-2 atoms (*versus* one in the shoulder domain). Other saccharides were assessed for their ability to bind the RBP in solution and it was suggested that phosphoglycerol, which is a component of lipoteichoic acid and teichoic acid, represented a potential receptor. This was the first direct proof of carbohydrate binding of a lactococcal 936-type phage RBP, confirming biological analyses that suggested the role of specific carbohydrates as receptors for 936-type phages [15,34].

The crystal structure of the RBP head domain of another 936 phage, bIL170, was solved and represents one of a number of phages that were not neutralised by the $V_{HH}5$ antibodies due to its variable C-terminal sequence [16]. The phylogeny of the RBPs of 936 phages has demonstrated clear links to host range and $V_{HH}5$ antibody-neutralisation, and it is well established that while the N-terminal region is well-conserved, it is the C-terminal (head) domain that determines the specific host interaction [15,45–47]. Interestingly, while there is little sequence conservation between the head domains of p2 and bIL170, their structure is strikingly similar [16]. Furthermore, it bears structural similarity to the head domain of the lactococcal P335 phage TP901–1 [16]. Akin to the RBP of p2, the head domain of bIL170 is proposed to be a homotrimer that binds saccharides near a tyrosine residue (Tyr226) whose side chain forms the upper wall of the crevice between the trimer components and is accessible to solvents [16].

Expression and purification of ORFs 15, 16, 17 and 18 of p2 by affinity chromatography revealed that the baseplate of p2 is composed of three proteins, ORFs 15, 16 and 18. The structural data generated from crystals, obtained from the *in vitro* expressed recombinant baseplate complex, was mapped onto the baseplate structure, generated from electron microscopy images of the virion, provided comparative data to verify the accuracy of the expressed baseplate complex structure [29]. ORF15 and 16 readily mapped onto the virion structural data while ORF18 displayed some variation. This variation was proven to be the result of conformational swapping of the RBP head domains either upward, facing the capsid in an inactive form when calcium is absent, or in the "activated" form in the presence of calcium (or strontium) in which case the RBP head domains rotate 200° downward. Calcium is required for infection by many phages and therefore it is not surprising that it may be involved in the initial stages of the phage-host interaction, however, the conformational swapping of the RBP head domains is the first report of any such baseplate activation for *Siphoviridae* phages [29]. Such studies highlight the importance of electron microscopy analysis to define the native virion baseplate structures because, while the use of camelid antibodies provided data on the baseplate/RBP in the activated state, the state of the free phage baseplate revealed a previously unknown mechanism. While calcium is known to be required for the activation of the baseplate, it is unclear as yet if there is also as role for calcium in the physiology and organization of the bacterial envelope, thereby endorsing phage infection or if the effect is limited to the activation of the phage baseplate.

The application of these "block cloning" strategies to the expression and purification of baseplate complexes of the lactococcal phage p2 and TP901–1 is an important tool to facilitate the study of lactococcal base plates [47,48]. This block cloning also provides discerning information that may define which proteins of an operon form part of the complex and this approach may be applied to any operon encoding structural components of heteromeric organelles. A model for the construction and assembly of lactococcal P335 phage tails has been proposed in which the tail tape measure (TMP), distal tail protein (Dit) and tail-associated lysin (Tal) converge to act as the "initiation complex" upon which the baseplate is assembled [40,41]. Mass spectrometry has recently been used as an alternative method to define the assembly pathway of the baseplates of p2 and TP901–1 [49]. Using this approach, the p2 baseplate is proposed to assemble through the interaction of a hexamer of ORF15 with a preassembled ORF16 trimer followed by the capture of six ORF18 trimers [49]. This approach was also applied to the P335 phage TP901–1 and an assembly pathway was proposed which is outlined below.

4.3. P335 Phage Baseplates

The possible assembly pathway for TP901–1 that was proposed involves the interaction of homotrimers of ORF48 (upper baseplate protein) and ORF49 (lower baseplate protein) [49]. These then form tripods comprised of 3×ORF48 and 9×ORF49, and after interaction between ORF48 with ORF46 (distal tail protein or Dit), oligomerisation of ORF46 may proceed with concomitant baseplate formation with a proposed final composition of 6×ORF46, 18×ORF48 and 54×ORF49 [49]. Furthermore, through electron microscopic analysis of the TP901–1 baseplate together with that of

TP901–1 mutants lacking either BppU (upper baseplate) or BppL (lower baseplate), the structure and composition of the TP901–1 baseplate have been determined [31]. Structural analysis of the RBP of TP901–1 has revealed the similarity of the modularity of the RBP of TP901–1 and the previously studied p2, pointing to common ancestral origins of such structures (Figure 2) [17]. This is consistent with the development of a chimeric RBP with the N-terminal portion of TP901–1 and the C-terminal head domain of p2 [50]. The resulting structure presented domains almost indistinguishable from the parental structures highlighting the modular conservation of these lactococcal phages of distinct species [49]. The head domain of TP901–1 RBP was also observed to bind glycerol and fluorescence quenching further consolidated the carbohydrate affinity of this phage protein, identifying affinity for glycerol and muramyl-dipeptide [17].

Unlike the 936-type phage p2, the TP901–1 baseplate is maintained in a so-called "infection-ready conformation" [31,51]. This observation highlights the bipartite baseplate conformations either maintained in an infection-ready state or in a state requiring activation. It is suggested that the signalling to induce DNA ejection after adsorption of TP901–1 may require a more subtle conformational change than that of p2 and the T4 Myoviruses [51]. In this study, two potential infection pathway scenarios were proposed: first is the possibility that the tail-associated lysin (Tal) may interact with the cell wall peptidoglycan resulting in triggering of DNA release through a so-called cascade effect channelled up through the tail tube in a similar manner proposed for SPP1 [14,51]. The second proposed injection pathway involves the binding of the RBP tripod complexes to the carbohydrate moiety on the cell surface with mechanical alterations leading to signalling to the lowermost ring of the major tail protein with ensuing "un-screwing" of the tail to allow DNA ejection to occur [51]. Furthermore, comparisons have been drawn between the activation systems of p2 and the Myophage T4 upon interaction with a carbohydrate moiety, which is an interesting analogy and further corroborates theories of common ancestral elements and conservation of systems and modular structures in the case of RBPs [46].

Tuc2009 is a temperate P335 phage that is closely related to TP901–1 [52], although its baseplate-encoding operon contains an additional gene designated *bppA* (accessory baseplate protein). This appears to be a non-essential structural element of the Tuc2009 baseplate that is possibly involved in the determination or extension of host range. A low resolution model of the Tuc2009 baseplate has been determined in which the baseplate was proposed to consist of six tripods, each comprising 3×BppU, 3×BppA and 3×BppL, although this model is currently being refined that might align Tuc2009 more closely to TP901–1 in terms of its baseplate structure [30]. Structural analysis of this baseplate may uncover a potential role for BppA, for which a functional assignment remains to be provided. The presence of homologues of BppA in other lactococcal P335 phages (ul36 ORF303, accession no. NP_663688.1; P335 ORF45, accession no. ABI54248.1), indicates that there may be a biological significance and ecological advantage to phages possessing the protein in their baseplate. Furthermore, the recent finding that Tuc2009 is one of the few sequenced members of the P335 species that requires calcium to produce plaques may highlight the subtleties that distinguish individual phages within a species [51]. Although significant data is emerging relating to the structure of P335 phage baseplates and individual components thereof, there remains a large gap between our knowledge of these phages and those of the model phages T4 and SPP1. While analogies may be drawn between these models and dairy phages it is imperative to also define relevant structural models within the dairy phage group.

Figure 2. A schematic representation of the distal tail region of Tuc2009, TP901–1 and p2. The stacks of tail rings composed of the major tail protein (MTP) (blue) with the distal tail (Dit) protein (orange) beneath, upon which the baseplate components are hinged (green and purple). In the case of Tuc2009 the green area represents the upper baseplate disc (BppU and BppA) while the TP901–1 upper disc is composed of BppU only. The lower baseplate (BppL) is presented in purple with a disc representation for TP901–1 and a so-called "petticoat" representation for Tuc2009. The protrusion from the baseplates of TP901–1 and Tuc2009 represent the tail associated lysin (Tal) (red). The schematic of the baseplate region of p2 highlights ORF15 (orange) at the base of the tail with ORF16 beneath (red) and the RBP in the active downward orientation (purple).

Tuc2009   TP901-1   p2

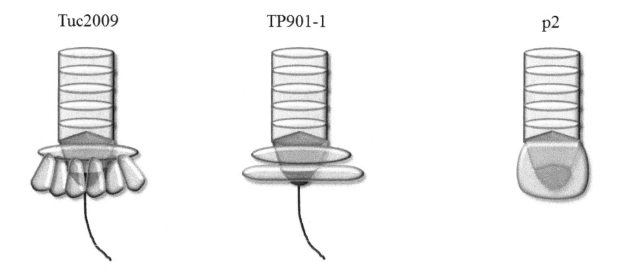

5. Conclusions and Future Perspectives

Structural studies relating to T4 in the 1990's and early 2000's paved the way for the structural analyses of many phage receptor binding proteins and their associated complexes and have permitted the detailed analysis of several phage-host interactions [9–12]. This, together with the isolation of the receptor for T5 and lambda in the 1990's and the 1970's prompted some of the early phage-host interaction studies for lactococcal phages [7,8,34]. Since then, numerous studies relating to the interactions of dairy phages and their hosts have been reported and have led to a surge of interest in structural analyses of these systems. The lactococcal phages p2, TP901–1 and Tuc2009 are emerging as model phages for the dairy *Siphoviridae* phages in terms of structural analyses. These data present a fresh perspective on studies attempting to define the host molecules with which dairy phages interact. The type of baseplate as defined by electron microscopy may be the initial indicator of the material that acts as receptor for the individual phage although this is merely a first glimpse. Structural analysis and/or detailed electron microscopic analysis may provide detailed information on the baseplate organisation and structure, which may be applied to phage-host interaction studies. The analysis of such structures is paramount to understanding how dairy phages bind to and subsequently infect their hosts, and an improved knowledge of phage-host interactions will aid in developing novel and effective anti-phage strategies such as those already explored through the use of camelid antibodies and designed ankyrin repeat proteins (DARPins) to inactivate phages [15,32].

Since structural data relating to these model dairy phages continue to accrue, it is inevitable that these will become the dominant model for the analysis of other dairy phages including those of *S. thermophilus* and *Lb. delbrueckii*, typical cultures used in dairy fermentations. It has been suggested that the receptor material for the *Lb. delbrueckii* phage LL-H is polyglycerophosphate type lipoteichoic acids [18], which is reminiscent of affinity of the 936 phage p2, further demonstrating the usefulness of the lactococcal phage data as a suitable model for other dairy phages. However, while structural analyses are essential to furthering our knowledge of phage-host interactions, it is equally important to apply biological and functional characterisation to verify conclusions drawn from structural analyses. The marriage of molecular microbiology with structural biology represents one of the most powerful modern tools to understand complex interactions and pathways involved in host recognition and infection, and may be the key to developing next generation anti-phage systems with potential application in the dairy industry.

Acknowledgments

D van Sinderen is a recipient of a Science Foundation Ireland (SFI) Principal Investigatorship award (Ref. No. 08/IN.1/B1909).

References

1. Brussow, H.; Hendrix, R.W. Phage genomics: Small is beautiful. *Cell* **2002**, *108*, 13–16.

2. Stern, A.; Sorek, R. The phage-host arms race: Shaping the evolution of microbes. *Bioessays* **2011**, *33*, 43–51.

3. Ackermann, H.W. 5500 phages examined in the electron microscope. *Arch. Virol.* **2007**, *152*, 227–243.

4. Baptista, C.; Santos, M.A.; Sao-Jose, C. Phage SPP1 reversible adsorption to *Bacillus subtilis* cell wall teichoic acids accelerates virus recognition of membrane receptor YueB. *J. Bacteriol.* **2008**, *190*, 4989–4996.

5. Valyasevi, R.; Sandine, W.E.; Geller, B.L. A membrane protein is required for bacteriophage c2 infection of *Lactococcus lactis* subsp. *lactis* c2. *J. Bacteriol.* **1991**, *173*, 6095–6100.

6. Lubbers, M.W.; Waterfield, N.R.; Beresford, T.P.; Le Page, R.W.; Jarvis, A.W. Sequencing and analysis of the prolate-headed lactococcal bacteriophage c2 genome and identification of the structural genes. *Appl. Environ. Microbiol.* **1995**, *61*, 4348–4356.

7. Randall-Hazelbauer, L.; Schwartz, M. Isolation of the bacteriophage lambda receptor from *Escherichia coli. J. Bacteriol.* **1973**, *116*, 1436–1446.

8. Boulanger, P.; le Maire, M.; Bonhivers, M.; Dubois, S.; Desmadril, M.; Letellier, L. Purification and structural and functional characterization of FhuA, a transporter of the *Escherichia coli* outer membrane. *Biochemistry* **1996**, *35*, 14216–14224.

9. Bartual, S.G.; Otero, J.M.; Garcia-Doval, C.; Llamas-Saiz, A.L.; Kahn, R.; Fox, G.C.; van Raaij, M.J. Structure of the bacteriophage T4 long tail fiber receptor-binding tip. *Proc. Natl. Acad. Sci. USA* **2010**, *107*, 20287–20292.

10. Trojet, S.N.; Caumont-Sarcos, A.; Perrody, E.; Comeau, A.M.; Krisch, H.M. The gp38 adhesins of the T4 Superfamily: A complex modular determinant of the phage's host specificity. *Genome Biol. Evol.* **2011**, *3*, 674–686.

11. Kostyuchenko, V.A.; Leiman, P.G.; Chipman, P.R.; Kanamaru, S.; van Raaij, M.J.; Arisaka, F.; Mesyanzhinov, V.V.; Rossmann, M.G. Three-dimensional structure of bacteriophage T4 baseplate. *Nat. Struct. Biol.* **2003**, *10*, 688–693.

12. Leiman, P.G.; Shneider, M.M.; Kostyuchenko, V.A.; Chipman, P.R.; Mesyanzhinov, V.V.; Rossmann, M.G. Structure and location of gene product 8 in the bacteriophage T4 baseplate. *J. Mol. Biol.* **2003**, *328*, 821–833.

13. Leiman, P.G.; Shneider, M.M.; Mesyanzhinov, V.V.; Rossmann, M.G. Evolution of bacteriophage tails: Structure of T4 gene product 10. *J. Mol. Biol.* **2006**, *358*, 912–921.

14. Plisson, C.; White, H.E.; Auzat, I.; Zafarani, A.; Sao-Jose, C.; Lhuillier, S.; Tavares, P.; Orlova, E.V. Structure of bacteriophage SPP1 tail reveals trigger for DNA ejection. *EMBO J.* **2007**, *26*, 3720–3728.

15. Tremblay, D.M.; Tegoni, M.; Spinelli, S.; Campanacci, V.; Blangy, S.; Huyghe, C.; Desmyter, A.; Labrie, S.; Moineau, S.; Cambillau, C. Receptor-binding protein of *Lactococcus lactis* phages: Identification and characterization of the saccharide receptor-binding site. *J. Bacteriol.* **2006**, *188*, 2400–2410.

16. Ricagno, S.; Campanacci, V.; Blangy, S.; Spinelli, S.; Tremblay, D.; Moineau, S.; Tegoni, M.; Cambillau, C. Crystal structure of the receptor-binding protein head domain from *Lactococcus lactis* phage bil170. *J. Virol.* **2006**, *80*, 9331–9335.

17. Spinelli, S.; Campanacci, V.; Blangy, S.; Moineau, S.; Tegoni, M.; Cambillau, C. Modular structure of the receptor binding proteins of *Lactococcus lactis* phages. The RBP structure of the temperate phage TP901–1. *J. Biol. Chem.* **2006**, *281*, 14256–14262.

18. Raisanen, L.; Schubert, K.; Jaakonsaari, T.; Alatossava, T. Characterization of lipoteichoic acids as *Lactobacillus delbrueckii* phage receptor components. *J. Bacteriol.* **2004**, *186*, 5529–5532.

19. Quiberoni, A.; Stiefel, J.I.; Reinheimer, J.A. Characterization of phage receptors in *Streptococcus thermophilus* using purified cell walls obtained by a simple protocol. *J. Appl. Microbiol.* **2000**, *89*, 1059–1065.

20. Vinga, I.; Baptista, C.; Auzat, I.; Petipas, I.; Lurz, R.; Tavares, P.; Santos, M.A.; Sao-Jose, C. Role of bacteriophage SPP1 tail spike protein gp21 on host cell receptor binding and trigger of phage DNA ejection. *Mol. Microbiol.* **2012**, *83*, 289–303.

21. Lhuillier, S.; Gallopin, M.; Gilquin, B.; Brasiles, S.; Lancelot, N.; Letellier, G.; Gilles, M.; Dethan, G.; Orlova, E.V.; Couprie, J.; *et al.* Structure of bacteriophage SPP1 head-to-tail connection reveals mechanism for viral DNA gating. *Proc. Natl. Acad. Sci. USA* **2009**, *106*, 8507–8512.

22. Veesler, D.; Blangy, S.; Lichiere, J.; Ortiz-Lombardia, M.; Tavares, P.; Campanacci, V.; Cambillau, C. Crystal structure of *Bacillus subtilis* SPP1 phage gp23.1, a putative chaperone. *Protein Sci.* **2010**, *19*, 1812–1816.

23. Veesler, D.; Robin, G.; Lichiere, J.; Auzat, I.; Tavares, P.; Bron, P.; Campanacci, V.; Cambillau, C. Crystal structure of bacteriophage SPP1 distal tail protein (gp19.1): A baseplate hub paradigm in gram-positive infecting phages. *J. Biol. Chem.* **2010**, *285*, 36666–36673.

24. Veesler, D.; Blangy, S.; Spinelli, S.; Tavares, P.; Campanacci, V.; Cambillau, C. Crystal structure of *Bacillus subtilis* SPP1 phage gp22 shares fold similarity with a domain of lactococcal phage p2 RBP. *Protein Sci.* **2010**, *19*, 1439–1443.

25. Goulet, A.; Lai-Kee-Him, J.; Veesler, D.; Auzat, I.; Robin, G.; Shepherd, D.A.; Ashcroft, A.E.; Richard, E.; Lichiere, J.; Tavares, P.; *et al.* The opening of the SPP1 bacteriophage tail, a prevalent mechanism in Gram-positive-infecting siphophages. *J. Biol. Chem.* **2011**, *286*, 25397–25405.

26. Chagot, B.; Auzat, I.; Gallopin, M.; Petitpas, I.; Gilquin, B.; Tavares, P.; Zinn-Justin, S. Solution structure of gp17 from the *Siphoviridae* bacteriophage SPP1: Insights into its role in virion assembly. *Proteins* **2012**, *80*, 319–326.

27. White, H.E.; Sherman, M.B.; Brasiles, S.; Jacquet, E.; Seavers, P.; Tavares, P.; Orlova, E.V. Capsid structure and its stability at the late stages of bacteriophage SPP1 assembly. *J. Virol.* **2012**, *86*, 6768–6777.

28. Auzat, I.; Droge, A.; Weise, F.; Lurz, R.; Tavares, P. Origin and function of the two major tail proteins of bacteriophage SPP1. *Mol. Microbiol.* **2008**, *70*, 557–569.

29. Sciara, G.; Bebeacua, C.; Bron, P.; Tremblay, D.; Ortiz-Lombardia, M.; Lichiere, J.; van Heel, M.; Campanacci, V.; Moineau, S.; Cambillau, C. Structure of lactococcal phage p2 baseplate and its mechanism of activation. *Proc. Natl. Acad. Sci. USA* **2010**, *107*, 6852–6857.

30. Sciara, G.; Blangy, S.; Siponen, M.; Mc Grath, S.; van Sinderen, D.; Tegoni, M.; Cambillau, C.; Campanacci, V. A topological model of the baseplate of lactococcal phage Tuc2009. *J. Biol. Chem.* **2008**, *283*, 2716–2723.

31. Bebeacua, C.; Bron, P.; Lai, L.; Vegge, C.S.; Brondsted, L.; Spinelli, S.; Campanacci, V.; Veesler, D.; van Heel, M.; Cambillau, C. Structure and molecular assignment of lactococcal phage TP901–1 baseplate. *J. Biol. Chem.* **2010**, *285*, 39079–39086.

32. Veesler, D.; Dreier, B.; Blangy, S.; Lichiere, J.; Tremblay, D.; Moineau, S.; Spinelli, S.; Tegoni, M.; Pluckthun, A.; Campanacci, V.; *et al.* Crystal structure and function of a darpin neutralizing inhibitor of lactococcal phage TP901–1: Comparison of DARPin and camelid VHH binding mode. *J. Biol. Chem.* **2009**, *284*, 30718–30726.

33. Deveau, H.; Labrie, S.J.; Chopin, M.C.; Moineau, S. Biodiversity and classification of lactococcal phages. *Appl. Environ. Microbiol.* **2006**, *72*, 4338–4346.

34. Geller, B.L.; Ngo, H.T.; Mooney, D.T.; Su, P.; Dunn, N. Lactococcal 936-species phage attachment to surface of *Lactococcus lactis*. *J. Dairy Sci.* **2005**, *88*, 900–907.

35. Dupont, K.; Janzen, T.; Vogensen, F.K.; Josephsen, J.; Stuer-Lauridsen, B. Identification of *Lactococcus lactis* genes required for bacteriophage adsorption. *Appl. Environ. Microbiol.* **2004**, *70*, 5825–5832.

36. Chapot-Chartier, M.P.; Vinogradov, E.; Sadovskaya, I.; Andre, G.; Mistou, M.Y.; Trieu-Cuot, P.; Furlan, S.; Bidnenko, E.; Courtin, P.; Pechoux, C.; *et al.* Cell surface of *Lactococcus lactis* is covered by a protective polysaccharide pellicle. *J. Biol. Chem.* **2010**, *285*, 10464–10471.

37. Ostergaard-Breum, S.; Neve, H.; Heller, K.J.; Vogensen, F.K. Temperate phages TP901–1 and phiLC3, belonging to the P335 species, apparently use different pathways for DNA injection in *Lactococcus lactis* subsp. *cremoris* 3107. *FEMS Microbiol. Lett.* **2007**, *276*, 156–164.

38. Dupont, K.; Vogensen, F.K.; Neve, H.; Bresciani, J.; Josephsen, J. Identification of the receptor-binding protein in 936-species lactococcal bacteriophages. *Appl. Environ. Microbiol.* **2004**, *70*, 5818–5824.

39. Duplessis, M.; Moineau, S. Identification of a genetic determinant responsible for host specificity in *Streptococcus thermophilus* bacteriophages. *Mol. Microbiol.* **2001**, *41*, 325–336.

40. Vegge, C.S.; Vogensen, F.K.; Mc Grath, S.; Neve, H.; van Sinderen, D.; Brondsted, L. Identification of the lower baseplate protein as the antireceptor of the temperate lactococcal bacteriophages TP901–1 and Tuc2009. *J. Bacteriol.* **2006**, *188*, 55–63.

41. Mc Grath, S.; Neve, H.; Seegers, J.F.; Eijlander, R.; Vegge, C.S.; Brondsted, L.; Heller, K.J.; Fitzgerald, G.F.; Vogensen, F.K.; van Sinderen, D. Anatomy of a lactococcal phage tail. *J. Bacteriol.* **2006**, *188*, 3972–3982.

42. Spinelli, S.; Desmyter, A.; Verrips, C.T.; de Haard, H.J.; Moineau, S.; Cambillau, C. Lactococcal bacteriophage p2 receptor-binding protein structure suggests a common ancestor gene with bacterial and mammalian viruses. *Nat. Struct. Mol. Biol.* **2006**, *13*, 85–89.

43. Ledeboer, A.M.; Bezemer, S.; de Hiaard, J.J.; Schaffers, I.M.; Verrips, C.T.; van Vliet, C.; Dusterhoft, E.M.; Zoon, P.; Moineau, S.; Frenken, L.G. Preventing phage lysis of *Lactococcus lactis* in cheese production using a neutralizing heavy-chain antibody fragment from llama. *J. Dairy Sci.* **2002**, *85*, 1376–1382.

44. De Haard, H.J.; Bezemer, S.; Ledeboer, A.M.; Muller, W.H.; Boender, P.J.; Moineau, S.; Coppelmans, M.C.; Verkleij, A.J.; Frenken, L.G.; Verrips, C.T. Llama antibodies against a lactococcal protein located at the tip of the phage tail prevent phage infection. *J. Bacteriol.* **2005**, *187*, 4531–4541.

45. Mahony, J.; Deveau, H.; Mc Grath, S.; Ventura, M.; Canchaya, C.; Moineau, S.; Fitzgerald, G.F.; van Sinderen, D. Sequence and comparative genomic analysis of lactococcal bacteriophages jj50, 712 and p008: Evolutionary insights into the 936 phage species. *FEMS Microbiol. Lett.* **2006**, *261*, 253–261.

46. Dupont, K.; Vogensen, F.K.; Josephsen, J. Detection of lactococcal 936-species bacteriophages in whey by magnetic capture hybridization pcr targeting a variable region of receptor-binding protein genes. *J. Appl. Microbiol.* **2005**, *98*, 1001–1009.

47. Castro-Nallar, E.; Chen, H.; Gladman, S.; Moore, S.C.; Seemann, T.; Powell, I.B.; Hillier, A.; Crandall, K.A.; Chandry, P.S. Population genomics and phylogeography of an Australian dairy factory derived lytic bacteriophage. *Genome Biol. Evol.* **2012**, *4*, 382–393.

48. Campanacci, V.; Veesler, D.; Lichiere, J.; Blangy, S.; Sciara, G.; Moineau, S.; van Sinderen, D.; Bron, P.; Cambillau, C. Solution and electron microscopy characterization of lactococcal phage baseplates expressed in *Escherichia coli. J. Struct. Biol.* **2010**, *172*, 75–84.

49. Shepherd, D.A.; Veesler, D.; Lichiere, J.; Ashcroft, A.E.; Cambillau, C. Unraveling lactococcal phage baseplate assembly by mass spectrometry. *Mol. Cell Proteom.* **2011**, *10*, M111.009787.

50. Siponen, M.; Sciara, G.; Villion, M.; Spinelli, S.; Lichiere, J.; Cambillau, C.; Moineau, S.; Campanacci, V. Crystal structure of orf12 from *Lactococcus lactis* phage p2 identifies a tape measure protein chaperone. *J. Bacteriol.* **2009**, *191*, 728–734.

51. Veesler, D.; Spinelli, S.; Mahony, J.; Lichiere, J.; Blangy, S.; Bricogne, G.; Legrand, P.; Ortiz-Lombardia, M.; Campanacci, V.; van Sinderen, D.; *et al.* Structure of the phage TP901–1 1.8 MDa baseplate suggests an alternative host adhesion mechanism. *Proc. Natl. Acad. Sci. USA* **2012**, *109*, 8954–8958.

52. Seegers, J.F.; Mc Grath, S.; O'Connell-Motherway, M.; Arendt, E.K.; van de Guchte, M.; Creaven, M.; Fitzgerald, G.F.; van Sinderen, D. Molecular and transcriptional analysis of the temperate lactococcal bacteriophage Tuc2009. *Virology* **2004**, *329*, 40–52.

8

Understanding Bacteriophage Specificity in Natural Microbial Communities

Britt Koskella * and Sean Meaden

BioSciences, University of Exeter, Cornwall Campus, Tremough, TR10 9EZ, UK;
E-Mail: sm341@exeter.ac.uk

* Author to whom correspondence should be addressed; E-Mail: B.L.Koskella@Exeter.ac.uk

Abstract: Studying the coevolutionary dynamics between bacteria and the bacteriophage viruses that infect them is critical to understanding both microbial diversity and ecosystem functioning. Phages can play a key role in shaping bacterial population dynamics and can significantly alter both intra- and inter-specific competition among bacterial hosts. Predicting how phages might influence community stability and apparent competition, however, requires an understanding of how bacteria-phage interaction networks evolve as a function of host diversity and community dynamics. Here, we first review the progress that has been made in understanding phage specificity, including the use of experimental evolution, we then introduce a new dataset on natural bacteriophages collected from the phyllosphere of horse chestnut trees, and finally we highlight that bacterial sensitivity to phage is rarely a binary trait and that this variation should be taken into account and reported. We emphasize that there is currently insufficient evidence to make broad generalizations about phage host range in natural populations, the limits of phage adaptation to novel hosts, or the implications of phage specificity in shaping microbial communities. However, the combination of experimental and genomic approaches with the study of natural communities will allow new insight to the evolution and impact of phage specificity within complex bacterial communities.

Keywords: coevolution; infection genetics; phage therapy; kill the winner

1. Introduction

Whether found in the soil [1,2], the leaf [3,4], the ocean [5,6], or the human body [7,8], microbial communities are proving to be more dynamic and diverse than could have been predicted. This incredible diversity is seen both within environments and among environments [7], but how it is generated and maintained is unclear. Early ideas about microbial diversity posited that "Everything is everywhere, but the environment selects [9]." However, this view of the microbial world remains hotly debated [10]. Much of the work testing this tenet has focused on the abiotic environment, such as salinity [11] or soil types [12], but the biotic environment is likely to be just as, if not more, important in shaping selection on microbial populations. Bacteria in any given environment face strong selection pressures from other microbes, predators, viruses, and in the case of bacteria living within another organism, the host immune response. In contrast to the abiotic environment, biotic "environments" have the potential to evolve in response to any changes in the microbial community, making them highly dynamic and capable of driving divergence among populations [13,14]. Bacteriophages (phages) represent perhaps the most ubiquitous of these biotic drivers [15–17]. To understand the role of phage-mediated selection in generating diversity, however, we need good insight into how specific phages are to their bacterial hosts.

For phages to alter the composition of a microbial community, there must exist a degree of specificity such that some hosts are more resistant to local phages than others or are better able to respond to phage-mediated selection. There is clear evidence that not all bacteria are infected by all phages, and indeed that most phages can only infect a subset of bacterial species (Table 1; [18]), but our understanding of phage host range is far from complete. Can phages easily adapt to infect new bacterial types as they become common? Can the same phage lineage shift from one bacterial species to another? These questions are far from new [19], but the development of recent techniques and the power of comparative genomics are moving us towards more satisfying answers. Experimental evolution (Box 1) provides one powerful approach to address these knowledge gaps, as the bacteria-phage interaction can be observed in the absence of other abiotic or biotic selection pressures; as such it has offered key advances in our understanding of the evolution of bacterial resistance to phages and reciprocal adaptations of phages to overcome such resistance. However, there are many reasons that the outcome of coevolution in a test tube might not be predictive of coevolution in nature, given the added biotic and abiotic complexity of most microbial ecosystems. For example, although experimental evolution studies have almost exclusively focused on phage adaptation within a population of one, or at most a few, bacterial species, most bacterial communities are highly rich, and therefore most bacterial species to which phage are adapting are rare. A comparison of culture-independent sequencing studies of microbial communities from the leaf surface, soil, atmosphere and the human body shows that the most dominant species in each given community represents a mere 2%–5% of sequences [20]. Given the heterogeneity and diversity of these microbial communities, it is unclear how a phage with narrow host range could evolve and be maintained.

For virulent bacteriophages, *i.e.*, those that reproduce within and then lyse their host cells, success depends on the chance event of encountering a susceptible host cell in the environment, and is most certainly reduced as a function of community diversity, dispersal, and exposure to the harsh conditions outside of a cell. It makes intuitive sense, therefore, that those phages with a larger host range should be at an inherent advantage. The data gathered so far, however, do not clearly support this intuition. First, many phages seem to be specific to a single bacterial species, and are often specific to only a few strains within that species [21–23] (Table 1). Second, there is building evidence that phages are "locally adapted" to their bacterial hosts [3,24], indicating a degree of specialization to common bacterial strains or species in a given population. Third, although phages do tend to increase their host range during the initial stages of coevolution, there is evidence that this expansion is short-lived [25]. The underlying mechanisms of phage infectivity and bacterial resistance are of course key to the evolution of phage host range [22,26], and have been the focus of extensive review elsewhere [21,27]. The data make it clear that phage infectivity is a complex function of adsorption [28], structural change of both host and phage [29], transport of nucleic acid into the bacterial cell, and avoidance of degradation once inside the cell [30], and is thus a result of both phage and bacterial phenotype. In addition, host susceptibility/resistance to phage can be determined by plasmids hosted by the bacterial cell [31,32]. This form of phage-plasmid interaction could lead to broad phage host range due to horizontal transfer of the plasmid among bacterial species within a community. A more thorough understanding of these interacting mechanisms will allow us to better predict the potential for host range expansion/contraction and therefore the effect of phages on microbial communities under both natural and therapeutic settings.

The more general question of why parasites specialize is of course not specific to bacteria-phage interactions, and we can apply much of the current coevolutionary theory to understanding the evolution of phage specificity. Many phages act as obligate parasites, as they are both unable to reproduce outside of their host cells and require cell lysis to transmit, thus killing their hosts. However, we acknowledge that, although virulent phages are obligate killers, other phages integrate into the host genome and their fitness relies on host reproduction. In these cases, the acquisition of a prophage can confer beneficial phenotypic change to the bacterial hosts, and therefore this latter relationship acts more synergistically than antagonistically. In either case, the question of host range for phages that are in the lytic cycle, and being transmitted among cells can be broken into two parts: first, the specificity of host resistance against infecting parasites; and second, the specificity of parasites on different hosts. It is often difficult to tease these two processes apart, but a recent review suggests that the failure to infect "nonhost" species (*i.e.*, those not considered to be hosts for the pathogen in question) may be the result of pathogen evolution leading to specialization on its own source host species and not the result of host evolution for resistance [33]. By reviewing studies across many host-parasite systems, the authors find a general trend towards decreasing parasite infection success on hosts of increasing genetic distance from the focal host. It remains to be determined whether this pattern is ubiquitous for bacteria-phage interactions, especially given the broad host ranges of some phages (Table 1).

The most supported evolutionary argument for why parasites specialize on given hosts, despite the clear advantage of a broad host range, is that there exists a trade-off between fitness and the breadth of parasite infectivity or host resistance [34]: In other words, the idea that "a jack of all trades is a master of none." This can be explained either by antagonistic pleiotropy, a situation where an adaptation that is advantageous in one host is deleterious in another, or else by selection for a less efficient but more general mechanism of infection. Support for this trade-off has been found for phage φ2, where individuals with broader host range within populations of its bacterial host, *Pseudomonas fluorescens*, were shown to pay a cost for this increased breadth relative to phages with narrow host ranges [35]. Specific evidence of antagonistic pleiotropy has also been found; during experimental host range expansion of phage φ6, spontaneous mutants able to infect novel hosts were found to be less infective to their native hosts in seven out of nine cases [36]. These trade-offs are also likely to be common in host populations. Indeed, recent results from experimental evolution (see Box 1) of *Prochlorococcus* hosts and their associated phages demonstrate that resistance to one phage genotype often came with the added cost of increased susceptibility to another phage genotype [37]. Similarly, experimental evolution of *P. syringae* in either single phage or multiple phage environments shows that bacteria evolved with multiple phages paid a higher cost of resistance than those evolving with single phages [38].

Box 1. Experimental evolution of phage specificity.

Experimental evolution of bacteriophage specificity has offered some key insights into the underlying process, evolutionary consequences, and fitness costs of host range expansion. The power of this method is that it allows replicate lines, started with genetically identical phages, to be passaged on homogeneous or heterogeneous hosts populations under a range of conditions (such as density and resources) for many thousands of generations (Figure 1). During this time, both the host bacterium and the phage can be frozen in time and resurrected at the end of the experiment, at which point [39] the fitness of evolved phages can be compared directly to both the ancestral types and phages experiencing a different selection regimen.

This approach has been used to demonstrate a number of key features of phage specificity, and has gone some way in explaining both when and how phage host range is likely to expand. First, in terms of range expansion within a host population (*i.e.*, the evolution of "generalist" phages capable of infecting more genotypes of a given bacterial species), there is evidence that phage φ2 is more likely to increase its host range during experimental coevolution with its bacterial host, *P. fluorescens*, than when the bacterial population is held constant [35]. Similar results were found in coevolving populations of phage SBW25φ2 and *P. fluorescens* [40]. Furthermore, the emergence of evolved "generalist" phenotypes of both bacteria and phages during experimental coevolution has been demonstrated both in a marine cyanobacteria and cyanophage system [5] and in a *P. fluorescens* and phage SBW25φ2 system [41]. Second, in terms of host range expansion to novel hosts, experimental evolution has provided evidence that phage φ6 populations are more likely to evolve expanded host range when there is strong competition for hosts (*i.e.*, when the focal host is rare in a population) [42].

It has also been shown that during the early stages of such a host shift, the likelihood of successful adaptation to a novel host is increased when contact with the native host is maintained, as this prevents extinction of the phage [43].

Figure 1. Illustrative example of experimental evolution of phage host range, where: (**A**) independent lines of genetically identical phage populations are propagated under different treatment regimens (e.g., different bacterial host species); and then (**B**) tested for infectivity on focal and alternate hosts. (**C**) Outcomes of these experiments might be a directional change towards increased host range over time (a), an initially increasing but then stable host range, perhaps indicative of coevolutionary response by the host population (b), or a decrease in host range associated with antagonistic pleiotropy during specialization on the focal host (c).

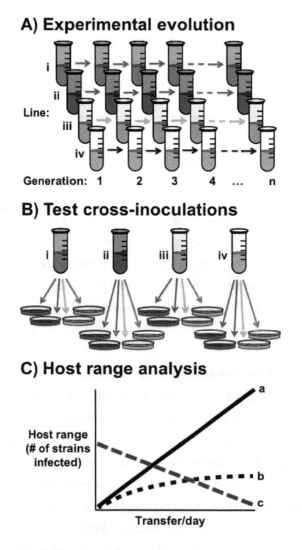

An experimental coevolution approach can also be taken to identify the mutations underlying gains or losses of host types. For example, host range expansion of phage φ6 to a novel host was found to be associated with a single nucleotide change [44]. Furthermore, phages experimentally coevolved with *P. fluorescens* hosts evolved increased host range over time and the phage genotypes with the broadest host ranges were found to have the most nonsynonymous amino acid changes, especially in the phage tail fiber gene [45]. Finally, experimental evolution of phage λ on populations of *Escherichia coli* that had lost the receptor used for phage attachment were found to evolve the ability to infect the bacterial host via a novel receptor following the spread of key precursor mutations, suggesting that phage host shifting can occur via entirely new innovations [46]. Clearly, the power of experimental evolution in understanding phage host range has not been fully exploited and moving forward this approach will offer further insight to the evolution of phage specificity in complex bacterial communities, fitness trade-offs between broad and narrow host ranges, and the potential limits of host shifting among phages.

1.1. The Structure of Bacteria-Phage Interaction Networks

Studies of phages in natural populations have confirmed that, like many other parasites, phages are well adapted to their local host populations [3,24,47]. However, whether individual phages are specialized on certain genotypes/species, the frequencies of which differ across space, or whether coevolution is driving among-population divergence remains to be determined. Importantly, phage specificity can only be meaningfully evaluated within a culturable reference set of host and/or parasite genotypes; and a different reference panel of host or pathogen genotypes may reveal different levels of specificity. Choosing an appropriate reference set is often challenging, as the interaction networks and species ranges of most phages is not known, but it is of central importance to choose a panel that is biologically meaningful and/or informative to the predictions being tested.

Those studies that have looked at host range of individual phages from the environment demonstrate large variation in specificity, both within and across bacterial species. In fact, some phages that appear to be "generalist," in the sense that they can infect bacteria spanning genera, fail to infect a subset of strains or species within genera (Table 1). This apparent inconsistency, which is likely the result of both specific phage adaptations and the subsequent evolution of bacterial resistance in some lineages, including via transfer of plasmids, makes it difficult to decipher specific rules regarding phage host range. For example, populations of *Candidatus* isolated from two separate sludge bioreactors were found to differ primarily in genomic regions encoding phage defense mechanisms, despite global dispersal of the strains among the two sites [48]. It is only when many of these studies are compared and datasets are compiled that clear patterns emerge. A recent meta-analysis of the statistical structure of bacteria-phage interactions shows that the infection networks are non-random and are typically nested [18]. This means that the highly resistant bacterial strains/species are only infected by broad host range phages, whereas the highly susceptible bacteria are infected by phages with narrow to broad host ranges. The observed lack of modularity may be suggestive of a true continuum of phage host range. Alternatively it might reflect the fact that most studies included in the analysis examine interactions either within a single bacterial species (*i.e.,* across multiple strains/genotypes) or between phages from one environment on bacteria from entirely different communities, habitats, and even continents. To determine the statistical structure of bacteria and phage communities in nature requires analysis of phage host range on representative bacterial hosts from the same local habitat.

1.2. The Evolutionary Implications of Phage Host Range

Given that phages are relatively specific, are capable of rapid adaptation, and are often obligate killers of their host cells, they can impose strong selection on bacterial populations and shape microbial communities. The "Kill the Winner" hypothesis posits that phages adapt to and preferentially infect the lineages of bacteria with the highest frequencies in the population, for example, those with higher metabolic fitness. Evidence for this hypothesis has been collected in a number of ways, including experimental coevolution of phages and bacteria in a test tube [49], monitoring of population change over time [50,51], and using metagenomic approaches [16]. Killing the winner is a form of negative frequency-dependent selection, as bacterial fitness is an indirect function of its frequency in the community. Specifically, bacterial species that are rare and free from phage attack will increase in frequency until the point at which an infective phage is introduced, either via mutation or migration, and spreads through the population. At this point, the common bacterial species will be at a relative disadvantage and may decrease in frequency.

Clearly, this type of dynamic is only possible if specificity underlying infection exists, but it also requires a time lag during which previously rare bacterial species can increase in frequency and remain free from phage attack. The length of this lag, *i.e.*, the time it takes for a phage mutant to arise by mutation or immigration, will dictate how common bacterial species can become before being targeted by coevolving phages. There is also likely to be a lag in the time it takes for a rare phage genotype to increase in frequency as its respective host becomes common, especially in populations where bacterial densities, and therefore rate of encounter, are low. Extending this theory to bacteria-phage interactions, we might predict that oscillatory dynamics should be more pervasive in relatively closed microbial communities, such as the human gut, than in highly connected communities, such as the ocean. This can be extended to predict that microbial communities with higher gene flow of bacteria and/or phages should show greater species evenness than closed communities, where frequencies are fluctuating over time. The data on phage infection of common bacterial species remains scarce, but the evidence we do have is in line with the above prediction. Estimates from marine communities suggest that cyanobacterial cell lysis by cyanophages ranges from a mere 0.005% to 3.2% per day, depending on the season [52]. Similarly, for bacterial isolates collected from the surface of tree leaves, a habitat that is open to constant immigration, only 3% of bacteria were found to be susceptible to local phages. This is in stark contrast to the interior of those same leaves, a more protected and closed microenvironment, where 45% of bacterial isolates were found to be susceptible to local phages [3]. It is important to note, however, that a number of differences exist among these habitats beyond the potential for immigration [53]. A comparable result was found in natural soil samples, where 33 to 40% of bacteria could be lysed by phages from the same sample [24]. Given the paucity of studies that have explored bacteria-phage dynamics in nature, the ubiquity of phage-mediated negative frequency-dependent selection remains unclear. Future studies exploring the natural prevalence of phage infection, coevolutionary dynamics of bacteria and phages over time, and the evolution of phage host range in either natural or experimental communities are still needed to predict how phages influence bacterial communities.

1.3. The Applied Implications of Phage Host Range

In addition to their potential role in shaping bacterial community composition, phages are key players in shaping the evolution of bacterial genomes [54]. As lytic bacteriophages reproduce within the host and reassemble, bacterial chromosomal DNA can be inadvertently packaged into the viral capsid along with the viral DNA. This mistake will lead to *generalized* transduction (in contrast to *specialized* transduction by prophages) and can move chromosomal DNA from one bacterial host to another. During this movement among hosts, phages can transfer genes encoding toxins or virulence factors, and thus critically alter the bacterial phenotype. For example, phage-mediated transfer of pathogenicity islands between *Listeria monocytogenes* and *Staphylococcus aureus* has been demonstrated in raw milk [55] and phage-mediated transfer of antibiotic resistance has been demonstrated among species of *Enterococcus* [56]. It is increasingly clear that transduction can occur across distantly related bacterial species, and even the seemingly highly conserved 16S rRNA gene has been found within the genome of a broad host range transducing phage [57]. Thus understanding phage specificity among bacterial strains and species is key to predicting potential movement of genes across bacterial species and habitats, and thus the potential emergence of novel pathogens.

An understanding of phage specificity is also central to predicting the success and consequences of phage therapy, *i.e.*, the use of phage or cocktails of phages to control the growth and/or virulence of pathogenic bacteria. The utility of phage therapy is often called into question because of the apparent specificity of phages [58]. However, this specificity is also a clear advantage of phage therapy over more general treatments, such as antibiotics, since the non-target bacterial populations should remain relatively undisturbed. The first steps in testing the benefits of a potentially therapeutic phage are to test (a) whether the phage is too specific to be effective against the standing strain variation of a pathogen in a host population and (b) the likelihood that the phage will affect other non-pathogenic bacteria, either immediately due to a large host range or over short evolutionary timescales as the phage evolves. For example, recent work from silage of dairy farms found a great deal of strain-to-strain variation in susceptibility of the food-borne pathogen, *L. monocytogenes*, to phages collected from silage. They tested the host range of 114 listeriaphages and found that 12% of these phages had narrow host ranges and could infect fewer than half of the strains tested, representing the nine major serotypes of *L. monocytogenes*. However, another 29% of the phages were capable of infecting nearly all of the strains tested, suggesting that these phages would be good candidates for therapeutic control of the pathogen [59]. Furthermore, given the ease of full genome sequencing, it is now possible to scan the phage genome for virulence factors and known toxin-encoding genes to ensure the phage will not act to increase the harm caused by a given pathogen.

2. Results and Discussion

2.1. Specificity within a Natural Phyllosphere Environment

The microbial community within eukaryotic hosts is a relatively closed system that holds the potential for long periods of uninterrupted bacteria-phage coevolution, especially if the hosts are long-lived. Recent work examining bacteria and phages from the horse chestnut phyllosphere (*i.e.*, the above-ground, aerial habitat of the plant) has demonstrated strong local adaptation of phages to bacteria collected from the same leaf [3]. Given that the culturable bacterial communities found within these leaves differed among the trees sampled, it is unclear whether this result demonstrates adaptation of multiple phages to common bacterial strains (which differ among populations) or whether it suggests phage adaptation to infect the common bacterial species within a given community. In other words the result could indicate species sorting according to infection success or it could suggest coevolution of bacteria and phages within each population. One way to tease these two possibilities apart would be to examine specific phage clones from each population and measure their host ranges both within the bacterial community from which they were isolated and from other communities.

Table 1. Examples of phage specificity from natural populations. Habitat refers to the environment from which the samples were selected, and host to the bacterial species used to first visualize the phage. For each study, the number of phages tested is reported and the host range of these phages is described depending on whether the phages were able to infect bacteria from multiple species and/or multiple genera. Finally, we report whether there was variability in phage infectivity on different strains/genotypes within a single bacterial species. In all cases, "n/a" is reported when the phages were not tested in a way that allowed for a given comparison. Two cases show both within-species specificity and an ability to infect multiple species, and these are highlighted in bold to emphasize the difficulty in describing a given phage as "generalist" versus "specialist." Note that this table is for illustrative purposes and is not exhaustive. For a formal meta-analysis of bacteria-phage infection networks, see recent review by Flores and coauthors [60].

			Host range			
Habitat	**Host**	**# Phages tested**	**Multi-species**	**Multi-genus**	**Within-species specificity**	**Reference**
Rhizosphere	Pseudomonas	5	4	0	n/a	Campbell *et al.* 1995 [61]
Sewage	Multiple hosts	11	n/a	11	n/a	Jensen *et al.* 1998 [62]
Industrial	Leuconostoc	6	0	0	Yes	Barrangou *et al.* 2002 [63]
Marine	Vibrio	13	**10**	n/a	**Yes**	Comeau *et al.* 2005 [64]
Soil	Burkholderia	6	**6**	n/a	**Yes**	Seed and Dennis 2005 [65]
Effluent	Salmonella	66	n/a	0	Yes	McLaughlin *et al.* 2006 [66]
Marine	Cellulophaga	46	0	0	Yes	Holmfeldt *et al.* 2007 [67]

As a first step towards this, we randomly selected 144 culturable bacterial isolates from across the eight trees sampled in the original experiment [3] and inoculated each isolate with a dilution series of the sympatric phage community (*i.e.*, filtered leaf homogenate, note that no enrichment procedure was used). For the 14 bacteria that were susceptible to their local phages, a single phage plaque was isolated and re-inoculated into an overnight culture of the bacteria it was able to infect. These co-cultures were then filtered, creating a high titer phage inoculum made up of a single phage clone. All 14 of these phages were cross-inoculated onto each of the 144 original bacterial host isolates, as well as 17 previously characterized *P. syringae* isolates representing nine different pathovars, to determine host range (Figure 2). In addition, we measured the susceptibility of each bacterium in the reference panel to ten phages that were previously collected from sewage and enriched on *P. syringae* pathovar tomato. It is important to note that the host range examined here represents the phage's *plaquing* host range (*i.e.,* the range of hosts a given phage can successfully infect and lyse in soft agar [21]) and may be an underrepresentation of its *productive* (*i.e.,* phage-producing) host range. We sequenced ≈800 bp of the 16S ribosomal RNA region from all hosts that were found to be susceptible to phage. Prior to sequencing, bacterial isolates were grown in KB broth overnight. These overnight cultures were then diluted 1:5 in PCR grade water and used as PCR template (5 µL) in reactions with universal 16S primers 515f (5-GTGCCAGCMGCCGCGGTAA-3) and 1492r (5'-GGTTACCTTGTTACGACTT-3') [68]. Diluted PCR products (50 ng/µL) were sequenced in both directions, and then aligned and compared against the NCBI database using Geneious (v2.5) software. Individual isolates were assigned to a given genera and species when possible according to the top BLAST hits associated with the sequence (with an e-score of 0.0). Geneious was used first to align the sequences (using MUSCLE, <u>mu</u>ltiple <u>s</u>equence <u>c</u>omparison by <u>log-e</u>xpectation [69]) and then a consensus neighbor-joining tree was assembled with pairwise distances calculated using the "Jukes-Cantor" formula.

We examined the resulting network of bacteria-phage interactions using a recently described method [70]. Relative to the null model, the network shows very little evidence of nestedness, in that the phages with broad host range do not tend to infect the more resistant hosts (the network is only 1% more nested than expected by chance). Instead, there is evidence that the network is highly modular; the network shows 82% of interactions occur between isolates from the same module relative to a null model of 66%. This is suggestive that the pattern of local adaptation observed previously [3] is indicative of many phages each coevolving with a subset of the bacterial host community. However, given that the phages and bacteria were collected from across eight separate microbial "populations," represented by eight different tree hosts [3], a larger analysis will be required to confirm that the modularity represents bacteria-phage coevolution at small spatial scales or whether it indeed indicates that these natural microbial communities are harboring phages with largely non-overlapping host ranges. The phages examined clearly fall across the continuum of "generalist" to "specialist" (note that none were restricted to a single bacterial isolate), but most are restricted to infecting fewer than one third of the bacterial isolates. Of the 13 phages isolated from the phyllosphere, 5 are capable of infecting both *Pseudomonas* and *Erwinia* species and most are capable of infecting multiple pathovars within *P. syringae* (Figure 2). Interestingly, the phages isolated from sewage had a relatively wider host range across the *Pseudomonas* isolates, infecting a mean of 10.9 (SD = 3.21) hosts, than did the phages from the phyllosphere, infecting a mean of 7.57 (SD = 3.18) hosts. This is not surprising, as the method of searching for phage in one environment using a host bacterium from another is likely to bias the resulting isolates towards more "generalist" phages. Finally, we found two phages collected from the horse chestnut leaf that are capable of infecting a previously characterized strain of *P. syringae* pathovar aesculi (Pae), the causal agent of bleeding canker disease in horse chestnut trees. These phages were not, however, capable of infecting the other three strains of Pae we tested, suggesting that there is variation in susceptibility of Pae strains to phages despite the relatively low genetic diversity typically found among isolates of this rapidly emerging pathogen [71].

2.2. The Importance of Dose in Measuring Specificity

In addition to taking into consideration the appropriate reference panel when measuring phage specificity, it is also necessary to take into account that bacterial sensitivity is likely to depend critically on both phage titer and test conditions. Most studies examining the specificity of parasites within their local communities [18,72–74] treat infectivity as a binary trait and examine bacterial sensitivity at a single phage titer and without taking into account environmental effects on the outcome of the interaction. This is often a necessary step given the large sample sizes of many studies. However, when environmental heterogeneity is taken into account it becomes clear that infection specificity is not simply a binary trait but rather that it can depend on local resources [75], temperature [76], experimental approach [21], and dose [67]. For example, examination of *Cellulophaga baltica* strains and their associated phages isolated from coastal waters showed differences of up to 6 orders of magnitude in bacterial sensitivity to the same titer of phage [67]. This variation can be critical to predicting the impact of phages on bacterial communities, as fine-scale differences among bacterial genotypes or species in sensitivity to the same phage would mean a fitness advantage of one over the other. Therefore, studies that treat infection success as binary would fail to predict this apparent competition. Similarly, if broad host range phages are able to infect more hosts simply because they reach higher prevalence within the environment (given the greater number of available hosts), this specificity can be considered context-dependent.

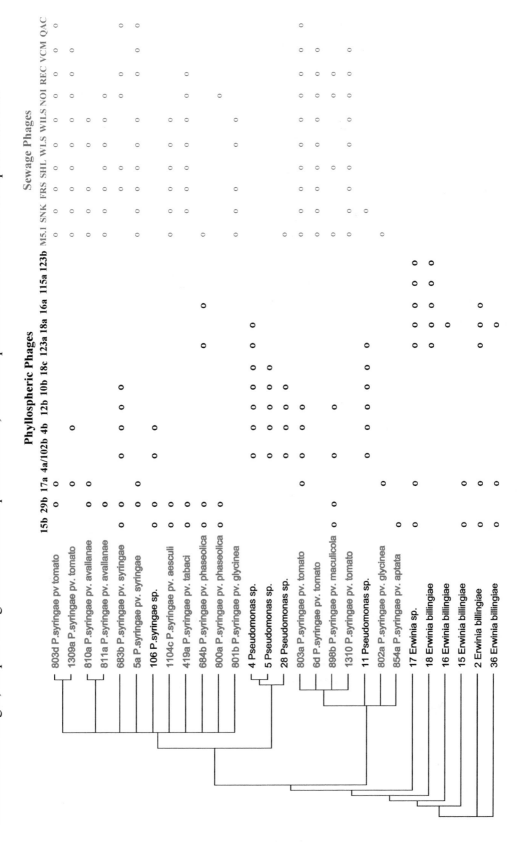

Figure 2. Neighbor-joining tree showing the phylogenetic relationships among bacteria used in this study and their susceptibility to bacteriophages from the phyllosphere (black) or from sewage (blue). Trees are based on 16S rRNA gene sequences (~800 bp). Bacterial isolates in red have been classified previously to the pathovar level. Phages 4a and 102b had identical host ranges, despite being isolated from separate leaves, and their profiles have thus been collapsed into one.

As an example, we measured whether the bacterial isolates used in our host range tests showed variation in sensitivity to the phages from the phyllosphere. To do this, we inoculated multiple hosts with a dilution series of the same phage inocula (and thus necessarily the same phage titer). In this way, we were able to compare the plaque forming units (PFUs) from each inoculum on the lawn of one bacterial host versus another. If the relationship between phage titer and phage infection success for a given inoculum were the same regardless of the host being tested, we would expect to find no significant difference in PFUs across hosts.

On the other hand, if different measures of PFUs are found for the same phage inoculum across different hosts, this would suggest that resistance is a quantitative trait, and therefore that host range may be dose-dependent. Overall, we found a significant interaction between phage inoculum and bacterial strain on the PFU per μL observed across 5 replicate dilution series (general linear model with log PFU as a response variable and phage and bacteria as explanatory variables; $F_{5, 156} = 3.25$, $p < 0.01$). Furthermore, when we included only the bacteria-phage combinations for which the test was fully reciprocal (*i.e.*, each of the two phages were infective to the same three hosts), we again found this interaction effect (Figure 3; $F_{1, 29} = 8.03$, $p < 0.01$). We later tested whether the density of bacterial cells in the soft agar overlay could explain this result and were able to rule out this possibility, as PFU was not correlated with bacterial density within the ranges used.

This suggests that direct competition between these three bacteria in the presence of either of these phages is likely to be biased towards the least susceptible, even though all three are within the host range of the phage, and reinforces the idea that infection is not a binary trait. However, given that all bacteria were susceptible overall, these results suggest that dose will only affect the binary outcome of host resistance when phage titer is very low.

Moving forward, researchers should consider whether bacterial resistance to phage should be considered a quantitative or qualitative trait with regard to their specific question of interest. For example, if experimental evolution lineages are being compared to a control treatment of the same bacterial strain, resistance might be treated as a binary trait to aid in statistical comparison. In this case, the only caution would be in interpreting a negative result such as lack of a cost of resistance, as the approach may have lumped different degrees of resistance into a single phenotype. On the other hand, resistance should be considered as a quantitative trait, varying for example across multiple environmental conditions or dose, if the goal is to make predictions regarding how phages will alter the competitive hierarchy of bacterial species in a community.

Figure 3. Results of the reciprocal cross-inoculation where the same phage inoculum was spotted in a dilution series onto lawns of each of three different *Erwinia sp.* bacterial isolates. The number of plaque forming units (PFUs) per microliter of inocula was measured for each cross, and the means across five replicates are shown on the Y-axis. Variation in PFU within a given inocula (*i.e.*, among the blue bars or among the yellow bars) represents variation in phage success across bacterial hosts. Note that the important comparison is within-phage variation, as between-phage variation reflects absolute differences in phage titer.

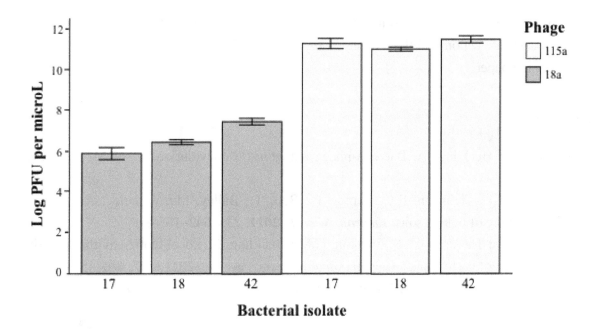

3. Conclusions

We set out to highlight the importance of understanding phage host range as a key factor in predicting (i) how phages shape microbial communities; (ii) when genes encoding virulence and toxins might be transferred among bacterial species; and (iii) the potential success of phages as therapeutic agents against bacterial pathogens. Recent advances in experimental evolution techniques and comparative genomics have given us important new insight, including that phage host range can be altered by single mutations [44], but is often the result of complex epistatic interactions among mutations [45], and that selection can lead to increased host range [42], but that this increase often carries a significant fitness cost [36,38,77]. Studies of the host ranges of natural phages have demonstrated a high variability of specificity, ranging from phages with extremely narrow ranges of hosts within a single species to those that can infect bacteria across genera. A key new insight comes from comparison across studies and systems, and suggests that bacteria-phage networks tend to be statistically nested [18,70]. However, our understanding of these networks remains limited, as very few whole community analyses have been completed and, importantly, estimates of phage host range are only as good as the reference panel against which they've been tested. Furthermore, phage infection is rarely if ever a binary trait, and therefore studies that measure infectivity at a single dose, and/or under a single environmental treatment, are likely to miss key aspects of phage specificity. Moving forward, data from natural populations in which the community of both bacteria and phages can be properly represented and tested in a way that captures variation in infection success beyond a simple "yes/no", will help elucidate the complex networks of interactions between phages and their bacterial host.

Acknowledgments

We would like to thank Timothée Poisot for help in examining the network structure of our bacteria-phage interaction and three anonymous reviewers for providing valuable feedback on earlier versions of the paper.

References

1. Gómez, P.; Buckling, A. Bacteria-phage antagonistic coevolution in soil. *Science* **2011**, *332*, 106–109.

2. Griffiths, R.I.; Thomson, B.C.; James, P.; Bell, T.; Bailey, M.; Whiteley, A.S. The bacterial biogeography of British soils. *Environ. Microb.* **2011**, *13*, 1642–1654.

3. Koskella, B.; Thompson, J.N.; Preston, G.M.; Buckling, A. Local biotic environment shapes the spatial scale of bacteriophage adaptation to bacteria. *Am. Nat.* **2011**, *177*, 440–451.

4. Lindow, S.E.; Brandl, M.T. Microbiology of the Phyllosphere. *Appl. Environ. Microbiol.* **2003**, *69*, 1875–1883.

5. Marston, M.F.; Pierciey, F.J.; Shepard, A.; Gearin, G.; Qi, J.; Yandava, C.; Schuster, S.C.; Henn, M.R.; Martiny, J.B.H. Rapid diversification of coevolving marine Synechococcus and a virus. *PNAS* **2012**, *109*, 4544–4549

6. Pommier, T.; Douzery, E.J.P.; Mouillot, D. Environment drives high phylogenetic turnover among oceanic bacterial communities. *Biol. Lett.* **2012**, *8*, 562–566.

7. Smillie, C.S.; Smith, M.B.; Friedman, J.; Cordero, O.X.; David, L.A.; Alm, E.J. Ecology drives a global network of gene exchange connecting the human microbiome. *Nature* **2011**, *480*, 241–244.

8. Hooper, L.V.; Littman, D.R.; Macpherson, A.J. Interactions Between the Microbiota and the Immune System. *Science* **2012**, *336*, 1268–1273.

9. De Wit, R.; Bouvier, T. "Everything is everywhere, but, the environment selects"; what did Baas Becking and Beijerinck really say? *Environ. Microb.* **2006**, *8*, 755–758.

10. O'Malley, M.A. "Everything is everywhere: but the environment selects": Ubiquitous distribution and ecological determinism in microbial biogeography. *Stud. Hist. Philos. Biol. Biomed. Sci.* **2008**, *39*, 314–325.

11. Lin, W.; Wang, Y.; Li, B.; Pan, Y. A biogeographic distribution of magnetotactic bacteria influenced by salinity. *ISME J.* **2011**, *6*, 475–479.

12. Chu, H.; Fierer, N.; Lauber, C.L.; Caporaso, J.G.; Knight, R.; Grogan, P. Soil bacterial diversity in the Arctic is not fundamentally different from that found in other biomes. *Environ. Microb.* **2010**, *12*, 2998–3006.

13. Buckling, A.; Rainey, P.B. The role of parasites in sympatric and allopatric host diversification. *Nature* **2002**, *420*, 496–499.

14. Childs, L.M.; Held, N.L.; Young, M.J.; Whitaker, R.J.; Weitz, J.S. Multiscale model of CRISPR-induced coevolutionary dynamics: Diversification at the interface of Larmarck and Darwin. *Evolution* **2012**, *66*, 2015–2029.

15. Suttle, C.A. Viruses in the sea. *Nature* **2005**, *437*, 356–361.

16. Rodriguez-Valera, F.; Martin-Cuadrado, A.-B.; Rodriguez-Brito, B.; Pasic, L.; Thingstad, T.F.; Rohwer, F.; Mira, A. Explaining microbial population genomics through phage predation. *Nat. Rev. Microbiol.* **2009**, *7*, 828–836.

17. Clokie, M.R.J.; Millard, A.D.; Letarov, A.V.; Heaphy, S. Phages in nature. *Bacteriophage* **2011**, *1*, 31–45.

18. Flores, C.O.; Meyer, J.R.; Valverde, S.; Farr, L.; Weitz, J.S. Statistical structure of host–phage interactions. *PNAS* **2011**, *108*, 288–297

19. Frisch, A.W.; Levine, P. Specificity of the Multiplication of Bacteriophage. *J. Immunol.* **1936**, *30*, 89–108.

20. Fierer, N.; Lennon, J.T. The generation and maintenance of diversity in microbial communities. *Am. J. Bot.* **2011**, *98*, 439–448.

21. Hyman, P.; Abedon, S.T. Bacteriophage host range and bacterial resistance. *Adv. Appl. Microbiol.* **2010**, *70*, 217–248.

22. Duplessis, M.; Moineau, S. Identification of a genetic determinant responsible for host specificity in Streptococcus thermophilus bacteriophages. *Mol. Microbiol.* **2001**, *41*, 325–336.

23. Miklič, A.; Rogelj, I. Characterization of lactococcal bacteriophages isolated from Slovenian dairies. *Int. J. Food Sci. Technol.* **2003**, *38*, 305‑311.

24. Vos, M.; Birkett, P.J.; Birch, E.; Griffiths, R.I.; Buckling, A. Local adaptation of bacteriophages to their bacterial hosts in soil. *Science* **2009**, *325*, 833.

25. Hall, A.R.; Scanlan, P.D.; Morgan, A.D.; Buckling, A. Host–parasite coevolutionary arms races give way to fluctuating selection. *Ecol. Lett.* **2011**, *14*, 635–642.

26. Riede, I.; Degen, M.; Henning, U. The receptor specificity of bacteriophages can be determined by a tail fiber modifying protein. *EMBO J.* **1985**, *4*, 2343.

27. Rakhuba, D.; Kolomiets, E.; Szwajcer Dey, E.; Novik, G. Bacteriophage receptors, mechanisms of phage adsorption and penetration into host cell. *Pol. J. Microbiol.* **2010**, *59*, 145–155.

28. Chatterjee, S.; Rothenberg, E. Interaction of Bacteriophage l with Its *E. coli* Receptor, LamB. *Viruses* **2012**, *4*, 3162–3178.

29. Mahony, J.; van Sinderen, D., Structural aspects of the interaction of dairy phages with their Host bacteria. *Viruses* **2012**, *4*, 1410–1424.

30. Richter, C.; Chang, J.T.; Fineran, P.C. Function and Regulation of Clustered Regularly Interspaced Short Palindromic Repeats (CRISPR)/CRISPR Associated (Cas) Systems. *Viruses* **2012**, *4*, 2291–2311.

31. De Vos, W.M.; Underwood, H.M.; Lyndon Davies, F. Plasmid encoded bacteriophage resistance in Streptococcus cremoris SK11. *FEMS Microbiol. Lett.* **1984**, *23*, 175–178.

32. Deng, Y.-M.; Harvey, M.L.; Liu, C.-Q.; Dunn, N.W. A novel plasmid-encoded phage abortive infection system from Lactococcus lactis biovar. diacetylactis. *FEMS Microbiol. Lett.* **1997**, *146*, 149–154.

33. Antonovics, J.; Boots, M.; Ebert, D.; Koskella, B.; Poss, M.; Sadd, B.M. The origins of specificity by means of natural selection: Evolved and nonhost resistance in host-pathogen interactions. *Evolution* **2012**, *6*, 1–9.

34. Straub, C.S.; Ives, A.R.; Gratton, C. Evidence for a trade-off between host-range breadth and host-use efficiency in aphid parasitoids. *Am. Nat.* **2011**, *177*, 389–395.

35. Poullain, V.; Gandon, S.; Brockhurst, M.A.; Buckling, A.; Hochberg, M.E. The evolution of specificity in evolving and coevolving antagonistic interactions between a bacteria and its phage. *Evolution* **2008**, *62*, 1–11.

36. Duffy, S.; Turner, P.E.; Burch, C.L. Pleiotropic costs of niche expansion in the RNA bacteriophage $\Phi6$. *Genetics* **2006**, *172*, 751-757.

37. Avrani, S.; Schwartz, D.A.; Lindell, D. Virus-host swinging party in the oceans: Incorporating biological complexity into paradigms of antagonistic coexistence. *Mob. Gen. Elem.* **2012**, *2*, 88–95.

38. Koskella, B.; Lin, D.M.; Buckling, A.; Thompson, J.N. The costs of evolving resistance in heterogeneous parasite environments. *Proc. R. Soc. B: Biol. Sci.* **2012**, *279*, 1896–1903.

39. Gaba, S.; Ebert, D. Time-shift experiments as a tool to study antagonistic coevolution. *TREE* **2009**, *24*, 226–232.

40. Hall, A.R.; Scanlan, P.D.; Buckling, A. Bacteria—Phage Coevolution and the Emergence of Generalist Pathogens. *Am. Nat.* **2011**, *177*, 44–53.

41. Buckling, A.; Rainey, P.B. Antagonistic coevolution between a bacterium and a bacteriophage. *Proc. R. Soc. B: Biol. Sci.* **2002**, *269*, 931–936.

42. Bono, L.M.; Gensel, C.L.; Pfennig, D.W.; Burch, C.L. Competition and the origins of novelty: Experimental evolution of niche-width expansion in a virus. *Biol. Let.* **2013**, *9*, doi:10.1098/rsbl.2012.0616.

43. Dennehy, J.J.; Friedenberg, N.A.; Holt, R.D.; Turner, P.E. Viral ecology and the maintenance of novel host use. *Am. Nat.* **2006**, *167*, 429–439.

44. Duffy, S.; Burch, C.L.; Turner, P.E. Evolution of host specificity drives reproductive isolation among RNA viruses. *Evolution* **2007**, *61*, 2614–2622.

45. Scanlan, P.D.; Hall, A.R.; Lopez-Pascua, L.D.C.; Buckling, A. Genetic basis of infectivity evolution in a bacteriophage. *Molec. Ecol.* **2011**, *20*, 981–989.

46. Meyer, J.R.; Dobias, D.T.; Weitz, J.S.; Barrick, J.E.; Quick, R.T.; Lenski, R.E. Repeatability and contingency in the evolution of a key innovation in phage lambda. *Science* **2012**, *335*, 428–432.

47. Held, N., Whitaker, R.J. Viral biogeography revealed by signatures in *Sulfolobus islandicus* genomes. *Environ. Microb.* **2009**, *11*, 457–466.

48. Kunin, V.; He, S.; Warnecke, F.; Peterson, S.B.; Garcia Martin, H.; Haynes, M.; Ivanova, N.; Blackall, L.L.; Breitbart, M.; Rohwer, F.; *et al.* A bacterial metapopulation adapts locally to phage predation despite global dispersal. *Gen. Res.* **2008**, *18*, 293–297.

49. Fuhrman, J.A.; Schwalbach, M. Viral Influence on Aquatic Bacterial Communities. *Biol. Bull.* **2003**, *204*, 192–195.

50. Rodriguez-Brito, B.; Li, L.; Wegley, L.; Furlan, M.; Angly, F.; Breitbart, M.; Buchanan, J.; Desnues, C.; Dinsdale, E.; Edwards, R.; *et al.* Viral and microbial community dynamics in four aquatic environments. *ISME J.* **2010**, *4*, 739–751.

51. Shapiro, O.H.; Kushmaro, A.; Brenner, A. Bacteriophage predation regulates microbial abundance and diversity in a full-scale bioreactor treating industrial wastewater. *ISME J.* **2010**, *4*, 327–36.

52. Waterbury, J.B.; Valois, F.W. Resistance to co-occurring phages enables marine Synechococcus communities to coexist with cyanophages abundant in seawater. *Appl. Environ. Microb.* **1993**, *59*, 3393.

53. Yu, X.; Lund, S.P.; Scott, R.A.; Greenwald, J.W.; Records, A.H.; Nettleton, D.; Lindow, S.E.; Gross, D.C.; Beattie, G.A. Transcriptional responses of Pseudomonas syringae to growth in epiphytic versus apoplastic leaf sites. *PNAS.* **2013**, *110*, 425–434

54. Fineran, P.C.; Petty, N.K.; Salmond, G.P.C. Transduction: Host DNA Transfer by Bacteriophages. In *The Encyclopedia of Microbiology*; Schaechter, M., Ed.; Elsevier, 2009.

55. Chen, J.; Novick, R.P. Phage-mediated intergeneric transfer of toxin genes. *Science* **2009**, *323*, 139–141.

56. Mazaheri Nezhad Fard, R.; Barton, M.; Heuzenroeder, M. Bacteriophage ⁻ mediated transduction of antibiotic resistance in enterococci. *Lett. Appl. Microb.* **2011**, *52*, 559–564.

57. Beumer, A.; Robinson, J.B. A Broad-Host-Range, Generalized Transducing Phage (SN-T) Acquires 16S rRNA Genes from Different Genera of Bacteria. *Appl. Environ. Microb.* **2005**, *71*, 8301–8304.

58. Loc-Carrillo, C.; Abedon, S.T. Pros and cons of phage therapy. *Bacteriophage* **2011**, *1*, 111–114.

59. Vongkamjan, K.; Switt, A.M.; den Bakker, H.C.; Fortes, E.D.; Wiedmann, M. Silage Collected from Dairy Farms Harbors an Abundance of Listeriaphages with Considerable Host Range and Genome Size Diversity. *Appl. Environ. Microb.* **2012**, *78*, 8666–8675.

60. Flores, C.O.; Meyer, J.R.; Valverde, S.; Farr, L.; Weitz, J.S. Statistical structure of host–phage interactions. *PNAS.* **2011**, *108*, 288–297.

61. Campbell, J.I.A.; Albrechtsen, M.; Sørensen, J. Large Pseudomonas phages isolated from barley rhizosphere. *FEMS Microb. Ecol.* **1995**, *18*, 63–74.

62. Jensen, E.C.; Schrader, H.S.; Rieland, B.; Thompson, T.L.; Lee, K.W.; Nickerson, K.W.; Kokjohn, T.A. Prevalence of Broad-Host-Range Lytic Bacteriophages of Sphaerotilus natans, Escherichia coli, andPseudomonas aeruginosa. *Appl. Environ. Microb.* **1998**, *64*, 575–580.

63. Barrangou, R.; Yoon, S.-S.; Breidt, J.F.; Fleming, H.P.; Klaenhammer, T.R. Characterization of Six Leuconostoc fallax Bacteriophages Isolated from an Industrial Sauerkraut Fermentation. *Appl. Environ. Microb.* **2002**, *68*, 5452–5458.

64. Comeau, A.M.; Buenaventura, E.; Suttle, C.A. A Persistent, Productive, and Seasonally Dynamic Vibriophage Population within Pacific Oysters (Crassostrea gigas). *Appl. Environ. Microb.* **2005**, *71*, 5324–5331.

65. Seed, K.D.; Dennis, J.J. Isolation and characterization of bacteriophages of the Burkholderia cepacia complex. *FEMS Microb. Lett.* **2005**, *251*, 273–280.

66. McLaughlin, M.R.; Balaa, M.F.; Sims, J.; King, R. Isolation of Salmonella Bacteriophages from Swine Effluent Lagoons. Journal article number J-10632 of the Mississippi Agricultural and Forestry Experiment Station. *J. Environ. Qual.* **2006**, *35*, 522–528.

67. Holmfeldt, K.; Middelboe, M.; Nybroe, O.; Riemann, L. Large Variabilities in Host Strain Susceptibility and Phage Host Range Govern Interactions between Lytic Marine Phages and Their Flavobacterium Hosts. *Appl. Environ. Microb.* **2007**, *73*, 6730–6739.

68. Turner, S.; Pryer, K.M.; Miao, V.P.W.; Palmer, J.D. Investigating Deep Phylogenetic Relationships among Cyanobacteria and Plastids by Small Subunit rRNA Sequence Analysis1. *J. Euk. Microb.* **1999**, *46*, 327–338.

69. Edgar, R.C. MUSCLE: multiple sequence alignment with high accuracy and high throughput, *Nucl. Acid. Res.* **2004**, *32*, 1792–1797.

70. Weitz, J.S.; Poisot, T.; Meyer, J.R.; Flores, C.O.; Valverde, S.; Sullivan, M.B.; Hochberg, M.E. Phage—bacteria infection networks. *Trends Microb.* **2012**, *21*, 82–91.

71. Green, S.; Laue, B.; Fossdal, C.G.; A'Hara, S.W.; Cottrell, J.E. Infection of horse chestnut (*Aesculus hippocastanum*) by *Pseudomonas syringae* pv. *aesculi* and its detection by quantitative real-time PCR. *Plant Path.* **2009**, *58*, 731–744.

72. Anderson, T.K.; Sukhdeo, M.V.K. Host Centrality in Food Web Networks Determines Parasite Diversity. *PLoS One* **2011**, *6*, e26798.

73. Poulin, R.; Moulillot, D. Parasite specialization from a phylogenetic perspective: A new index of host specificity. *Parasitol.* **2003**, *126*, 473–480.

74. Luijckx, P.; Ben-Ami, F.; Mouton, L.; Du Pasquier, L.; Ebert, D. Cloning of the unculturable parasite Pasteuria ramosa and its Daphnia host reveals extreme genotype–genotype interactions. *Ecol. Lett.* **2011**, *14*, 125–131.

75. Poisot, T.; Lepennetier, G.; Martinez, E.; Ramsayer, J.; Hochberg, M.E. Resource availability affects the structure of a natural bacteria–bacteriophage community. *Biology Letters* **2011**, *7*, 201–204.

76. Seeley, N.D.; Primrose, S.B. The Effect of Temperature on the Ecology of Aquatic Bacteriophages. *J. Gen. Virol.* **1980**, *46*, 87–95.

77. Lennon, J.; Khatana, S.; Marston, M.; Martiny, J. Is there a cost of virus resistance in marine cyanobacteria? *ISME J.* **2007**, *1*, 300.

9

Interaction of Bacteriophage λ with Its *E. coli* Receptor, LamB

Sujoy Chatterjee and Eli Rothenberg *

Department of Biochemistry and Molecular Pharmacology, NYU Medical School, 550 First Avenue, New York, NY 10016, USA; E-Mail: sujoy.chatterjee@nyumc.org

* Author to whom correspondence should be addressed; E-Mail: Eli.Rothenberg@nyumc.org

Abstract: The initial step of viral infection is the binding of a virus onto the host cell surface. This first viral-host interaction would determine subsequent infection steps and the fate of the entire infection process. A basic understating of the underlining mechanism of initial virus-host binding is a prerequisite for establishing the nature of viral infection. Bacteriophage λ and its host *Escherichia coli* serve as an excellent paradigm for this purpose. λ phages bind to specific receptors, LamB, on the host cell surface during the infection process. The interaction of bacteriophage λ with the LamB receptor has been the topic of many studies, resulting in wealth of information on the structure, biochemical properties and molecular biology of this system. Recently, imaging studies using fluorescently labeled phages and its receptor unveil the role of spatiotemporal dynamics and divulge the importance of stochasticity from hidden variables in the infection outcomes. The scope of this article is to review the present state of research on the interaction of bacteriophage λ and its *E. coli* receptor, LamB.

Keywords: bacteriophage; lambda phage; LamB receptor; protein J; single-virus tracking

1. Introduction

Viral infections are initiated through a binding process, which involves a specific interaction between the virus and the host cell surface [1–3]. The nature of initial virus host interaction greatly varies amongst different systems, and is mediated by various cell surface receptors, co-receptors and other molecules, depending on the specific virus and host [4–6]. Nevertheless, the underlining mechanism of this interaction relies on a diffusion limited viral-receptor finding process. This viral receptor-finding process is crucial for the propagation of the infection process as it determines the fate of infection [4,6]. Here we review the studies of this interaction focused on the *Escherichia coli* bacterium and its virus, bacteriophage λ as a virus-host model system.

Since the seminal discovery of bacteriophage λ, it has drawn paramount interest by the molecular biologists and served as an excellent genetic tool for studying fundamental principles of biology such as gene regulation, DNA replication, homologous and site-specific recombination [7–9]. Even today, the λ system continues to yield new insights into its gene regulatory circuits. The best-characterized feature of phage λ biology is the genetic switch that decides whether a phage propagates by cell lysis, or integrates into host genomes to become a prophage [10–12]. Other well studied aspects of phage λ biology is the adsorption of the phages and subsequent DNA delivery into the host *E. coli* cell [13–16]. Phage λ uses the bacterial maltose pore LamB (λ-receptor) for delivery of its genome into the bacterial cell. The interaction between λ and its receptor has been extensively studied both biochemically using purified components and genetically by either mutated phage or bacterial receptor [17–25]. However, the quantitative understanding of the dependence of viral target-finding on virus-receptor interactions and cellular architecture came from recent single virus tracking studies [26,27]. In this review we focus on the interaction of phage λ and its receptor LamB. We will summarize previous biochemical and structural studies of phage λ and its receptor (LamB), and provide an overview of a recent study of virus target searching mechanisms at single molecule level.

2. Structure of the LamB Receptor and Interaction with Bacteriophage λ

Bacteriophage λ is one of the well studied models in molecular biology. Although a vast amount of information is available about the gene regulation network of λ [10–12], the quantitative depiction of host cell infection was determined only recently [26]. Bacteriophage λ consists of an icosahedrally symmetric (5,3,2 rotational symmetries) head of diameter 60 nm encapsulating the 48,502 bp double strand DNA molecule and a flexible tail through which the viral DNA expel during infection (Figure 1) [28,29]. Following infection, the invading phage DNA can either replicate within the host, forming new phages and propagate by lysing the host cell (lytic pathway), or it can become a prophage by integrating its DNA into the host chromosome, which then replicate as a part of host chromosome (lysogenic pathway). However, a switch from lysogenic to lytic pathway can be induced where the prophages can replicate independently, assemble the head and tail, forming new viruses and promoting lysis for further propagation [10–12]. The *in vivo* head and tail assembly are complex process and beyond the limit of present discussion [29,30]. Early genetic experiments showed that most of the *E. coli* K-12 mutations resistant to λ phage are located in two genetic regions *malA* and *malB* [14,16,21,22]. This *malB* region contains a gene *lamB* whose product, LamB, involves in the λ receptor synthesis, a component of *E. coli* outer membrane. However, a recent study shows that mutant form of bacteriophage λ can target alternative receptor. When phage cI26 (a strictly lytic derivative of phage λ) was cultured with *E. coli* in condition that suppressed the expression of LamB, mutant phage changed their specificity from LamB to a new receptor, OmpF [31]. A combination of four mutants in phage tail protein J were required for targeting this new receptor. It is noteworthy that some host mutations prevented phage from evolving this new function, demonstrating the complexity of interactions in a co-evolving population. The adsorption of λ phages onto bacterial surface is the first step in the infection process [15]. At this stage, phages can either dissociate from the host cell, known as desorption or alternatively bind irreversibly to the host cell [14]. However, once the phage irreversibly bind to the cell surface it triggers a series of poorly understood events and finally delivers its DNA into the bacterial cytoplasm through the channel formed by its tail, leaving the phage protein capsid behind. The tail fibers of bacteriophages are also important to make specific contacts with receptor molecules on the surface of the bacterial cell. The common laboratory strain of bacteriophage λ, so called λ wild type carries a frameshift mutation in *stf* gene relative to Ur-λ, the original isolate. The Ur-λ phages have thin tail fibers which are absent in λ wild type and the Ur-λ has expanded receptor specificity and adsorbs to host cells more rapidly, suggesting the importance of tail fibers [32]. The process between phage adsorption and DNA injection can be sub divided in three steps: 'lag', 'trigger' and 'uptake' [33]. It has been shown that the free tail can itself adsorb to the cell surface, however, the head attachment is required for the lag reaction [9,29,34].

Figure 1. A cartoon representation of λ phage.

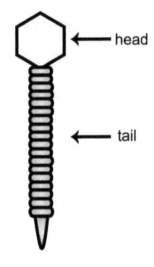

Gram-negative bacteria have two membranes [35]; the outer monolayer is composed of lipopolysaccharide (LPS) as its major lipid, and the inner leaflet contains mostly phosphatidylethanolamine, small amounts of phosphatidylglycerol and cardiolipin [36]. The outer membranes of Gram negative bacteria protect cells from harmful agents by slowing down their penetration while allowing uptake of nutrients [37–39]. The major proteins of outer membrane are β-barrel proteins, which play crucial roles in adhesion and virulence. In Gram negative bacteria β-barrel proteins are found exclusively in the outer membranes and contributed ~2%–3% of the Gram negative proteome [40]. However, their occurrence is not restricted to Gram negative bacteria and also found in Mycobacteria, mitochondria and chloroplasts [41–44]. Transmembrane β-barrels also found in several microbial toxins such as heptameric pore-forming α-haemolysin from Staphylococcus aureus, toxin aerolysin and the anthrax-protective antigen [45–47].

Well characterized constitutive outer membrane proteins are porins. They form a general diffusion pore with a defined exclusion limit for hydrophilic solutes within the outer membrane. In addition to the constitutive porins, the outer membrane may contain porins that are induced only under special growth conditions. The most abundant porins in *Escherichia coli*, the OmpF and OmpC porins, are called general porins, allowing passage of hydrophilic solutes up to a size limit determined by the constrictions of their channels [37]. The biogenesis of β-barrel proteins is complex owing to their intrinsic structure. The insertion of individual β-strands into the hydrophobic core of the bilayer is thermodynamically restricted as they cannot assemble from previously inserted transmembrane segments due to the lack of hydrogen bonds of individual β-strands of β-barrel proteins. Henceforth, membrane insertion must be orchestrated with acquisition of both secondary and tertiary structure to produce the contiguous membrane-spanning barrel held by inter-strand hydrogen bonds [37–39]. Furthermore, the porins must also attain their quaternary structure, presumably by associations formed in the membrane.

The LamB protein is a well-characterized example of a porin, termed as a maltoporin, because it is required for growth on limiting concentrations of maltose. The protein coded by gene *lamB* of the maltose operon also serves as receptors for several phages, such as λ, K10 and TP1 [48–50]. The molecular weight of LamB is 135.6 kDa, looking like a half-open tulip, formed by 3 identical subunits, each one having a molecular weight of 45.9 kDa [51–53]. A major contribution to understanding the molecular basis of the λ phage interaction with LamB receptor has come from determination of the crystal structure of LamB [54]. This study showed that each subunit of the trimeric protein formed by an 18-stranded antiparallel β-barrel, which form a wide channel with a diameter of about 2.5 nm (Figure 2A,B). Loops are found at the end of barrel. Three loops, L1, L3 and L6 (in Figure 2C, they are colored as green, red and yellow respectively) interact with the Loop 2 from an adjacent subunit and packed against the inner wall of the barrel and line the channel. The other loops form a compact structure at the cell surface. Six aromatic residues (Y6, Y41, W74, W358, W420 and F227) lining up the channel interior, forms the 'greasy slide', and actively participate in carbohydrate (maltooligosaccharides are in apolar van der Waals contact with the "greasy slide") transport. Tyrosine 118 (Y118), located opposite of the greasy slide has a major impact on ion and carbohydrate transport through LamB [54,55]. When all six residues of the greasy slide are mutated to alanine, the mutation Y118W is sufficient to confer to LamB maltopentaose transport *in vivo* and maltopentaose binding *in vitro*. How the mutation in LamB could correlate with the resistance to phage λ infection? We have summarized the mutants of LamB that conferred resistance for phage λ in Table 1 [56]. Only about half of the phage λ resistant mutants are surface exposed, and are labeled in yellow in Figure 2D [54]. The remaining mutations (labeled in red in Figure 2D) might have indirect effect. They may alter or the cause structural change on the surface, or they might alter the dynamic behavior of the loops. For instance, one mutation, G18V, is known to affects the stability of trimers and may thus have long-range effects [56]. Similar to the LamB receptor, a number of studies had been carried to understand how λ utilizes the LamB receptor to inject its DNA inside the host cell [24,25]. Phage λ tail contains a hollow tube that consists of 32-stacked disks, where each disk is formed by six subunits of the major tail protein gpV, arranged such that each disk has a central, 3 nm hole. Phage λ uses this channel to eject its DNA, though the tail by itself (without the head) can attach to the host cell. Genetic evidence indicates that gpJ directly interacts with the outer membrane protein LamB during the attachment of the bacteriophage to the surface of the cell. When the J gene from phage λ was substituted with the tail fiber gene from a closely related bacteriophage 434, the resulting phage was found to bind to a different membrane receptor OmpC [16], which phage 434 uses for infection [57]. Further studies indicate that the C-terminal of gpJ protein determines the host specificity of the phage [24,25]. Electron microscopy imaging of soluble LamB receptor and LamB protein incorporated into liposomes revealed the presence of two different types of bacteriophage λ—LamB complexes. In one type of the complexes binding occurs near the end of the tail fiber (J protein), while in the other type of complexes the distal end of the tail tube was directly and irreversibly attached to the receptor particles or the liposome. Genetic studies showed that the mutations in λ phages that have compensatory effect with the LamB mutants tightly blocks the phage λ adsorption are located in the C-terminal portions of J [24]. In fact, using a chimeric protein comprising the last 249 amino acids of J in fusion to the C-terminal end of the carrier maltose binding protein (MBP) could bind to LamB trimers and inhibited recognition by anti-LamB antibody [17]. Electron microscopy study showed that this chimeric J protein could also bind to the LamB at the cell surface and this interaction prevented λ adsorption. In this same study, when this chimeric protein was reconstituted with either LamB or the loop deletion mutant LamB 4+6+9v, both of them showed similar blocking of ion current in the lipid bilayer experiment, which indicated that the phage λ binding includes not only the extracellular loops.

Table 1. Mutation in LamB receptor that confers resistance to lambda phage infection (From [56]).

Residue(s)	Substitution
18	Gly → Val
148	Glu → Lys
151	Gly → Asp
152	Ser → Phe
154	Ser → Phe
155	Phe → Ser
163	Tyr → Asp
164	Thr → Pro
245	Gly → Arg
245	Gly → Val
247	Ser → Leu
249	Gly → Asp
250	Ser → Phe
259	Phe → Val
259	Phe → Val
382	Gly → Asp
382	Gly → Val
401	Gly → Asp

Figure 2. Crystal structure of 18 antiparallel β-strands barrel trimer LamB. (**A**) Side view and (**B**) Top view. Three monomers are shown by different colors (red, yellow and green). (**C**) Interaction of loops in barrel. Three loops L1 (green), L3 (red) and L6 (yellow) interact with L2 from an adjacent subunit and packed against the inner wall of the barrel. (**D**) The mutation(s) of LamB confer resistance to phage infection. The surface exposed mutant shown in yellow and the others in red. Mutation G18V is known to affect the stability of trimer and shown in green. The Adopted and modified after [54,56].

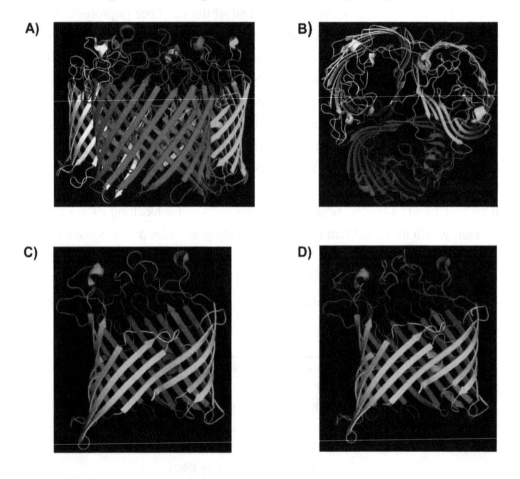

3. The λ Phage LamB Target Finding Process

Numerous genetic and biochemical studies had been carried out to provide the nature of interaction of the LamB receptor and protein J. Recently, a study of the spatiotemporal dynamics by which a phage λ, initially diffusing in bulk, arrives at a specific LamB site on the cell surface, provided a fundamental understanding of the initial viral-host target finding process.

The classical problem of diffusion limited target finding on a cell surface was first addressed by the concept of reaction rate enhancement by dimensional reduction (RREDR), introduced by Adam and Delbrück [58]. To explain the enhancement of adsorption rates they proposed a two-stage capture model , where free phages diffuse in three dimensions (3D), but once in contact with the cell surface this 3D motion is replaced by a two-dimensional (2D) 'random walk' on the cell surface, until it is captured by the receptor. This model of reduction in spatial dimension, from 3D to 2D, implementing a 3D+2D searching strategy initially explained the accelerated rate of the process of target finding. In later years, Berg and Purcell showed that to make this 2D diffusion to be advantageous, the adsorption energy has to be strong enough to keep the phage on the surface of the bacterial cell while weak enough to allow the 2D diffusion [59]. Nonetheless, this model was also the first to predict the adsorption rate k as a function of receptors N on a bacterium and the calculated maximum adsorption rate (k_{max}) based on this model was noticeably less than experimental findings, which leads to hypothesize the possible contribution of bacterial swimming to the search process. Further understanding of the adsorption process comes from a recent work [60], which showed that upon incubation of phage λ with *E. coli* strain Ymel (this strain carries wild type λ receptor) in a solution of 10 mM $MgSO_4$ (pH = 7.4), the decreases of population of free phage in bulk obeys a double-exponential function with a fast and a slow decay time. Both the fast and slow processes are specific to interactions between phage λ and its receptor and henceforth the interaction is an on-and-off process followed by an irreversible binding. It is also noteworthy that the reversible and the irreversible binding rate in nearly independent of temperature, suggesting the entropic nature of phage retention by the receptor. However, this simplified model is based on uniform distribution of receptor sites in the cell surface of the bacteria, while studies using fluorescently labeled λ tail showed that spatial distribution of LamB in the outer membrane is not uniform, and rather LamB accumulated in irregular and spiral patterns that are dynamic and depend on cell length [61]. In another study, fluorescently labeled λ phages adsorbed on individual cells revealed a preferential binding of phages to the bacterial poles rather than cover the cell surface uniformly [62]. Interestingly, this preference for polar sites is not restricted to λ. When similar experiments were carried with other *E. coli* phages such as P1 [63], virulent coliphages T4 [64] and T7 [65] or with other bacteria than *E. coli* K-12, like *Yersinia pseudotuberculosis* (and with T7-like Yersinia phage A1122) [66] and *Vibrio Cholera* (with T4 like vibrio phage KVP40) [67], similar preference to bacterial poles had been documented (See Table 2) [62]. Taken together, these findings indicated that a model based on uniform 2D diffusion search process is lacking.

Table 2. Localization of different phages on cell surface. QDots label phages are compared for their localization on cell surface (Adopted and modified from [62]).

Phage: Host	Foci at the pole and mid-cell (%)	Foci in other location (%)
λ: *E. coli*	69	31
T7: *E. coli*	71	29
P1: *E. coli*	78	22
T4: *E. coli*	95	5
λØ80: *E. coli*	68	32
KVP40: *V. cholera*	73	27
ØA1122: *Y. pseudotuberculosis*	68	32

An accurate depiction of the search process was recently established using single-particle tracking experiments of fluorescently labeled phages. This experimental approach allowed to monitor the early stages of infection in live bacteria at single-virus:single-cell level with nanometer localization accuracy and ~30-ms time resolution [26]. In this study live *E. coli* cells were first attached to the surface of a microfluidic chamber followed by addition of fluorescent viruses enabling to monitor their adsorption onto cells. The motion of individual phages was categorized in three different modes, free diffusion, motion on the surface of the host cells, and attachment (Figure 3A,B). Similar to the classical pictures [58], free phages initial diffuse in 3D until they encounter a bacterial cell, hence their motion will transition into a 2D diffusion on the cell surface. The phages will continue to diffuse on the surface of the cell until they either irreversibly bind to a receptor, or fall off and continue their free 3D motion. It is noteworthy that the diffusion coefficient value for each mode of motion is strikingly different (Figure 3D). Both free viruses and viruses moving on the cell surface followed normal Fick's laws of diffusion but with an order of magnitude difference in their diffusion coefficient. On the other hand, viruses that were attached to the cells exhibited slow local motion. However, in contrary to the classic view of uniform 2D movement of phages on the cell surface, phages exhibited a distinctly anisotropic motion pattern, with a tendency to move along the short axis of the cell (Figure 3C). Their 2D motion on cell surface was also spatially heterogeneous, showed a spatial focusing along the cell. The bound phages also showed a distinct preference for the poles of the cells (Figure 3F).

These observations were hypothesized to be linked to an ordered pattern of phage receptors, LamB, on the surface of the cell. This hypothesis is also in good agreement with the previous results [61] which showed, using fluorescently labeled phage tails that a spatial arrangement of LamB receptors on the *E. coli* surface was reminiscent of the helices and rings found for other bacterial surface proteins [68,69]. To test that notion, the arrangement of LamB receptors on the cell surface was examined using an *E. coli* strain (S2188:pLO16) with an inducible expression for a modified biotinylated version of the LamB protein [70], enabling to specifically label LamB receptors using SA-conjugated fluorophores. Labeling multiple receptors with a high concentration of QDs (10 nM) resolved the spatial organization of LamB on the cell surface, clearly showing various striped patterns reminiscent of rings and helices (Figure 4A). The distribution of receptors along individual cells showed distinct peaks corresponding to the observed rings and helices with high receptor concentration around the cell pole. The striking resemblance of organization of the LamB receptors with the features exhibited by viruses moving on the cell surface, suggest a unique virus-receptor interaction resulting in an increased viral residence in receptor rich regions. This idea is further supported by: (1) a similar angular distribution for viral trajectories and LamB bands (Figure 4C); (2) an increased viral affinity for polar localization; and (3) distinct LamB bands at poles (Figure 4B). Further proof that the viral motion of phages on the cell surface is governed by interaction with LamB receptors was provided by comparing the dwell times of viruses on wild-type cells and receptor-deleted cells (Figure 5B) [70,71]. Cells lacking receptors exhibited a >15-fold decrease in dwell time, indicating that LamB receptors are required for prolonged interaction between viruses and cell surface. Along this line of evidence, co localization experiments of moving viruses and receptors (Figure 5A,C) showed that the viruses spent a mean of 73.6 ± 3.7% of their total trajectory time in receptor-rich regions.

Figure 3. The target-finding process of individual viruses. (**A**) Time-lapse images of a single fluorescently label λ-phage virus (green spot) diffusing near and on an *E. coli* host cell (black) (images scale: height = 10 μm, width = 11 μm). (**B**) Cartoon of the observed stages for virus receptor-finding process: (I) Virus initially diffuses freely until it encounters a cell, followed by (II) motion on the host cell and (III) binding to a receptor (or detachment from the host cell and continued free diffusion). (**C**) Four representative single-virus trajectories plotted in normalized bacterial coordinates (X_L, long axis: length normalized coordinate; X_T, short axis: width normalized coordinate [26]), showing a tendency for motion along the short axis X_T (perpendicular to the long axis). (**D**) Normalized distribution of the angle between the momentary displacement vector and the short axis, X_T, (138 viral trajectories). The histogram shows a predominant inclination for motion along the short axis (X_T) with a mean angle = 29.6 ± 0.4° (mean ± SE, SD = 24.38 degrees). (**E**) The calculated MSD as a function of lag time for individual viral trajectories in each of the observed regimes, forming distinct groups with more than an order of magnitude separation in MSD values: off-cell diffusion trajectories (green), on-cell diffusion (red), and attachment (blue). Trajectories for both off-cell diffusion and on-cell diffusion yielded a log-log slope of ~1, indicative of normal diffusion. Viruses bound to the host showed either small, confined movement or no movement, with a 0.5–0 slope range. (**F**) Distributions of virus positions along the cell (X_L) throughout the spatial focusing process. Error bars: mean ± SE. (i) Initial random point of encounter (top panel, 47 viral trajectories). (ii) Unbound viruses moving on the surface of the host (middle panel, 138 viral trajectories). (iii) Distribution of final infection sites with a clear trend for polar localization (bottom panel, 59 viral trajectories). The area of all three distributions was normalized to facilitate comparison between the different distributions. Adopted and modified after [26].

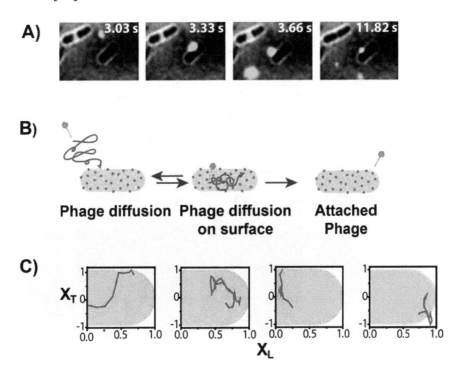

Figure 3. *Cont.*

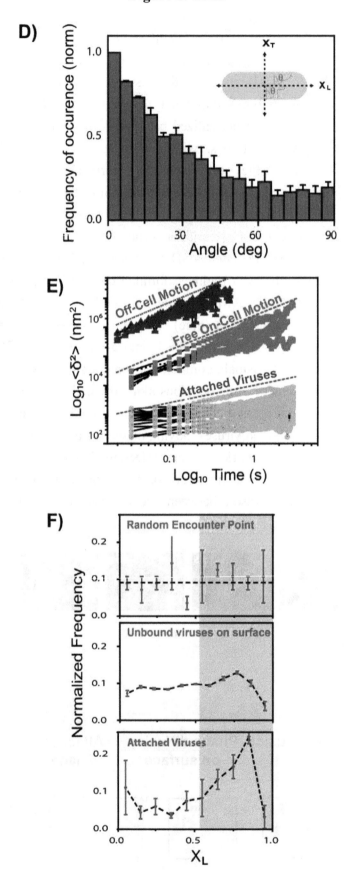

Figure 4. Arrangement of the LamB receptors network on cell surface. (**A**) One cell is shown. Rendered 3D image obtained by sectioning epifluorescence microscopy of cells with labeled receptors, showing nearly continuous helices (scale bar = 2 μm). Quantum dots (green) were used to label the receptor and cell outline is in red. The pattern is described by two helices in blue. (**B**) Normalized LamB distribution along X_L averaged for 50 cells. A clear peak at the cellular poles indicates of a high concentration of receptors. (**C**) Angular distribution of LamB bands from 98 cells, showing a tendency for band orientation along the short axis X_T (perpendicular to the long axis). Adopted and modified after [26].

A)

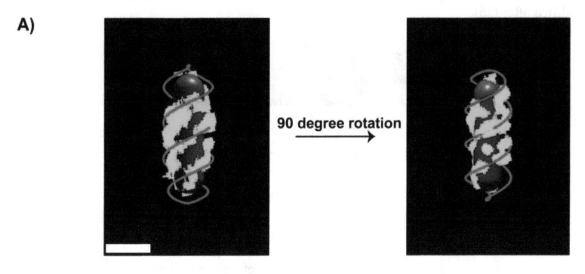

B)   **C)**

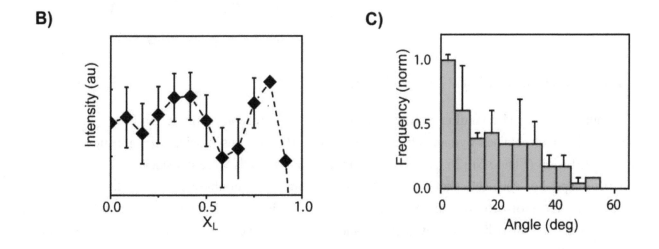

Figure 5. Viral motion on the cell surface is dominated by the interaction with LamB receptors. (**A**) Time-lapse images of the colocalization of a free virus (green) moving on a cell (blue) with LamB receptor bands (purple). The virus is observed to be predominantly moving on the LamB receptor bands (scale bar = 2 μm). (**B**) The mean dwell times of viruses on host cell with (+LamB) and without (−LamB) LamB. (**C**) Colocalization dwell-time analysis of 70 unbound viruses moving on cells with labeled bio LamB, showing the distribution of the fraction of a trajectory the viruses spent on LamB regions, with strong prevalence for residing in highly labeled, receptor-rich regions. Inset: Distribution of the cell area (fraction of entire area) with labeled receptors. Adopted and modified after [26].

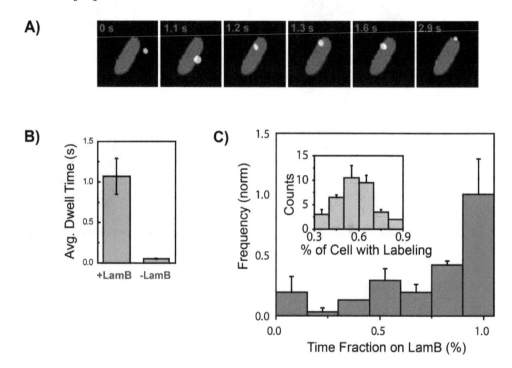

4. Theoretical Model for Viral Target Finding

A quantitative understanding of the dynamics of viral receptor-finding process came from a theoretical model obtained by simulating the motion of 10,000 viruses using an algorithm for curved surface diffusion (Figure 6) [26,72,73]. In this model, *E. coli* cells were considered as cylinders with spherical caps. For each cell, the cell size defined the pitch of a double helix. This double helix was defined as a receptor-rich area, and the rest of the cell surface was defined as receptor-free. A virus located in a receptor-rich zone may diffuse randomly within the receptor-rich area, it may become attached to the surface, or it may move into a receptor-free area, whereas a virus located in the receptor-free area may diffuse within this area, diffuse into the receptor-rich area, or fall off the cell surface. The resulting spatial focusing features generated by the model and the resulting trajectories showed to be in agreement with the experiments. Viruses arriving at random places along the cell followed by gradually concentrate at receptor-rich regions and the final attachment sites show a pronounced preference for the receptor-rich areas, and especially the cell poles.

Figure 6. Theoretical model of phage target finding. (**A**) A schematic description of the model, showing a typical phage trajectory (green) on the surface of an *E. coli* cell. Cell shape was modeled as a cylinder with hemispherical caps (radius R_c = 0.4 μm). The cell surface was divided into receptor-free (transparent pink) and receptor-rich (solid red) areas. The receptors form a 50-nm-thick double helical or multi-ring pattern along the cell surface. The zoomed-in area shows the kinetic scheme in detail. A phage located in a receptor-rich zone can either move within the receptor area with a stepsize consistent with the diffusion coefficient D_{RR}, become attached with a probability P_{Att} (chosen to match the experimentally observed dwell time to attachment), or move into a receptor-free zone. This move can be rejected with a probability $1 - P_{NR \rightarrow R}$, in which case the move will be repeated. On the other hand, a phage located in a receptor-free area can move with a diffusion coefficient D_{NR}. If the phage attempts to enter the receptor zone, the move will always be allowed. The phage can also fall off the cell with a probability $P_{Fall} = \tau sim/<\tau_{Fall}>$, where τ_{sim} is the time step of the simulation and $<\tau_{Fall}>$, is the experimentally observed dwell time to fall-off in a LamB-*E. coli* strain. (**B**) A simplified kinetic scheme of the model, shown as a four-state, discrete-time Markov chain. The nonreceptor (NR) and receptor (R) states are transient, and the attachment (A) and fall-off (F) states are absorbing. (**C**) Generating population heterogeneity. Cell lengths were randomly chosen from a log-normal distribution spanning the experimentally observed cell sizes. The cell size defines the pitch of receptor double helix according to the relationship obtained experimentally. For short cells with helix angles of <10, a multi-ring pattern was used instead of a double helix. (**D**) Representative 2D projections of phage trajectories in normalized units (green lines). The initial landing site is shown as a gray circle. Panels in the first two rows display trajectories ending in attachment (blue circle), and panels in the last two rows display trajectories ending with the phage falling off the cell (at a position denoted with a yellow circle). Adopted and modified after [26].

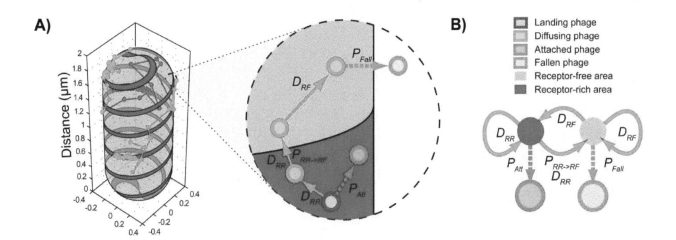

Figure 6. *Cont.*

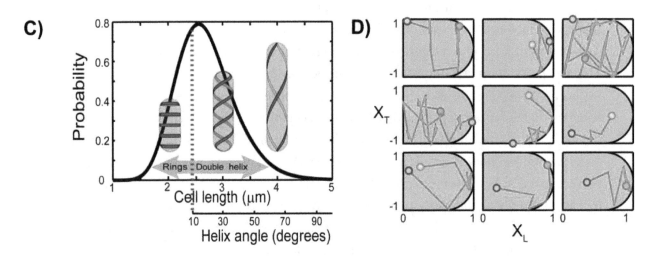

5. Summary

The mechanisms by which molecules, viruses and cells find their targets in biological systems are crucial for a fundamental understanding all biological processes. The interaction of bacteriophage λ with its cell surface receptor, LamB, and its receptor finding process serves an excellent model for this purpose. A plethora of theoretical and experimental studies had been done to address this problem [14–16,18–26]. The classic RREDR model [58], a reduction of dimensionality from 3D motion in bulk to 2D motion on cell surface, was initially considered to be in good agreement with the experimental observation [14,15,58–60,74]. However, finding that the interaction of viruses with a localized network of receptors on the cell surface limits their sampling motion to a fraction of the surface, in effect rendering it quasi-one-dimensional, indicates that the searching process is more complex than what was previously thought and paved the way for further modification of the existing model. We are concluding our present discussion by speculating that the connection between target finding processes and the spatial distribution of the target may be universal, and could be better addressed by considering fractal dimensional motion.

References

1.   Marsh, M.; Helenius, A. Virus entry: Open sesame. *Cell* **2006**, *124*, 729–740.
2.   Mudhakir, D.; Harashima, H. Learning from the viral journey: How to enter cells and how to overcome intracellular barriers to reach the nucleus. *AAPS J.* **2009**, *11*, 65–77.
3.   Smith, A.E.; Helenius, A. How viruses enter animal cells. *Science* **2004**, *304*, 237–242.
4.   Dimitrov, D.S. Virus entry: Molecular mechanisms and biomedical applications. *Nat. Rev. Microbiol.* **2004**, *2*, 109–122.
5.   Mercer, J.; Helenius, A. Virus entry by macropinocytosis. *Nat. Cell Biol.* **2009**, *11*, 510–520.

6. Skehel, J.J.; Wiley, D.C. Receptor binding and membrane fusion in virus entry: The influenza hemagglutinin. *Annu. Rev. Biochem.* **2000**, *69*, 531–569.

7. Friedman, D.I.; Court, D.L. Bacteriophage lambda: Alive and well and still doing its thing. *Curr. Opin. Microbiol.* **2001**, *4*, 201–207.

8. Hendrix, R.W. Bacteriophage genomics. *Curr. Opin. Microbiol.* **2003**, *6*, 506–511.

9. Hendrix, R.; Roberts, J.; Stahl, F.; Wesberg, R. *Lambda II*; Cold Spring Harbor Laboratory Press: Plainview, NY, USA, 1983.

10. Ptasne, M. *A Genetic Switch, Gene Control and Phage Lambda*; Blackwell Scientific and Cell Press: Palo Alto, CA, USA, 1986.

11. Ptasne, M. *A Gentic Switch: Phage Lambda Revisited*; Cold Spring Harbor: New York, NY, USA, 2004.

12. Oppenheim, A.B.; Kobiler, O.; Stavans, J.; Court, D.L.; Adhya, S. Switches in bacteriophage lambda development. *Annu. Rev. Genet.* **2005**, *39*, 409–429.

13. Randall-Hazelbauer, L.; Schwartz, M. Isolation of the bacteriophage lambda receptor from *Escherichia coli*. *J. Bacteriol.* **1973**, *116*, 1436–1446.

14. Schwartz, M. Reversible interaction between coliphage lambda and its receptor protein. *J. Mol. Biol.* **1975**, *99*, 185–201.

15. Schwartz, M. The adsorption of coliphage lambda to its host: Effect of variations in the surface density of receptor and in phage-receptor affinity. *J. Mol. Biol.* **1976**, *103*, 521–536.

16. Schwartz, M. *Interaction of Phages with Their Receptor Proteins*; Chapman and Hall: London, UK, 1980.

17. Berkane, E.; Orlik, F.; Stegmeier, J.F.; Charbit, A.; Winterhalter, M.; Benz, R. Interaction of bacteriophage lambda with its cell surface receptor: An *in vitro* study of binding of the viral tail protein gpJ to LamB (Maltoporin). *Biochemistry* **2006**, *45*, 2708–2720.

18. Charbit, A.; Clement, J.M.; Hofnung, M. Further sequence analysis of the phage lambda receptor site. Possible implications for the organization of the lamB protein in *Escherichia coli* K12. *J. Mol. Biol.* **1984**, *175*, 395–401.

19. Charbit, A.; Werts, C.; Michel, V.; Klebba, P.E.; Quillardet, P.; Hofnung, M. A role for residue 151 of LamB in bacteriophage lambda adsorption: Possible steric effect of amino acid substitutions. *J. Bacteriol.* **1994**, *176*, 3204–3209.

20. Clement, J.M.; Lepouce, E.; Marchal, C.; Hofnung, M. Genetic study of a membrane protein: DNA sequence alterations due to 17 lamB point mutations affecting adsorption of phage lambda. *EMBO J.* **1983**, *2*, 77–80.

21. Hofnung, M.; Hatfield, D.; Schwartz, M. malB region in *Escherichia coli* K-12: Characterization of new mutations. *J. Bacteriol.* **1974**, *117*, 40–47.

22. Hofnung, M.; Jezierska, A.; Braun-Breton, C. lamB mutations in E. coli K12: Growth of lambda host range mutants and effect of nonsense suppressors. *Mol. Gen. Genet.* **1976**, *145*, 207–213.

23. Scandella, D.; Arber, W. An *Escherichia coli* mutant which inhibits the injection of phage lambda DNA. *Virology* **1974**, *58*, 504–513.

24. Wang, J.; Hofnung, M.; Charbit, A. The C-terminal portion of the tail fiber protein of bacteriophage lambda is responsible for binding to LamB, its receptor at the surface of *Escherichia coli* K-12. *J. Bacteriol.* **2000**, *182*, 508–512.

25. Wang, J.; Michel, V.; Hofnung, M.; Charbit, A. Cloning of the J gene of bacteriophage lambda, expression and solubilization of the J protein: First *in vitro* studies on the interactions between J and LamB, its cell surface receptor. *Res. Microbiol.* **1998**, *149*, 611–624.

26. Rothenberg, E.; Sepulveda, L.A.; Skinner, S.O.; Zeng, L.; Selvin, P.R.; Golding, I. Single-virus tracking reveals a spatial receptor-dependent search mechanism. *Biophys. J.* **2011**, *100*, 2875–2882.

27. Golding, I. Decision making in living cells: Lessons from a simple system. *Annu. Rev. Biophys.* **2011**, *40*, 63–80.

28. Sanger, F.; Coulson, A.R.; Hong, G.F.; Hill, D.F.; Petersen, G.B. Nucleotide sequence of bacteriophage lambda DNA. *J. Mol. Biol.* **1982**, *162*, 729–773.

29. Hendrix, R. *Lambda II*; Cold Spring Harbor Laboratory: New York, NY, USA, 1985.

30. Weigle, J. Assembly of phage lambda *in vitro*. *Proc. Natl. Acad. Sci. USA* **1966**, *55*, 1462–1466.

31. Meyer, J.R.; Dobias, D.T.; Weitz, J.S.; Barrick, J.E.; Quick, R.T.; Lenski, R.E. Repeatability and contingency in the evolution of a key innovation in phage lambda. *Science* **2012**, *335*, 428–432.

32. Hendrix, R.W.; Duda, R.L. Bacteriophage lambda PaPa: Not the mother of all lambda phages. *Science* **1992**, *258*, 1145–1148.

33. Roa, M.; Scandella, D. Multiple steps during the interaction between coliphage lambda and its receptor protein *in vitro*. *Virology* **1976**, *72*, 182–194.

34. Weigle, J. Studies on head-tail union in bacteriophage lambda. *J. Mol. Biol.* **1968**, *33*, 483–489.

35. Smit, J.; Kamio, Y.; Nikaido, H. Outer membrane of Salmonella typhimurium: Chemical analysis and freeze-fracture studies with lipopolysaccharide mutants. *J. Bacteriol.* **1975**, *124*, 942–958.

36. Kamio, Y.; Nikaido, H. Outer membrane of Salmonella typhimurium: Accessibility of phospholipid head groups to phospholipase c and cyanogen bromide activated dextran in the external medium. *Biochemistry* **1976**, *15*, 2561–2570.

37. Nikaido, H. Porins and specific channels of bacterial outer membranes. *Mol. Microbiol.* **1992**, *6*, 435–442.

38. Nikaido, H. Molecular basis of bacterial outer membrane permeability revisited. *Microbiol. Mol. Biol. Rev.* **2003**, *67*, 593–656.

39. Nikaido, H.; Vaara, M. Molecular basis of bacterial outer membrane permeability. *Microbiol. Rev.* **1985**, *49*, 1–32.

40. Zhai, Y.; Saier, M.H., Jr. The beta-barrel finder (BBF) program, allowing identification of outer membrane beta-barrel proteins encoded within prokaryotic genomes. *Protein Sci.* **2002**, *11*, 2196–2207.

41. Engelhardt, H.; Heinz, C.; Niederweis, M. A tetrameric porin limits the cell wall permeability of *Mycobacterium smegmatis*. *J. Biol. Chem.* **2002**, *277*, 37567–37572.

42. Heinz, C.; Engelhardt, H.; Niederweis, M. The core of the tetrameric mycobacterial porin MspA is an extremely stable beta-sheet domain. *J. Biol. Chem.* **2003**, *278*, 8678–8685.

43. Model, K.; Prinz, T.; Ruiz, T.; Radermacher, M.; Krimmer, T.; Kuhlbrandt, W.; Pfanner, N.; Meisinger, C. Protein translocase of the outer mitochondrial membrane: Role of import receptors in the structural organization of the TOM complex. *J. Mol. Biol.* **2002**, *316*, 657–666.

44. Schleiff, E.; Eichacker, L.A.; Eckart, K.; Becker, T.; Mirus, O.; Stahl, T.; Soll, J. Prediction of the plant beta-barrel proteome: A case study of the chloroplast outer envelope. *Protein Sci.* **2003**, *12*, 748–759.

45. Parker, M.W.; Buckley, J.T.; Postma, J.P.; Tucker, A.D.; Leonard, K.; Pattus, F.; Tsernoglou, D. Structure of the Aeromonas toxin proaerolysin in its water-soluble and membrane-channel states. *Nature* **1994**, *367*, 292–295.

46. Petosa, C.; Collier, R.J.; Klimpel, K.R.; Leppla, S.H.; Liddington, R.C. Crystal structure of the anthrax toxin protective antigen. *Nature* **1997**, *385*, 833–838.

47. Song, L.; Hobaugh, M.R.; Shustak, C.; Cheley, S.; Bayley, H.; Gouaux, J.E. Structure of staphylococcal alpha-hemolysin, a heptameric transmembrane pore. *Science* **1996**, *274*, 1859–1866.

48. Hancock, R.E.; Reeves, P. Lipopolysaccharide-deficient, bacteriophage-resistant mutants of *Escherichia coli* K-12. *J. Bacteriol.* **1976**, *127*, 98–108.

49. Roa, M. Interaction of bacteriophage K10 with its receptor, the lamB protein of *Escherichia coli*. *J. Bacteriol.* **1979**, *140*, 680–686.

50. Wandersman, C.; Schwartz, M. Protein Ia and the lamB protein can replace each other in the constitution of an active receptor for the same coliphage. *Proc. Natl. Acad. Sci. USA* **1978**, *75*, 5636–5639.

51. Ishii, J.N.; Okajima, Y.; Nakae, T. Characterization of lamB protein from the outer membrane of *Escherichia coli* that forms diffusion pores selective for maltose-maltodextrins. *FEBS Lett.* **1981**, *134*, 217–220.

52. Nakae, T.; Ishii, J.N. Molecular weights and subunit structure of LamB proteins. *Annales de Microbiologie* **1982**, *133A*, 21–25.

53. Neuhaus, J.M. The receptor protein of phage lambda: Purification, characterization and preliminary electrical studies in planar lipid bilayers. *Annales de Microbiologie* **1982**, *133A*, 27–32.

54. Schirmer, T.; Keller, T.A.; Wang, Y.F.; Rosenbusch, J.P. Structural basis for sugar translocation through maltoporin channels at 3.1 A resolution. *Science* **1995**, *267*, 512–514.

55. Dutzler, R.; Wang, Y.F.; Rizkallah, P.; Rosenbusch, J.P.; Schirmer, T. Crystal structures of various maltooligosaccharides bound to maltoporin reveal a specific sugar translocation pathway. *Structure* **1996**, *4*, 127–134.

56. Charbit, A.; Gehring, K.; Nikaido, H.; Ferenci, T.; Hofnung, M. Maltose transport and starch binding in phage-resistant point mutants of maltoporin. Functional and topological implications. *J. Mol. Biol.* **1988**, *201*, 487–496.

57. Fuerst, C.R.; Bingham, H. Genetic and physiological characterization of the J gene of bacteriophage lambda. *Virology* **1978**, *87*, 437–458.

58. Adam, G.; Delbruck, M. Reduction of dimensionality in biological diffusion processes. In *Structural Chemistry and Molecular Biology*; W.H. Freeman & Company: San Francisco, NC, USA, 1968.

59. Berg, H.C.; Purcell, E.M. Physics of chemoreception. *Biophys. J.* **1977**, *20*, 193–219.

60. Moldovan, R.; Chapman-McQuiston, E.; Wu, X.L. On kinetics of phage adsorption. *Biophys. J.* **2007**, *93*, 303–315.

61. Gibbs, K.A.; Isaac, D.D.; Xu, J.; Hendrix, R.W.; Silhavy, T.J.; Theriot, J.A. Complex spatial distribution and dynamics of an abundant *Escherichia coli* outer membrane protein, LamB. *Mol. Microbiol.* **2004**, *53*, 1771–1783.

62. Edgar, R.; Rokney, A.; Feeney, M.; Semsey, S.; Kessel, M.; Goldberg, M.B.; Adhya, S.; Oppenheim, A.B. Bacteriophage infection is targeted to cellular poles. *Mol. Microbiol.* **2008**, *68*, 1107–1116.

63. Kaiser, D.; Dworkin, M. Gene transfer to myxobacterium by *Escherichia coli* phage P1. *Science* **1975**, *187*, 653–654.

64. Hashemolhosseini, S.; Holmes, Z.; Mutschler, B.; Henning, U. Alterations of receptor specificities of coliphages of the T2 family. *J. Mol. Biol.* **1994**, *240*, 105_110.

65. Steven, A.C.; Trus, B.L.; Maizel, J.V.; Unser, M.; Parry, D.A.; Wall, J.S.; Hainfeld, J.F.; Studier, F.W. Molecular substructure of a viral receptor-recognition protein. The gp17 tail-fiber of bacteriophage T7. *J. Mol. Biol.* **1988**, *200*, 351–365.

66. Garcia, E.; Elliott, J.M.; Ramanculov, E.; Chain, P.S.; Chu, M.C.; Molineux, I.J. The genome sequence of Yersinia pestis bacteriophage phiA1122 reveals an intimate history with the coliphage T3 and T7 genomes. *J. Bacteriol.* **2003**, *185*, 5248–5262.

67. Miller, E.S.; Heidelberg, J.F.; Eisen, J.A.; Nelson, W.C.; Durkin, A.S.; Ciecko, A.; Feldblyum, T.V.; White, O.; Paulsen, I.T.; Nierman, W.C.; *et al.* Complete genome sequence of the broad-host-range vibriophage KVP40: Comparative genomics of a T4-related bacteriophage. *J. Bacteriol.* **2003**, *185*, 5220–5233.

68. Osborn, M.J.; Rothfield, L. Cell shape determination in *Escherichia coli*. *Curr. Opin. Microbiol.* **2007**, *10*, 606–610.

69. Vats, P.; Shih, Y.L.; Rothfield, L. Assembly of the MreB-associated cytoskeletal ring of Escherichia coli. *Mol. Microbiol.* **2009**, *72*, 170–182.

70. Oddershede, L.; Dreyer, J.K.; Grego, S.; Brown, S.; Berg-Sorensen, K. The motion of a single molecule, the lambda-receptor, in the bacterial outer membrane. *Biophys. J.* **2002**, *83*, 3152–3161.

71. Brown, S. Engineered iron oxide-adhesion mutants of the *Escherichia coli* phage lambda receptor. *Proc. Natl. Acad. Sci. USA* **1992**, *89*, 8651–8655.

72. Golding, I.; Cox, E.C. Physical nature of bacterial cytoplasm. *Phys. Rev. Lett.* **2006**, *96*, 098102.

73. Holyst, R.; Plewczynski, D.; Aksimentiev, A.; Burdzy, K. Diffusion on curved, periodic surfaces. *Phys. Rev. E* **1999**, *60*, 302–307.

74. Axelrod, D.; Wang, M.D. Reduction-of-dimensionality kinetics at reaction-limited cell surface receptors. *Biophys. J.* **1994**, *66*, 588–600.

10

The Staphylococci Phages Family

Marie Deghorain * and Laurence van Melderen *

Laboratoire de Génétique et Physiologie Bactérienne, Faculté de Sciences, IBMM, Université Libre de Bruxelles (ULB), Gosselies B-6141, Belgium

* Author to whom correspondence should be addressed; E-Mails: lvmelder@ulb.ac.be (L.V.M.); mdeghora@ulb.ac.be (M.D.)

Abstract: Due to their crucial role in pathogenesis and virulence, phages of *Staphylococcus aureus* have been extensively studied. Most of them encode and disseminate potent staphylococcal virulence factors. In addition, their movements contribute to the extraordinary versatility and adaptability of this prominent pathogen by improving genome plasticity. In addition to *S. aureus*, phages from coagulase-negative *Staphylococci* (CoNS) are gaining increasing interest. Some of these species, such as *S. epidermidis*, cause nosocomial infections and are therefore problematic for public health. This review provides an overview of the staphylococcal phages family extended to CoNS phages. At the morphological level, all these phages characterized so far belong to the *Caudovirales* order and are mainly temperate *Siphoviridae*. At the molecular level, comparative genomics revealed an extensive mosaicism, with genes organized into functional modules that are frequently exchanged between phages. Evolutionary relationships within this family, as well as with other families, have been highlighted. All these aspects are of crucial importance for our understanding of evolution and emergence of pathogens among bacterial species such as *Staphylococci*.

Keywords: bacteriophages; *Staphylococcus*; horizontal transfer; virulence

1. Introduction

The vast majority of bacteria contain prophages, either integrated into their chromosome or as extra-chromosomal elements, accounting for substantial genetic variability. Not only do phages shape bacterial genome architecture, they also constitute major vehicles for horizontal gene transfer [1,2]. In addition, they contribute to virulence by encoding numerous virulence or fitness factors and by their movements within genomes (see below, section 2.2.) [2–5]. These mobile elements are responsible for gene disruption, provide anchor region for genomic rearrangements, protect bacteria from lytic infections or, in contrast, provoke cell lysis through prophage induction [2]. Thus, phages play essential roles in bacterial evolution and adaptation.

Phages are widespread in *Staphylococcus aureus* and have been extensively studied [1,3,6]. They were firstly used for the typing of clinical *S. aureus* isolates [7,8]. *S. aureus* is a major human and animal pathogen that causes both nosocomial and community-acquired infections. It colonizes skin and mucous membranes, with the anterior nares being the primary niche in humans. While found in healthy carriers, *S. aureus* is also responsible for a wide range of diseases, from mild skin infections to severe life-threatening infections, such as sepsis or endocarditis [6]. The number of prophages in *S. aureus* genome is generally high. All *S. aureus* genome sequenced so far do contain at least one prophage, and many strains carry up to four [1]. These encode numerous staphylococcal toxins responsible for pathogenesis [1,2,6].

Staphylococci also comprise coagulase-negative species (coagulase-negative *Staphylococci*, CoNS), which are distinguishable from *S. aureus* by the lack of coagulase-encoding gene. In contrast to *S. aureus*, which is only found in part of the population, these species belong to the commensal flora of healthy humans. Some species are associated to specific niches, and others appear to be more 'generalist' and are generally found on the body surface [9–12]. CoNS include human opportunistic pathogens often associated with medical devices. *S. epidermidis* is referred to as a frequent cause of nosocomial infections [9–12]. In addition, 'true' pathogens that are not associated with medical devices may also be problematic for public health. As an example, *S. saprophyticus* is considered as a frequent pathogen responsible for uncomplicated urinary tract infections [11,13]. Pathogenesis of CoNS species relies on factors required for their commensal mode of life or fitness (e.g., factors involved in adhesion, in biofilm formation and in persistence) and not on toxins, as observed for *S. aureus* [10,11]. As a consequence, less attention has been paid to these phages.

During the past decade, sequencing of *Staphylococci* genomes and extensive comparative genomic analyses have significantly increased the number of staphylococcal phages identified. Up to now, more than 68 *Staphylococci* phages and prophages sequences, mainly from *S. aureus*, are found in the [14]. In addition, 268 *Staphylococci* genomes are available on the PATRIC server [15] and offer a remarkable source of novel prophage sequences for further studies (see below).

In this review, we provide an overview of *Staphylococci* phages with a focus on their contribution to pathogenesis. A special interest is placed on the classification methods, as well as on the evolutionary relationships connecting staphylococcal phages. Phage classification is often problematic, due to the modular organization of phage genomes. Relationships between *Staphylococci* phages and phages from other species are also discussed in an evolutionary perspective. Finally, the potential use of staphylococcal phages for bio-technological and medical applications is briefly addressed.

2. The Phages of *S. aureus*

2.1. Global Features of S. aureus *Phages*

2.1.1. Morphological Families

As the vast majority of phages, the *S. aureus* phages known so far are double-stranded DNA phages belonging to the *Siphoviridae* family of the *Caudovirales* order (Table S1) (reviewed in [2,4,6]). In general, they are temperate phages detected as prophage inserted in the chromosome, some of them being lytic due to mutations in the lysogeny functions (e.g., phiIPLA35 and phiIPLA88; [16]; or SA11; [17]). According to the morphological classification previously proposed by Ackermann [18], staphylococcal *Siphoviridae* are composed of an icosahedral capsid and a non-contractile tail ended by a base-plate structure. Capsids may adopt elongated or isometric shapes, and tail length varies from short (130 nm) to long (400 nm). A small number of *S. aureus Podoviridae* and *Myoviridae* phages, also belonging to the *Caudovirales* order, were described (Table S1). *Podoviridae,* such as the recently identified SAP-2 phage [19], are composed of a small icosahedral capsid and a short, non-flexible, non-contractile tail. *Myoviridae* phages, such as the well-known phage K [20], are characterized by an icosahedral capsid and a long contractile tail.

2.1.2. Genomic Characteristics of *S. aureus* Phages

A comparative study of *S. aureus* phage genomes performed by Pelletier and co-workers [3] revealed several key genomic features, which are globally applicable to all staphylococcal phages described so far. The analysis encompassed 27 genomes from *S. aureus* phages and prophages belonging to the three morphological families described above.

Genome size extends from less than 20 kb up to more than 125 kb [3]. In contrast to phages from *Mycobacterium* [21] or *Pseudomonas aeruginosa* [22], genome sizes are not uniformly distributed, and three categories can be established and used to classify *Staphylococci* phages (class I: < 20 kb; class II: ≈ 40 kb, class III: >125 kb; see section 4.1) [3]. Interestingly, genome size categories correlate with the morphological classification, *Podoviridae* harboring the smallest genomes (class I), *Myoviridae* the largest ones (class III), and *Siphoviridae* showing intermediate sizes (class II).

Coding regions are tightly packed with very few and small intergenic regions and a high gene density (1.67 genes/kb in average) [3]. Their GC content is similar to that of the host. *S. aureus* phages provide an impressive, mainly unexplored source of genetic diversity. On the 2,170 predicted proteins from the 27 phages analyzed in the study of Pelletier and co-workers, a function could be assigned using BLAST to 35% of the ORFs. No match was detected for 44% of these ORFS in Bacteria and Phages Gene Bank databases [3].

Genomes of *S. aureus Siphoviridae* display the typical structure of the morphological family (Figure 1a) [2,4,23]. Five functional modules are arranged as follows: lysogeny, DNA metabolism, DNA packaging and capsid morphogenesis, tail morphogenesis and host cell lysis. The DNA metabolism module can be divided into replication and regulation functions. When present, virulence factors are generally encoded downstream of the lysis module [2,4]. In some cases, they are inserted between the lysogeny and DNA metabolism modules as reported for phiNM1 to four prophages found in the *S. aureus* Newman strain [24]. Genes are generally transcribed on the same strand, except for small clusters, such as genes involved in host genome integration [3,4].

While harboring a modular structure as well, organization of *Podoviridae* genomes is different (Figure 1b) [3,19,25]. One major distinction resides in a smaller number of ORFs, as indicated by a smaller size (20 to 32 ORFs) [3,19,25]. Functional modules encoding DNA packaging and capsid morphogenesis, tail morphogenesis and lysis were identified, in addition to genes of unknown function. In contrast to *Siphoviridae*, modules are not well defined and tail and lysis genes are overlapping. In addition, the lysogeny module is absent as expected for lytic phages.

Genome organization of staphylococcal *Myoviridae* (e.g., phages K, G1, Twort) [3,20,26] is similar to *E. coli* T4 phage, the paradigm for *Myoviridae* [27]. Genomes are organized into functional modules of conserved genes (replication, structural elements), interrupted by large plastic regions encoding mainly genes of unknown function (Figure 1c). Structural modules found in staphylococcal *Myoviridae* phages are more closely related (in terms of gene content and organization) to modules found in staphylococcal *Siphoviridae* phages than to modules of *Myoviridae* found in different bacterial species.

Figure 1. Modular organization of *Staphylococci* phages genomes (**a**) *Siphoviridae* genomes. Colored boxes represent the five functional modules found in *Siphoviridae* genomes. Red: lysogeny, yellow: DNA metabolism, green: DNA packaging and capsid morphogenesis, blue: tail morphogenesis, pink: cell host lysis. Virulence genes (purple) are generally found downstream the lysis module, or inserted between the lysogeny and DNA metabolism module. A closer view of the DNA packaging and capsid morphogenesis shows the structural genes pattern typical of the Sfi21- and Sfi11- like phages genera (see text for details); (**b**) *Podoviridae* genomes. Lysis module (pink) and tail morphogenesis modules (blue) are overlapping. DNA metabolism genes (*i.e.*, single-strand DNA binding protein and DNA polymerase; green) are located in a region of genes of unknown function (gray), upstream to the tail module. An encapsidation protein is encoded next to the DNA polymerase in the staphylococcal *Podoviridae* genomes described so far. (**c**) *Myoviridae* genomes. The phage Twort genome is represented as an example. A large region encodes genes of unknown function (gray). DNA metabolism genes (yellow) are distributed in two distinct modules (known as replication modules), as well as lysis genes (pink) that are found upstream and downstream to DNA packaging and capsid (green) and tail modules (blue). TerS: small subunit terminase, TerL: large subunit terminase, Port: portal protein,

Prot: protease, MHP: major capsid protein, H: capsid morphogenesis protein, mHP: minor capsid protein, SS-DNA binding: single strand DNA binding protein.

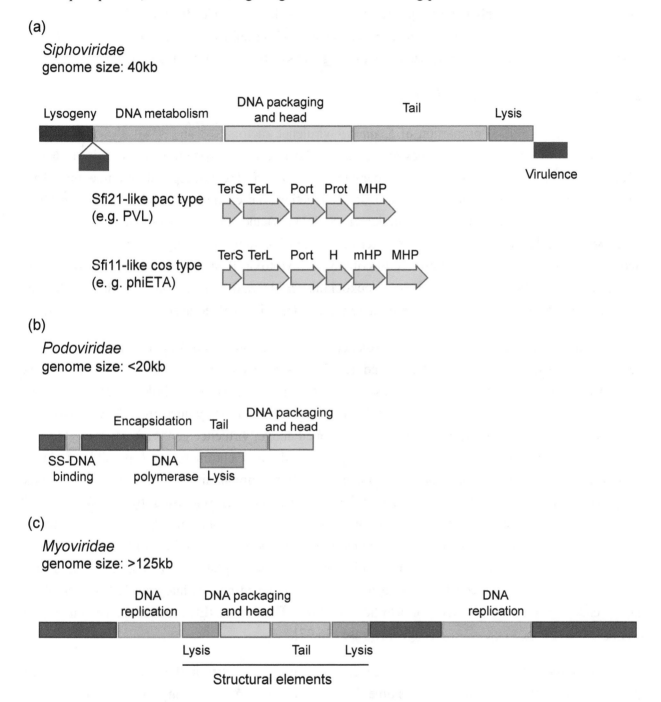

2.2. Role of Phages in S. aureus Pathogenesis

The *S. aureus* genome is mainly composed of a backbone of genes that are conserved among the different strains, both in terms of sequence and synteny [28,29]. These genes constitute the 'core genome' in opposition to the 'accessory genome', which is variable between strains and constituted by integrated plasmids, transposons, genomic islands, pathogenicity islands (saPIs) and prophages. This 'accessory genome' may represent up to 25% of *S. aureus* genome and largely contributes to the high genetic and phenotypic plasticity of the pathogen. Indeed, one of the remarkable characteristics of *S. aureus* is represented by its versatility and ability to adapt to diverse and hostile environments [1,6, 28,30,31], and phages are playing an essential role in this phenomenon.

2.2.1. Phages-encoded Virulence Factors

Phages encode a large proportion of *S. aureus* virulence factors and provide the pathogen with a large variety of toxins, mainly allowing escaping host immune system (reviewed in [1,6,31]). Many factors have been described and characterized, such as the widespread immune modulator staphylokinase (*sak*) responsible for host tissue destruction, the chemotaxis inhibitory protein CHIP (*chp*), the staphylococcal inhibitor of complement SCIN (*scn*) and several superantigens (*sea, seg, sek, sek2, sep, seq*). These superantigens are enterotoxins causing food poisoning, toxic shock syndrome and necrotizing fasciitis. In addition, the bi-component cytotoxin Panton-Valentine leukocidin (PVL, encoded by *lukF-PV, lukS-PV*) and related leukocidins (*lukM, lukF*)) form pores into leukocytes and cause necrotic infections. Finally, the exfoliative toxin A (*eta*) is involved in severe skin infections.

In general, phages are carrying single virulence factor genes, although some exceptions have been reported. For examples, phiSa3 phages and relatives, such as phiN315, may encode up to five virulence factors, which form an immune escape complex (IEC) [5,31–33]. Virulence factor genes are not strictly associated to a specific phage and appear to be exchanged by horizontal gene transfer and recombination [2,6,31] (see section 2.2.3). As mentioned above, virulence genes are often located near the attachment site (*att*) of the prophage, *i.e.*, adjacent to the host chromosome. A possible origin for their acquisition by phages might be aberrant excision events from an ancestral bacterial chromosome [34]. They might also derive from mobile genetic elements, as suggested by the presence of transposase genes flanking toxin genes as in phiPV83 [35]. Thus, phages clearly impact virulence by positive lysogenic conversion, since they provide novel functions and activities to the host. Negative conversion also occurs as prophage insertion might inactivate genes [2,4,31]. In most of the cases, both phenomena occur simultaneously. As an example, phiSa3-related phages integrate in beta-hemolysin gene, rendering lysogens defective for beta-hemolyse but effective for IEC production. The integration site relies on the specificity of the phage integrase [36,37].

Expression of phage-encoded virulence genes is maximal upon entry in the lytic cycle, since latent promoters are activated and phage genome is replicated, leading to an increase in genome copy number [5,38,39], although some expression is detected during lysogeny [34].

2.2.2. Phage-mediated Mobilization of Virulence Factors: SaPIs Pathogenicity Islands

Phages are the primary vehicles for horizontal transfer between *S. aureus* strains. They spread chromosomally-encoded virulence determinants through generalized transduction. In addition, they are also responsible for the mobilization of SaPIs, which encodes major toxin genes, such as the toxic shock syndrome toxin 1 and other superantigens (reviewed in [40]). SaPIs are widespread in *S. aureus* genomes. They are discrete chromosomal DNA segments that have been acquired by horizontal transfer. They are not mobile by themselves and rely on a helper phage for moving. SaPIs are replicated and mobilized either in response to SOS-induced excision of a helper prophage present in the same strain, either following the infection by a helper phage or by the joint entry of SaPI and helper phage [40]. The underlying molecular mechanism for induction is the specific interaction of a SaPI repressor and a de-repressor encoded by the helper phage. Different proteins of a particular helper phage may be involved in induction of different SaPIs. In particular, phi80alpha is able to mobilize at least five different SaPIs [40–42]. Hence, SaPIs mobilization represents a remarkable example of evolutionary adaptation involving pathogenicity islands and phages [42].

2.2.3. Phage Dynamics Contribute to *S. aureus* Evolution and Pathogenesis

Additionally to gene transfer, phages contribute to genetic alterations during infection, providing the species with broad genetic variations [5,38,43,44,45]. In addition to phage acquisition or excision (lysogenic conversion); duplication, ectopic integration and stable extra-chromosomal form of phages have been reported, increasing genetic diversity within bacterial populations [36,44]. Hence, generating heterogeneity within a population upon infection offers different virulence potentials and provides the pathogen with the ability to develop a flexible response to host defenses. It was shown that phage mobilization and atypical genomic integration are favored in pathogenic strains upon infection conditions, compared to colonizing strains in healthy carriers [5,36,44]. Factors causing phage induction *in situ* are environmental conditions that lead to bacterial DNA damage, such as antibiotics treatments or reactive oxygen species released by macrophages. Phage-mediated phenotypic diversification acts in concert with other mechanisms for genetic variations, such as recombination and mutations, and is under the influence of external factors, such as the presence of co-infecting species [43,45].

3. What about Phages in Non-*aureus Staphylococci*?

Comparison of CoNS genomes with *S. aureus* genomes revealed the inter-species conservation of both sequence and synteny of a large proportion of genes (core genes) with variable regions carrying species-specific genes [13,46–49]. Phage-encoded virulence factors responsible for *S. aureus* pathogenesis are absent in CoNS, in correlation with the difference in pathogenesis-mediated by CoNS. However, toxin and antibiotic resistance genes have been identified in several of these species. Although mobilization of virulence genes by phages has not been demonstrated, it is conceivable that phages might play a role in pathogenesis and evolution of CoNS, such as observed for *S. aureus*. Accordingly, phages were shown to impact genetic variability in *S. epidermidis* [50]. Although phage prevalence in clinical isolates might be underestimated [51], CoNS genomes described so far contain only few prophages or genomic islands, if any.

Most of the CoNS phages belong to the *Siphoviridae* family, while several virulent *Podoviridae* phages have been recently isolated directly from human anterior nares [52]. As for *S. aureus* phages, the first interest brought to phage infecting these species relied on their use for clinical isolate typing [53,54]. Several studies report the isolation and characterization of phages from *S. epidermidis* and *S. saprophyticus* [52,55–57], but only nine phages and prophages have been sequenced and studied, both at the molecular and physiological levels (Table S2). Among these, five phages and prophages are from *S. epidermidis* [51,58,59], one prophage from *S. carnosus* [48], two phages from *S. hominis* [37] and one phage from *S. capitis* [37]. The *S. epidermidis* vB_SepiS_phiIPLA5 (Table S2) is strictly lytic due to a defective lysogeny module [51,57], and the *S. carnosus* phiTM300 prophage (Table S2) appears to have lost its mobility, as it could not be induced upon mitomycin C treatment [48]. These CoNS phages show the general genomic features described for the *S. aureus* phages (see above, section 2.1.2; Table S2). Their genome size range falls into the class II proposed by the Pelletier group [3], with high gene density and a typical genomic organization into five functional modules (lysogeny, DNA metabolism, DNA packaging and capsid morphogenesis, tail morphogenesis and lysis). As for *S. aureus* phages, a function could be only assigned to a small proportion of the ORFs (from 29% to 53%, depending on the phage). Known virulence determinants were not found in these genomes. Comparative analysis of genome sequences revealed that *S. epidermidis* vB_SepiS_phiIPLA5, vB_SepiS_phiIPLA7, phiPH15 and phiCNPH82 are closely related, while phi909 showed a high similarity to *S. aureus* phages [48,51]. Interestingly, we also detected close relationships between *S. aureus* and CoNS phages (see below, section 4) [37].

Genomic islands related to SaPIs elements were identified in *S. haemolyticus* [46] and *S. saprophyticus* [13], although they usually lack superantigen-encoding genes. The first CoNS superantigen-bearing genomic island was recently described in *S. epidermidis* [59]. SePI-I encodes the staphylococcal enterotoxin C3 (SEC3) and enterotoxin-like toxin L (SEIL). Interestingly, the *seil* gene is homologous to those of *S. aureus*, indicating horizontal transfer events between *Staphylococci* species.

4. Classification and Evolution of the Staphylococcal Phage Family

4.1. Classification of Staphylococci Phages, a Long-Term Challenge

Early classifications proposed for *S. aureus* phages were based on their lytic properties, serotypes, morphology, on number and size of virion proteins and on genome size and organization as revealed through DNA hybridization or endonuclease restriction patterns ([18,60–63]; reviewed in [26]). More recently, progress in genomics and bio-informatics have allowed alternative classification methods.

Comparative genomic studies led to the subdivision of the morphological families into sub-families and genera. Following a classification proposed by Brussow and Desiere (although still not recognized by ICTV (International Comity on Taxonomy of Viruses)), *Siphoviridae* phages of low GC gram-positive bacteria, including *Staphylococci,* are categorized as Sfi21- or Sfi11-like phages by several authors [64] (Tables S1 and S2). This distinction is based on the capsid genes pattern as reported for *Streptococcus thermophilus* phages [65]. Sfi21-like phages share characteristic features over the capsid region with *E. coli* HK97 phage and use a similar cos-site based strategy for DNA packaging (Figure 1a). In *S. aureus*, Sfi21-like genus is subdivided into three groups: the first being represented by phiPVL-phiPV83-phi13-phiSa3mw phages, the second by phiSLT-phiSa2mw-phi12 and the third by phiMu50A-phiN315 [4] (Table S1). While showing different capsid morphology (isometric or elongated), these phages encode typical capsid gene pattern of Sfi21-like phages (portal protein-protease-major capsid protein). Sfi11-like *pac*-type phages are related to the *B. subtilis* SPP1 phage and differ from Sfi21-like phages, notably by the lack of the protease-encoding gene (Figure 1a) [64, 65].

The two other families of the *Caudovirales* order were recently reassessed on the basis of protein similarities [66,67]. Among *Podoviridae*, *S. aureus* phages constitute a novel genus called 44AHJD-like. This genus belongs to the *Picovirinae* sub-family that also includes the phi29-like genus represented by the well-described *Bacillus* phi29 [67]. Among *Myoviridae*, the Twort phage is a representative of the Twort-like genus within the *Spounaviridae* sub-family [66]. This sub-family also includes the *Bacillus subtilis* SPO1 phage and relatives, as well as the *Lactobacillus plantarum* LP65 phage [66,68].

Other classification approaches based on specific marker genes found within *S. aureus Siphoviridae* genomes have been proposed [29,63,69,70]. They rely on PCR detection of genes representative of different phage types or categories. In addition to providing classification schemes, these methods are useful for *S. aureus* prophage detection, which is of great interest for epidemiological studies. The markers encompass genes coding for structural components, such as tail fibers, capsid proteins [63] or integrase genes [29,36]. In the latter case, classification correlates with distinct integration sites into the host chromosome. A good correlation between the type of integrase and virulence determinants has also been reported [36]. However, these methods do not provide information about mosaic structure, although detection of representatives of each functional module might provide some clues about mosaicism [70].

Recently, the group of Pelletier [3] proposed a classification taking into account the genome size, the gene organization, in addition to comparative nucleotide and protein sequence analysis. Using comparative genomic, Class II (*Siphoviridae*) was divided into three clades (A-C), and a new clade (D) was added later on by the Fischetti group after including *S. epidermidis* and additional *S. aureus* phage sequences [58]. Our group has refined this classification by extending the analysis to 85 phage and prophage genomes, among which 15 originated from CoNS [37]. The approach was based on the similarity of protein repertoires using tree-like and network-like methods [71,72]. Both methods established nine distinct clusters (data obtained with the network-like method is shown in Figure 2). Seven of these clusters (1–6, 9) are composed of *Siphoviridae* and constitute class II, according to the Pelletier classification. Clusters 7 and 8 constitute class III and I and are composed of *Myoviridae* and *Podoviridae*, respectively. They are unrelated to the other clusters. Within class II, our analysis generates seven related clusters, instead of the four clades previously proposed by the Pelletier group. Most importantly, one cluster (cluster 9) is composed exclusively of non-*S. aureus* phages and constitutes an entirely new clade, as compared to the earlier classification. Four of the seven clusters are composed exclusively of *S. aureus* phages (clusters 2, 4, 5 and 6). Interestingly, the two last clusters (clusters 1 and 3) are composed of *S. aureus* and non-*S. aureus* phages, revealing close relationships between phages of different *Staphylococci* species.

Figure 2. Network representation of relationships between *Staphylococci* phages based on protein content (adapted from [37]). Circles represent the nine different clusters defined by Markov cluster algorithm (MCL). The color indicates the host species (magenta: *S. aureus*; purple: *S. aureus* and CoNS; blue: CoNS). The number of genomes is indicated into brackets. Cluster 8 corresponds to the class I (*Podoviridae*), clusters 1–6 and 9 to class II (*Siphoviridae*) and cluster 7 to class III (*Myoviridae*). Cluster 1 corresponds to clade A, cluster 4 to clade B, and clades C and D were split into two sub-clades (clusters 2 and 3, and 5 and 6, respectively). Cluster 9 constitutes a new clade. In this schematic representation, gray lines between distinct clusters indicate that at least 30% of homologous proteins are shared between at least two phage genomes. Following this analysis, two previously unclassified phages (2638A and 187) were included in cluster 6 and 1, respectively. PT1028 was not included in this analysis.

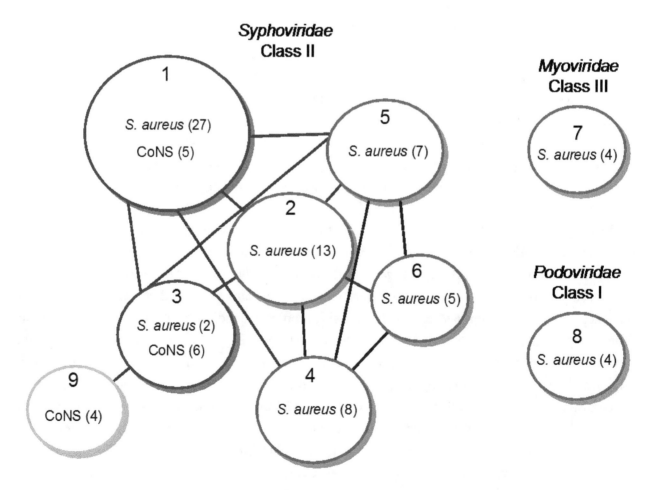

4.2. Modular Evolution of Staphylococci Phages

4.2.1. Extensive Genome Mosaicism in Staphylococci Phages

The mosaic gene organization is consistent with the theory of modular evolution based on module exchanges by horizontal transfer and recombination events [3,4,73]. This mosaicism can be viewed at either the nucleotide or amino acid level. At the nucleotide level, genome comparisons reveal exchange events that are likely to have occurred recently. Comparisons at the protein level identify homologous proteins that are shared by distantly related phages. They clearly derived from a common ancestor, but have diverged with time, and similarity at the nucleotide level is undetectable [73]. *S. aureus* phages are highly mosaic, indicating that gene exchange is common within this phage group [3,33,35,37,74–79]. Exchanges not only concern single genes, but also protein domains or a group of genes, such as functional modules [3,73]. Interestingly, *S. aureus* phages often share large, highly similar sequences with at least two other phages and with different phages along their genome. As an example, more than 80% of the ROSA genome is covered by large identical sequences from at least five other *S. aureus* phage genomes, with remarkable co-linearity. Exchange events are favored between phages of the same genome size range, which correspond to different morphological families (see above section 2.1.2) [79].

Different molecular mechanisms have been proposed to explain how these exchanges occur [23,73]. A first model invokes homologous recombination events between conserved sequences, which are generally found at gene borders, such as detected in lambdoid *E. coli* phages [80,81]. Recombination events require large homologous regions that are recognized either by host- or phage-encoded recombinases [80]. A second model proposes alternative recombination mechanisms, occurring randomly or between short sequences [82]. It is likely that these two mechanisms contribute to mosaicism generation. For instance, PVL-related phage genomes are composed of long regions that are shared between different phages, flanked by conserved sequences, called junctions, suitable for homologous recombination events. However, shuffling might occur at other regions, since short regions appear to be shared by phage genomes [35,74–76]. Even though these mechanisms certainly lead to a high percentage of non-viable phages, it is considered as a highly creative process that might provide countermeasures against anti-phage mechanisms evolved by bacteria, such as CRISPR sequences, toxin-antitoxin systems and restriction-modification [82].

4.2.2. Evolutionary Relationships within Staphylococci Phages

An interesting aspect of mosaicism is that comparison of a large number of genomes allows the establishment of phylogenetic relationships of specific regions instead of entire phages. Phylogenetic trees of representative genes revealed distinct evolutionary histories for different modules, thereby highlighting extensive mosaicism. This observation thus provides a tool to investigate evolution of *S. aureus* phages.

General clustering approaches, as used by our group to extend the classification of *Staphylococci* phages to CoNS species [37], are also useful to represent relationships among phage groups. In order to study the modular evolution of the CoNS StB12, StB27 and StB20 phages, families of homologous proteins were defined and the relative proportion of homologs in the nine clusters shown in Figure 2 was calculated. As previously shown [79], phages belonging to different morphological families do not share homologs. Our study confirms that morphogenesis module genes are in general less prone to horizontal swapping [3,73], although genes involved in host-phage interactions are notable exceptions. Clustering of *Staphylococcus Siphoviridae* phages relies mainly on structural features, indicating that within the overall similarities of virion structure, subtle variations allow differentiation. This is in accordance with a modular mechanism of evolution in which differentiation processes mainly rely on the exchange of a restricted number of genes or protein domains within structural modules [37]. Importantly, mosaicism encompasses genomes from both *Siphoviridae* phages from *S. aureus* and CoNS species, suggesting inter-species gene exchanges, which might be of crucial importance for *Staphylococci* pathogenesis.

4.3. Evolutionary Relationships between Staphylococci Phages and Other Species

Among the numerous approaches developed to outline evolutionary and functional relationships between phages [58,66,67,71,79,83–87], an original method developed by Lima-Mendez et al., [72,87] allowed them to study phage modular evolution on a large scale. The analysis was based on the clustering of phage proteins families in evolutionary conserved modules (ECMs) and the establishment of their distribution among phages. ECMs are defined as groups of protein that have a similar phylogenetic profile, meaning that they co-occur in genomes. In virulent phages, they are generally larger than functional modules, while in temperate phages, they tend to correspond to functional modules [87]. The clustering method allowed classification of phages into groups distinguished by different combination of ECMs. On the other hand, ECMs specific to particular phage groups can be identified. As an example, some modules were strictly associated with *Staphylococci* phages [72,87]. These ECMs encode virulence determinants, capsid and tail morphogenesis genes, or genes of unknown functions. The latter case is of special interest, since these ORFs of unknown functions might potentially be involved in phage/host interaction and may therefore include novel virulence factors or other proteins of medical relevance [72]. Other ECMs associated with *Staphylococci* phages were shared with *Streptococci* phages (modules involved in capsid and long non-contractile tail morphogenesis, phage integration/excision, replication or regulation functions), indicating an evolutionary link between these species.

Development of such methods is helpful for understanding phage evolution in general. These studies also complete the ICTV classification system based on morphology and specific molecular markers.

5. Use of *Staphylococci* Phages for Phage Therapy and Other Bio-Technological Applications

Due to the renewed interest for phage therapy, an increasing number of phages have been isolated and characterized for their potential use against *Staphylococci* infections, both in humans and animals (see notably [26,51,88–91]). Besides phage therapy, staphylococcal phages are also attractive candidates for food preservation [92–95]. Most phages selected for phage therapy or food preservation are strictly lytic because of the complications caused by lysogeny (e. g., resistance of lysogen strains to phage infection and unexpected transduction of host genes) [2,88,89]. The direct use of lysins is often an alternative for the use of entire phages (see notably [19,96,97]).

Phage candidates are mainly virulent phages belonging to *Myoviridae* (e. g., phage K, [97,98] or the recently identified phiStau2A, [99]) and *Podoviridae* (e. g., phiSAP-2, [19]). They were isolated from diverse environments, including dairy products, farm environments (milk from cow infected mastitis), humans and medical devices implanted into patients (see notably [16,19,100]).

Another important characteristic of phages selected for therapeutic and bio-technological applications is their narrow host range [89]. As an example, phage K appears to be specific to particular *S. aureus* clinical isolates, as well as to particular CoNS strains, while phiStau2 shows a larger host range among *S. aureus* clinical isolates, but is inefficient against the CoNS strains tested [99]. Therefore, to set up an efficient treatment, it is mandatory to precisely determine the bacterial species/isolate responsible for the infection, as well as disposing of a battery of phages of known host range. Traditionally, phage cocktails are used to overcome these issues [89,101]. On the other hand, this property allows specific treatment against pathogens without affecting the commensal flora.

Nevertheless, phage therapy remains promising as a large number of phages—more than 10^{31} phage particles are estimated in the biosphere—await discovery, opening the perspectives for novel therapeutic approaches [89].

6. Concluding Remarks

The large number of *Staphylococci* phages sequenced and characterized so far revealed an extensive mosaicism, which indicates gene shuffling between phages of *Staphylococci* species, including *S. aureus* and CoNS. An important aspect of *Staphylococci* phages is their pivotal role in *S. aureus* pathogenesis. Horizontal transfer of phages is an efficient way to rapidly disseminate virulence determinants among pathogens. Such transfers have been reported between *S. aureus* clinical isolates during infections [6]. CoNS pathogenesis is less understood, and virulence genes appear to be absent in CoNS phage genomes. However, CoNS phage diversity remains to be explored, and their function in CoNS pathogenesis is likely to be underestimated. Identification and characterization of novel phages from pathogenic and non-pathogenic CoNS strains is therefore of crucial importance to further understand the evolutionary relationships connecting *S. aureus* phages with phages found in non-pathogenic commensal *Staphylococci*.

Acknowledgments

Work in MD, LVM lab is supported by the European Space Agency (MISSEX project ESA AO-2004: Prodex C90255), the Fonds Jean Brachet and the Van Buuren Fondation. MD is Chargé de Recherches at the Fonds de la Recherche Scientifique, Belgium (FNRS).

References

1.   Lindsay, J.A. Genomic variation and evolution of Staphylococcus aureus. *Int. J. Med. Microbiol.* **2010**, *300*, 98–103.

2.   Brussow, H.; Canchaya, C.; Hardt, W.D. Phages and the evolution of bacterial pathogens: From genomic rearrangements to lysogenic conversion. *Microbiol. Mol. Biol. Rev.* **2004**, *68*, 560–602.

3.   Kwan, T.; Liu, J.; DuBow, M.; Gros, P.; Pelletier, J. The complete genomes and proteomes of 27 Staphylococcus aureus bacteriophages. *Proc. Natl. Acad. Sci. USA* **2005**, *102*, 5174-5179.

4.   Canchaya, C.; Proux, C.; Fournous, G.; Bruttin, A.; Brussow, H. Prophage Genomics. *Microbiol. Mol. Biol. Rev.* **2003**, *67*, 238–276.

5.   Goerke, C.; Wirtz, C.; Fluckiger, U.; Wolz, C. Extensive phage dynamics in Staphylococcus aureus contributes to adaptation to the human host during infection. *Mol. Microbiol.* **2006**, *61*, 1673–1685.

6.   Feng, Y.; Chen, C.J.; Su, L.H.; Hu, S.; Yu, J.; Chiu, C.H. Evolution and pathogenesis of Staphylococcus aureus: lessons learned from genotyping and comparative genomics. *FEMS Microbiol. Rev.* **2008**, *32*, 23–37.

7.   Wentworth, B.B. Bacteriophage Typing of the Staphylococci. *Bacteriol. Rev.* **1963**, *27*, 253–272.

8.   Rosenblum, E.D.; Tyrone, S. Serology, Density, and Morphology of Staphylococcal Phages. *J. Bacteriol.* **1964**, *88*, 1737–1742.

9.   Otto, M. Molecular basis of Staphylococcus epidermidis infections. *Semin. Immunopathol.* **2012**, *34*, 201–214.

10. Otto, M. Staphylococcus epidermidis--the 'accidental' pathogen. *Nat. Rev. Microbiol.* **2009**, *7*, 555–567.

11. Von Eiff, C.; Peters, G.; Heilmann, C. Pathogenesis of infections due to coagulase-negative staphylococci. *Lancet Infect. Dis.* **2002**, *2*, 677–685.

12. Frank, K.L.; Del Pozo, J.L.; Patel, R. From clinical microbiology to infection pathogenesis: How daring to be different works for Staphylococcus lugdunensis. *Clin. Microbiol. Rev.* **2008**, *21*, 111–133.

13. Kuroda, M.; Yamashita, A.; Hirakawa, H.; Kumano, M.; Morikawa, K.; Higashide, M.; Maruyama, A.; Inose, Y.; Matoba, K.; Toh, H.; *et al.* Whole genome sequence of Staphylococcus saprophyticus reveals the pathogenesis of uncomplicated urinary tract infection. *Proc. Natl. Acad. Sci. USA* **2005**, *102*, 13272–13277.

14. EMBL phage database. Genomes pages – Phages. Available online: http://www.ebi.ac.uk/genomes/phage.html (accessed on 30 October 2012).

15. Patric database. Staphylococcus page. Available online: http://www.patricbrc.org/portal/portal/patric/Taxon?cType=taxon&cId=1279 (accessed on accessed on 30 October 2012).

16. Garcia, P.; Martinez, B.; Obeso, J.M.; Lavigne, R.; Lurz, R.; Rodriguez, A. Functional genomic analysis of two Staphylococcus aureus phages isolated from the dairy environment. *Appl. Environ. Microbiol.* **2009**, *75*, 7663–7673.

17. Kim, M.S.; Myung, H. Complete Genome of Staphylococcus aureus Phage SA11. *J. Virol.* **2012**, *86*, 10232.

18. Ackermann, H.W. Tailed bacteriophages: the order caudovirales. *Adv. Virus Res.* **1998**, *51*, 135–201.

19. Son, J.S.; Lee, S.J.; Jun, S.Y.; Yoon, S.J.; Kang, S.H.; Paik, H.R.; Kang, J.O.; Choi, Y.J. Antibacterial and biofilm removal activity of a podoviridae Staphylococcus aureus bacteriophage SAP-2 and a derived recombinant cell-wall-degrading enzyme. *Appl. Microbiol. Biotechnol.* **2010**, *86*, 1439–1449.

20. O'Flaherty, S.; Coffey, A.; Edwards, R.; Meaney, W.; Fitzgerald, G.F.; Ross, R.P. Genome of Staphylococcal Phage K: a New Lineage of Myoviridae Infecting Gram-Positive Bacteria with a Low G+C Content. *J. Bacteriol.* **2004**, *186*, 2862–2871.

21. Pedulla, M.L.; Ford, M.E.; Houtz, J.M.; Karthikeyan, T.; Wadsworth, C.; Lewis, J.A.; Jacobs-Sera, D.; Falbo, J.; Gross, J.; Pannunzio, N.R.; Brucker, W.; Kumar, V.; Kandasamy, J.; Keenan, L.; Bardarov, S.; Kriakov, J.; Lawrence, J.G.; Jacobs, W.R., Jr.; Hendrix, R.W.; Hatfull, G.F. Origins of highly mosaic mycobacteriophage genomes. *Cell* **2003**, *113*, 171–182.

22. Kwan, T.; Liu, J.; Dubow, M.; Gros, P.; Pelletier, J. Comparative genomic analysis of 18 Pseudomonas aeruginosa bacteriophages. *J. Bacteriol.* **2006**, *188*, 1184–1187.

23. Hatfull, G.F. Bacteriophage genomics. *Curr. Opin. Microbiol.* **2008**, *11*, 447–453.

24. Bae, T.; Baba, T.; Hiramatsu, K.; Schneewind, O. Prophages of Staphylococcus aureus Newman and their contribution to virulence. *Mol. Microbiol.* **2006**, *62*, 1035–1047.

25. Vybiral, D.; Takac, M.; Loessner, M.; Witte, A.; von Ahsen, U.; Blasi, U. Complete nucleotide sequence and molecular characterization of two lytic Staphylococcus aureus phages: 44AHJD and P68. *FEMS Microbiol. Lett.* **2003**, *219*, 275–283.

26. Lobocka, M.; Hejnowicz, M.S.; Dabrowski, K.; Gozdek, A.; Kosakowski, J.; Witkowska, M.; Ulatowska, M.I.; Weber-Dabrowska, B.; Kwiatek, M.; Parasion, S.; *et al.* Genomics of staphylococcal Twort-like phages--potential therapeutics of the post-antibiotic era. *Adv. Virus Res.* **2012**, *83*, 143–216.

27. Krisch, H.M.; Comeau, A.M. The immense journey of bacteriophage T4--from d'Herelle to Delbruck and then to Darwin and beyond. *Res. Microbiol.* **2008**, *159*, 314–324.

28. Holden, M.T.; Feil, E.J.; Lindsay, J.A.; Peacock, S.J.; Day, N.P.; Enright, M.C.; Foster, T.J.; Moore, C.E.; Hurst, L.; Atkin, R.; *et al.* Complete genomes of two clinical Staphylococcus aureus strains: evidence for the rapid evolution of virulence and drug resistance. *Proc. Natl. Acad. Sci. USA* **2004**, *101*, 9786–9791.

29. Lindsay, J.A.; Holden, M.T. Staphylococcus aureus: superbug, super genome? *Trends Microbiol.* **2004**, *12*, 378–385.

30. Suzuki, H.; Lefebure, T.; Bitar, P.P.; Stanhope, M.J. Comparative genomic analysis of the genus Staphylococcus including Staphylococcus aureus and its newly described sister species Staphylococcus simiae. *BMC Genomics* **2012**, *13*, 38.

31. Malachowa, N.; DeLeo, F.R. Mobile genetic elements of Staphylococcus aureus. *Cell. Mol. Life Sci.* **2010**, *67*, 3057–3071.

32. Coleman, D.C.; Sullivan, D.J.; Russell, R.J.; Arbuthnott, J.P.; Carey, B.F.; Pomeroy, H.M. Staphylococcus aureus bacteriophages mediating the simultaneous lysogenic conversion of beta-lysin, staphylokinase and enterotoxin A: molecular mechanism of triple conversion. *J. Gen. Microbiol.* **1989**, *135*, 1679–1697.

33. van Wamel, W.J.; Rooijakkers, S.H.; Ruyken, M.; van Kessel, K.P.; van Strijp, J.A. The innate immune modulators staphylococcal complement inhibitor and chemotaxis inhibitory protein of Staphylococcus aureus are located on beta-hemolysin-converting bacteriophages. *J. Bacteriol.* **2006**, *188*, 1310–1315.

34. Wagner, P.L.; Waldor, M.K. Bacteriophage control of bacterial virulence. *Infect. Immun.* **2002**, *70*, 3985–3993.

35. Zou, D.; Kaneko, J.; Narita, S.; Kamio, Y. Prophage, phiPV83-pro, carrying panton-valentine leukocidin genes, on the Staphylococcus aureus P83 chromosome: comparative analysis of the genome structures of phiPV83-pro, phiPVL, phi11, and other phages. *Biosci. Biotechnol. Biochem.* **2000**, *64*, 2631–2643.

36. Goerke, C.; Pantucek, R.; Holtfreter, S.; Schulte, B.; Zink, M.; Grumann, D.; Broker, B.M.; Doskar, J.; Wolz, C. Diversity of prophages in dominant Staphylococcus aureus clonal lineages. *J. Bacteriol.* **2009**, *191*, 3462–3468.

37. Deghorain, M.; Bobay, L.M.; Smeesters, P.R.; Bousbata, S.; Vermeersch, M.; Perez-Morga, D.; Dreze, P.A.; Rocha, E.P.; Touchon, M.; van Melderen, L. Characterization of novel phages isolated in coagulase-negative Staphylococci reveals evolutionary relationships with *S. aureus* phages. *J. Bacteriol.* **2012**, *194*, 5829–5839.

38. Goerke, C.; Koller, J.; Wolz, C. Ciprofloxacin and trimethoprim cause phage induction and virulence modulation in Staphylococcus aureus. *Antimicrob. Agents Chemother.* **2006**, *50*, 171–177.

39. Sumby, P.; Waldor, M.K. Transcription of the toxin genes present within the Staphylococcal phage phiSa3ms is intimately linked with the phage's life cycle. *J. Bacteriol.* **2003**, *185*, 6841–6851.

40. Novick, R.P.; Christie, G.E.; Penades, J.R. The phage-related chromosomal islands of Gram-positive bacteria. *Nat. Rev. Microbiol.* **2010**, *8*, 541–551.

41. Lindsay, J.A.; Ruzin, A.; Ross, H.F.; Kurepina, N.; Novick, R.P. The gene for toxic shock toxin is carried by a family of mobile pathogenicity islands in Staphylococcus aureus. *Mol. Microbiol.* **1998**, *29*, 527–543.

42. Tormo-Mas, M.A.; Mir, I.; Shrestha, A.; Tallent, S.M.; Campoy, S.; Lasa, I.; Barbe, J.; Novick, R.P.; Christie, G.E.; Penades, J.R. Moonlighting bacteriophage proteins derepress staphylococcal pathogenicity islands. *Nature* **2010**, *465*, 779–782.

43. Goerke, C.; Wolz, C. Adaptation of Staphylococcus aureus to the cystic fibrosis lung. *Int. J. Med. Microbiol.* **2010**, *300*, 520–525.

44. Goerke, C.; Wolz, C. Regulatory and genomic plasticity of Staphylococcus aureus during persistent colonization and infection. *Int. J. Med. Microbiol.* **2004**, *294*, 195–202.

45. McAdam, P.R.; Holmes, A.; Templeton, K.E.; Fitzgerald, J.R. Adaptive evolution of Staphylococcus aureus during chronic endobronchial infection of a cystic fibrosis patient. *PLoS One* **2011**, *6*, c24301.

46. Takeuchi, F.; Watanabe, S.; Baba, T.; Yuzawa, H.; Ito, T.; Morimoto, Y.; Kuroda, M.; Cui, L.; Takahashi, M.; Ankai, A.; *et al.* Whole-genome sequencing of staphylococcus haemolyticus uncovers the extreme plasticity of its genome and the evolution of human-colonizing staphylococcal species. *J. Bacteriol.* **2005**, *187*, 7292–7308.

47. Gill, S.R.; Fouts, D.E.; Archer, G.L.; Mongodin, E.F.; Deboy, R.T.; Ravel, J.; Paulsen, I.T.; Kolonay, J.F.; Brinkac, L.; Beanan, M.; *et al.* Insights on evolution of virulence and resistance from the complete genome analysis of an early methicillin-resistant Staphylococcus aureus strain and a biofilm-producing methicillin-resistant Staphylococcus epidermidis strain. *J. Bacteriol.* **2005**, *187*, 2426–2438.

48. Rosenstein, R.; Nerz, C.; Biswas, L.; Resch, A.; Raddatz, G.; Schuster, S.C.; Gotz, F. Genome analysis of the meat starter culture bacterium Staphylococcus carnosus TM300. *Appl. Environ. Microbiol.* **2009**, *75*, 811–822.

49. Rosenstein, R.; Gotz, F. Genomic differences between the food-grade Staphylococcus carnosus and pathogenic staphylococcal species. *Int. J. Med. Microbiol.* **2010**, *300*, 104–108.

50. Lina, B.; Bes, M.; Vandenesch, F.; Greenland, T.; Etienne, J.; Fleurette, J. Role of bacteriophages in genomic variability of related coagulase-negative staphylococci. *FEMS Microbiol. Lett.* **1993**, *109*, 273–277.

51. Gutierrez, D.; Martinez, B.; Rodriguez, A.; Garcia, P. Genomic characterization of two Staphylococcus epidermidis bacteriophages with anti-biofilm potential. *BMC Genomics* **2012**, *13*, 228.

52. Aswani, V.; Tremblay, D.M.; Moineau, S.; Shukla, S.K. Staphylococcus epidermidis bacteriophages from the anterior nares of humans. *Appl. Environ. Microbiol.* **2011**, *77*, 7853–7855.

53. Boussard, P.; Pithsy, A.; Devleeschouwer, M.J.; Dony, J. Phage typing of coagulase-negative staphylococci. *J. Clin. Pharm. Ther.* **1992**, *17*, 165–168.

54. Barcs, I.; Herendi, A.; Lipcsey, A.; Bognar, C.; Hashimoto, H. Phage pattern and antibiotic resistance pattern of coagulase-negative staphylococci obtained from immunocompromised patients. *Microbiol. Immunol.* **1992**, *36*, 947–959.

55. Bes, M.; Ackermann, H.W.; Brun, Y.; Fleurette, J. Morphology of Staphylococcus saprophyticus bacteriophages. *Res. Virol.* **1990**, *141*, 625–635.

56. Bes, M. Characterization of thirteen Staphylococcus epidermidis and S. saprophyticus bacteriophages. *Res. Virol.* **1994**, *145*, 111–121.

57. Gutierrez, D.; Martinez, B.; Rodriguez, A.; Garcia, P. Isolation and characterization of bacteriophages infecting Staphylococcus epidermidis. *Curr. Microbiol.* **2010**, *61*, 601–608.

58. Daniel, A.; Bonnen, P.E.; Fischetti, V.A. First complete genome sequence of two Staphylococcus epidermidis bacteriophages. *J. Bacteriol.* **2007**, *189*, 2086–2100.

59. Madhusoodanan, J.; Seo, K.S.; Remortel, B.; Park, J.Y.; Hwang, S.Y.; Fox, L.K.; Park, Y.H.; Deobald, C.F.; Wang, D.; Liu, S.; *et al.* An Enterotoxin-Bearing Pathogenicity Island in Staphylococcus epidermidis. *J. Bacteriol.* **2011**, *193*, 1854–1862.

60. Ackermann, H.W.; DuBow, M.S.; Jarvis, A.W.; Jones, L.A.; Krylov, V.N.; Maniloff, J.; Rocourt, J.; Safferman, R.S.; Schneider, J.; Seldin, L.; *et al.* The species concept and its application to tailed phages. *Arch. Virol.* **1992**, *124*, 69–82.

61. Lee, J.S.; Stewart, P.R. The virion proteins and ultrastructure of Staphylococcus aureus bacteriophages. *J. Gen. Virol.* **1985**, *66*, 2017–2027.

62. Stewart, P.R.; Waldron, H.G.; Lee, J.S.; Matthews, P.R. Molecular relationships among serogroup B bacteriophages of Staphylococcus aureus. *J. Virol.* **1985**, *55*, 111–116.

63. Pantucek, R.; Doskar, J.; Ruzickova, V.; Kasparek, P.; Oracova, E.; Kvardova, V.; Rosypal, S. Identification of bacteriophage types and their carriage in Staphylococcus aureus. *Arch. Virol.* **2004**, *149*, 1689–1703.

64. Brussow, H.; Desiere, F. Comparative phage genomics and the evolution of Siphoviridae: Insights from dairy phages. *Mol. Microbiol.* **2001**, *39*, 213–222.

65. Le Marrec, C.; van Sinderen, D.; Walsh, L.; Stanley, E.; Vlegels, E.; Moineau, S.; Heinze, P.; Fitzgerald, G.; Fayard, B. Two groups of bacteriophages infecting Streptococcus thermophilus can be distinguished on the basis of mode of packaging and genetic determinants for major structural proteins. *Appl. Environ. Microbiol.* **1997**, *63*, 3246–3253.

66. Lavigne, R.; Darius, P.; Summer, E.J.; Seto, D.; Mahadevan, P.; Nilsson, A.S.; Ackermann, H.W.; Kropinski, A.M. Classification of Myoviridae bacteriophages using protein sequence similarity. *BMC Microbiol.* **2009**, *9*, 224.

67. Lavigne, R.; Seto, D.; Mahadevan, P.; Ackermann, H.W.; Kropinski, A.M. Unifying classical and molecular taxonomic classification: analysis of the Podoviridae using BLASTP-based tools. *Res. Microbiol.* **2008**, *159*, 406–414.

68. Chibani-Chennoufi, S.; Dillmann, M.L.; Marvin-Guy, L.; Rami-Shojaei, S.; Brussow, H. Lactobacillus plantarum bacteriophage LP65: a new member of the SPO1-like genus of the family Myoviridae. *J. Bacteriol.* **2004**, *186*, 7069–7083.

69. Grossi, P.A. Early appropriate therapy of Gram-positive bloodstream infections: the conservative use of new drugs. *Int. J. Antimicrob. Agents* **2009**, *34*, Suppl. 4, S31–34.

70. Kahankova, J.; Pantucek, R.; Goerke, C.; Ruzickova, V.; Holochova, P.; Doskar, J. Multilocus PCR typing strategy for differentiation of Staphylococcus aureus siphoviruses reflecting their modular genome structure. *Environ. Microbiol.* **2010**, *12*, 2527–2538.

71. Rohwer, F.; Edwards, R. The Phage Proteomic Tree: a genome-based taxonomy for phage. *J. Bacteriol.* **2002**, *184*, 4529–4535.

72. Lima-Mendez, G.; Toussaint, A.; Leplae, R. A modular view of the bacteriophage genomic space: identification of host and lifestyle marker modules. *Res. Microbiol.* **2011**, *162*, 737–746.

73. Hatfull, G.F.; Hendrix, R.W. Bacteriophages and their genomes. *Curr. Opin. Virol.* **2011**, *1*, 298–303.

74. Ma, X.X.; Ito, T.; Kondo, Y.; Cho, M.; Yoshizawa, Y.; Kaneko, J.; Katai, A.; Higashiide, M.; Li, S.; Hiramatsu, K. Two different Panton-Valentine leukocidin phage lineages predominate in Japan. *J. Clin. Microbiol.* **2008**, *46*, 3246–3258.

75. Ma, X.X.; Ito, T.; Chongtrakool, P.; Hiramatsu, K. Predominance of clones carrying Panton-Valentine leukocidin genes among methicillin-resistant Staphylococcus aureus strains isolated in Japanese hospitals from 1979 to 1985. *J. Clin. Microbiol.* **2006**, *44*, 4515–4527.

76. Narita, S.; Kaneko, J.; Chiba, J.; Piemont, Y.; Jarraud, S.; Etienne, J.; Kamio, Y. Phage conversion of Panton-Valentine leukocidin in Staphylococcus aureus: molecular analysis of a PVL-converting phage, phiSLT. *Gene* **2001**, *268*, 195–206.

77. Iandolo, J.J.; Worrell, V.; Groicher, K.H.; Qian, Y.; Tian, R.; Kenton, S.; Dorman, A.; Ji, H.; Lin, S.; Loh, P.; *et al.* Comparative analysis of the genomes of the temperate bacteriophages phi 11, phi 12 and phi 13 of Staphylococcus aureus 8325. *Gene* **2002**, *289*, 109–118.

78. Zhang, M.; Ito, T.; Li, S.; Jin, J.; Takeuchi, F.; Lauderdale, T.L.; Higashide, M.; Hiramatsu, K. Identification of the third type of PVL phage in ST59 methicillin-resistant Staphylococcus aureus (MRSA) strains. *FEMS Microbiol. Lett.* **2011**, *323*, 20–28.

79. Belcaid, M.; Bergeron, A.; Poisson, G. Mosaic graphs and comparative genomics in phage communities. *J. Comput. Biol.* **2010**, *17*, 1315–1326.

80. Susskind, M.M.; Botstein, D. Molecular genetics of bacteriophage P22. *Microbiol. Rev.* **1978**, *42*, 385–413.

81. Clark, A.J.; Inwood, W.; Cloutier, T.; Dhillon, T.S. Nucleotide sequence of coliphage HK620 and the evolution of lambdoid phages. *J. Mol. Biol.* **2001**, *311*, 657–679.

82. Hendrix, R.W. Bacteriophages: evolution of the majority. *Theor. Popul. Biol.* **2002**, *61*, 471–480.

83. Hatfull, G.F.; Pedulla, M.L.; Jacobs-Sera, D.; Cichon, P.M.; Foley, A.; Ford, M.E.; Gonda, R.M.; Houtz, J.M.; Hryckowian, A.J.; Kelchner, V.A.; *et al.* Exploring the mycobacteriophage metaproteome: phage genomics as an educational platform. *PLoS Genet.* **2006**, *2*, e92.

84. Glazko, G.; Makarenkov, V.; Liu, J.; Mushegian, A. Evolutionary history of bacteriophages with double-stranded DNA genomes. *Biol. Direct.* **2007**, *2*, 36.

85. Huson, D.H.; Bryant, D. Application of phylogenetic networks in evolutionary studies. *Mol. Biol. Evol.* **2006**, *23*, 254–267.

86. Lawrence, J.G.; Hatfull, G.F.; Hendrix, R.W. Imbroglios of viral taxonomy: genetic exchange and failings of phenetic approaches. *J. Bacteriol.* **2002**, *184*, 4891–4905.

87. Lima-Mendez, G.; Van Helden, J.; Toussaint, A.; Leplae, R. Reticulate representation of evolutionary and functional relationships between phage genomes. *Mol. Biol. Evol.* **2008**, *25*, 762–777.

88. Mann, N.H. The potential of phages to prevent MRSA infections. *Res. Microbiol.* **2008**, *159*, 400–405.

89. Lu, T.K.; Koeris, M.S. The next generation of bacteriophage therapy. *Curr. Opin. Microbiol.* **2011**, *14*, 524–531.

90. Matsuzaki, S.; Yasuda, M.; Nishikawa, H.; Kuroda, M.; Ujihara, T.; Shuin, T.; Shen, Y.; Jin, Z.; Fujimoto, S.; Nasimuzzaman, M.D.; *et al.* Experimental protection of mice against lethal Staphylococcus aureus infection by novel bacteriophage phi MR11. *J. Infect. Dis.* **2003**, *187*, 613–624.

91. Kwiatek, M.; Parasion, S.; Mizak, L.; Gryko, R.; Bartoszcze, M.; Kocik, J. Characterization of a bacteriophage, isolated from a cow with mastitis, that is lytic against Staphylococcus aureus strains. *Arch. Virol.* **2012**, *157*, 225–234.

92. Hagens, S.; Loessner, M.J. Bacteriophage for biocontrol of foodborne pathogens: calculations and considerations. *Curr. Pharm. Biotechnol.* **2010**, *11*, 58–68.

93. Bueno, E.; Garcia, P.; Martinez, B.; Rodriguez, A. Phage inactivation of Staphylococcus aureus in fresh and hard-type cheeses. *Int. J. Food Microbiol.* **2012**, *158*, 23–27.

94. Garcia, P.; Madera, C.; Martinez, B.; Rodriguez, A.; Evaristo Suarez, J. Prevalence of bacteriophages infecting Staphylococcus aureus in dairy samples and their potential as biocontrol agents. *J. Dairy Sci.* **2009**, *92*, 3019–3026.

95. Garcia, P.; Martinez, B.; Obeso, J.M.; Rodriguez, A. Bacteriophages and their application in food safety. *Lett. Appl. Microbiol.* **2008**, *47*, 479–485.

96. Fischetti, V.A. Bacteriophage endolysins: a novel anti-infective to control Gram-positive pathogens. *Int. J. Med. Microbiol.* **2010**, *300*, 357–362.

97. O'Flaherty, S.; Coffey, A.; Meaney, W.; Fitzgerald, G.F.; Ross, R.P. The recombinant phage lysin LysK has a broad spectrum of lytic activity against clinically relevant staphylococci, including methicillin-resistant Staphylococcus aureus. *J. Bacteriol.* **2005**, *187*, 7161–7164.

98. O'Flaherty, S.; Ross, R.P.; Meaney, W.; Fitzgerald, G.F.; Elbreki, M.F.; Coffey, A. Potential of the polyvalent anti-Staphylococcus bacteriophage K for control of antibiotic-resistant staphylococci from hospitals. *Appl. Environ. Microbiol.* **2005**, *71*, 1836–1842.

99. Hsieh, S.E.; Lo, H.H.; Chen, S.T.; Lee, M.C.; Tseng, Y.H. Wide host range and strong lytic activity of Staphylococcus aureus lytic phage Stau2. *Appl. Environ. Microbiol.* **2011**, *77*, 756–761.

100. O'Flaherty, S.; Ross, R.P.; Flynn, J.; Meaney, W.J.; Fitzgerald, G.F.; Coffey, A. Isolation and characterization of two anti-staphylococcal bacteriophages specific for pathogenic Staphylococcus aureus associated with bovine infections. *Lett. Appl. Microbiol.* **2005**, *41*, 482–486.

101. Merabishvili, M.; Pirnay, J.P.; Verbeken, G.; Chanishvili, N.; Tediashvili, M.; Lashkhi, N.; Glonti, T.; Krylov, V.; Mast, J.; van Parys, L.; *et al.* Quality-controlled small-scale production of a well-defined bacteriophage cocktail for use in human clinical trials. *PLoS One* **2009**, *4*, e4944.

11

Lysogenic Conversion and Phage Resistance Development in Phage Exposed *Escherichia coli* Biofilms

Pieter Moons [†]**, David Faster and Abram Aertsen** *

Laboratory of Food Microbiology, Department of Microbial and Molecular Systems (M^{2}S), Faculty of Bioscience Engineering, Katholieke Universiteit Leuven, Kasteelpark Arenberg 22, Leuven 3001, Belgium; E-Mails: pieter.moons@ua.ac.be (P.M.); david.faster@biw.kuleuven.be (D.F.)

[†] Present address: Laboratory of Medical Microbiology, Vaccine & Infectious Disease Institute, Universiteit Antwerpen, Universiteitsplein 1, S6.21, Antwerpen 2610, Belgium.

* Author to whom correspondence should be addressed; E-Mail: abram.aertsen@biw.kuleuven.be

---

Abstract: In this study, three-day old mature biofilms of *Escherichia coli* were exposed once to either a temperate Shiga-toxin encoding phage (H-19B) or an obligatory lytic phage (T7), after which further dynamics in the biofilm were monitored. As such, it was found that a single dose of H-19B could rapidly lead to a near complete lysogenization of the biofilm, with a subsequent continuous release of infectious H-19B particles. On the other hand, a single dose of T7 rapidly led to resistance development in the biofilm population. Together, our data indicates a profound impact of phages on the dynamics within structured bacterial populations.

Keywords: *Escherichia coli*; biofilm; lysogenic conversion; resistance development; Shiga toxin; phage

---

1. Introduction

Many bacterial populations are organized as biofilms, which consist of cells attached to a surface embedded in a matrix of variable composition [1]. These biofilms structurally vary from flat layers of cells to complex structured communities, consisting of tower or mushroom shaped micro-colonies interspersed with water channels that allow access of nutrients and removal of metabolites [2–4]. Bacterial biofilms present a medical hazard, since they confer increased resistance against antimicrobials and the host immune system [5], and are linked to persistent infections [6–8].

Apart from being a predominant species among the facultative anaerobic bacteria in the gastrointestinal tract [9], *E. coli* is considered a major zoonotic food-borne pathogen [10] and is one of the main causes of nosocomial infections [11], such as those associated with biofilm formation on urinary catheters [12]. Based on their disease-associated virulence factors, the species is divided into pathotypes, which are known to cause diarrhoeal disease, extra-intestinal and urinary tract infections, sepsis and meningitis [13,14]. Virulence characteristics include a variety of adhesion and colonization factors, the formation of attaching and effacing lesions, the production of toxins and the presence of antibiotic resistance genes [13,15,16]. Among diarrheagenic *E. coli*, those producing the potent Shiga (or Vero) toxins (Stxs) are the most virulent and can cause the potentially fatal haemolytic uremic syndrome [17,18]. Genes encoding Stxs are typically harbored by temperate phages (Stx-phages) integrated in the genome of such pathogenic *E. coli* strains, which makes them subject to lateral gene transfer [19,20].

The worldwide emergence of multi-drug resistant pathogens and empty antibiotic development pipelines reduce medical treatment options and necessitate the research into alternative therapies. One such option is phage therapy, which makes use of lytic phages as natural bacterial enemies whose narrow host range minimizes their impact on the normal flora [21–24]. In fact, phages are well suited to affect biofilms, since infection leads to local enrichment of viral particles. Moreover, many bacteriophages possess enzymes capable of bacterial lysis and less common biofilm matrix degradation [22,25,26], aiding in the accessibility of biofilm cells toward viral particles [27,28] and leading to biofilm dispersal [29–31].

In this study, we examined the impact of a temperate or lytic phage on mature *E. coli* biofilms, with particular interest in the lateral transfer of virulence determinants and phage resistance development, respectively.

2. Results and Discussion

2.1. Lysogenic Conversion of a Mature E. coli Biofilm with an Stx-Encoding Phage

In order to examine to which extent phage encoded virulence factors could be captured and spread within an existing *E. coli* biofilm, we decided to make acquisition of the naturally *stx1*-encoding temperate H-19B phage [32] readily detectable through selective plating by equipping it with an antibiotic resistance marker. Since only a small fragment of the H-19B genome sequence is currently known [32], a random transposon mutagenesis procedure was followed to tag this (pro)phage. More specifically, *E. coli* MG1655 was first lysogenized with H-19B, after which a random

Tn10-transposon library of ca. 10,000 clones was constructed in this lysogen using the λNK1324 hop protocol described by Kleckner *et al.* [33]. Assuming some of the Tn10 insertions to be located within the H-19B prophage, mitomycin C was subsequently used to induce the prophage in the obtained pool of transposon mutants. To isolate H-19B::Tn10 mutants within the corresponding phage lysate, it was first plaqued on *E. coli* MG1655. Since the Tn10 transposon codes for the *cat* gene and confers chloramphenicol resistance, lysogens arising in the middle of turbid plaques were scored for the presence or absence of chloramphenicol resistance. As such, nine chloramphenicol resistant lysogens were obtained, which were further confirmed to simultaneously have acquired the *stx1* operon of H-19B. The main advantage of this procedure is that it automatically disregards phage mutants compromised in lytic or lysogenic development. Moreover, in contrast to phage recombineering protocols [34], this protocol can be applied without prior knowledge of the phage's genome sequence and resembles the method used previously by Acheson *et al.* [35] to look for phage encoded exported proteins.

From one of the obtained H-19B::Tn10 lysogens, the Tn10 insertion site was determined and found to map within a gene bearing homology to the *nleG* virulence genes. NleG proteins are effectors of the type 3 secretion system that are thought to mimic eukaryotic E3 ubiquitin ligases [36], and are generally found in the late region of phage genomes [37]. This insertion underscores the presence of additional virulence genes to be present in H-19B, supports the observation that neither lytic nor lysogenic behaviour is affected in H-19B::Tn10 and due to the uptake of an additional piece of DNA (*i.e,.* the transposon) adds evidence to the plasticity of Stx genomes [38]. Subsequently, a mature three-day old *E. coli* biofilm was only once exposed to a small number (*i.e.*, 150 viral particles spread over one hour) of the corresponding H-19B::Tn10 derivative in order to mimic an accidental exposure. Subsequently, the spread of this virulence conferring phage genome throughout the biofilm population was tracked. More specifically, the effluent of the biofilm was examined for the presence of (i) cells lysogenized with H-19B::Tn10, and (ii) free H-19B::Tn10 phage (Figure 1). Interestingly, from this analysis it became clear that lysogenic conversion of the biofilm proceeded very rapidly, with the emergence of *ca.* 10^6 CFU/mL of H-19B::Tn10 lysogens on a total effluent cell count of circa 10^9 CFU/mL (*i.e.*, 0.06% conversion) after 24 h, and an above 50% conversion reached after five days. Moreover, at the end of the experiment the ratio between total and lysogenized cells within the actual biofilm itself corresponded to that observed in the effluent, demonstrating that, at least at the end of the experiment, lysogenic conversion had occurred in the entire biofilm Furthermore, during the first two days after phage exposure, the concentration of free H-19B::Tn10 phage in the effluent quickly rose from the applied 50 PFU/mL to 10^5–10^6 PFU/mL, after which it remained stable at this level. It can be anticipated that part of the biofilm is being lysed, leading to a local enrichment in H-19B::Tn10 concentration. Nevertheless, it remains unclear exactly to what extent the emergence of H-19B::Tn10 lysogens is the result of *de novo* lysogenic conversion of pre-existing wild-type cells or of clonal enrichment of the first converted cells.

Figure 1. Numbers of viable cells (CFU/mL) and phages (PFU/mL) in the effluent of a mature *E. coli* biofilm exposed for 1 hour to 50 PFU/mL (total of 150 PFU per biofilm) of H-19B::Tn10 phage at day 0, after sampling the effluent for initial total cell count. Dark blue bars represent total cell concentration, while light blue bars represent the concentration of H-19B::Tn10 lysogens. White markers represent free H-19B::Tn10 phage. Dark and light grey bars represent total cell counts and lysogen cell counts, respectively, in the attached biofilm obtained after dismantling the flow cell setup. The data represent the average and standard deviations of three biological replicates.

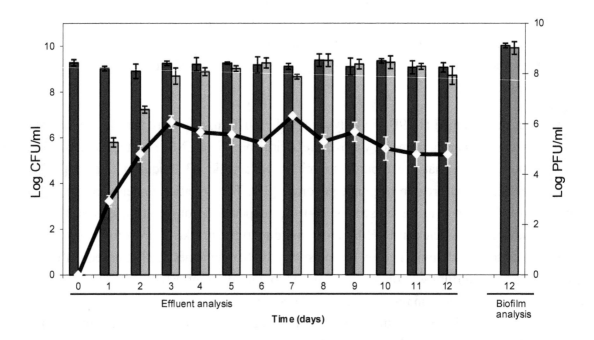

2.2. Resistance Development of E. coli *biofilms Against a Lytic Phage*

In a second approach, biofilm dynamics were examined upon single exposure of a 3-day old mature *E. coli* biofilm to *ca.* 10^8 particles per milliliter of the obligatory lytic T7 phage. The high phage titer selected in this approach was considered to best reflect the clinical application of phage in combatting biofilm-related infections. After three additional days of incubation, cells within the biofilm were harvested and enumerated. A comparison between the number of viable cells within the T7-exposed or -unexposed biofilm revealed an almost 100-fold reduction (*p*-value in two tailed homoscedastic T-test <0.05) in the phage exposed biofilm (Figure 2). However, in between the normal colonies observed during plating, colonies with a mucoid appearance were observed in biofilms treated with T7 phages (Figure 3A), but not in untreated biofilms. This phenotype was earlier shown to be correlated with T7 resistance due to a mutation resulting in the production of excess capsular polysaccharide preventing T7 adhesion [39] due to physical blocking of the phage binding site [40], and actual T7 resistance of these colonies could be confirmed as in Figure 3B. After examining *ca.* 400 colonies of both types, phage resistant clones with a non-mucoid phenotype or phage sensitive clones with a mucoid phenotype were not observed. In turn, this enabled us to accurately determine the number of resistant cells in the biofilm. However, a large variation in the number of resistant cells was found between biological replicates (Table 1), ranging from circa 0.05% to over 28%. As in Luria-Delbruck fluctuation experiments, this variation likely stemmed from clonal enrichment of stochastically pre-existing T7-resistant mutants that spontaneously arise within a population. Since 3 days after initial phage exposure T7 particles still remained present in the effluent and the biofilm, on-going selection for T7-resistance development seems warranted upon longer incubation of the biofilm.

Figure 2. Number of viable cells (CFU) within a mature 3-day old *E. coli* biofilm exposed once, for one hour, to either water (control; dark bars) or *ca.* 10^8 PFU/mL of T7 phage (light bars) followed by an additional three days of biofilm development. The data represent the average and standard deviations of three biological replicates.

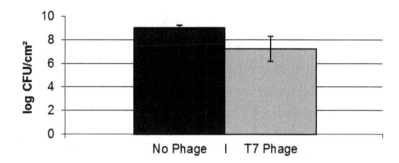

Figure 3. (A) mucoid (bottom) and non mucoid (top) colony of *E. coli*. (B) Phage resistance of a bacterial colony was demonstrated by streaking a line of *E. coli* cells through a perpendicularly streaked line of T7 solution. T7 phage resistant cells (bottom) were capable of growth past this line, while T7 sensitive cells (top) were not.

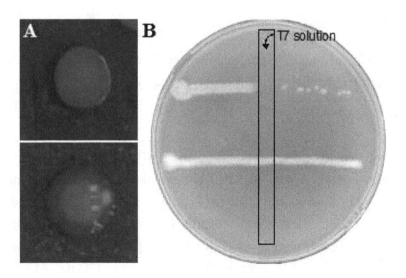

Table 1. Viable cell counts within three mature three-day old *E. coli* biofilms each exposed once for one hour to *ca.* 10^8 PFU/mL of T7 phage, followed by three additional days of growth before plating.

Cell counts (log CFU/cm²)	Biofilm 1	Biofilm 2	Biofilm 3
Total cell count	7.31	5.82	7.91
T7 resistant cells	6.76	2.49	7.05
T7 sensitive cells	7.17	5.82	7.85

3. Experimental Section

3.1. Strains, Standard Culture Conditions and Chemicals

E. coli MG1655 [41], MG1655 *lacZ::Tc* [42], and MG1655 H-19B (*i.e.*, lysogenized with *stx1* encoding phage H-19B; [43]) were used in this study, and were cultured at 37 °C under shaking conditions in Lysogeny Broth (LB). In order to determine the number of bacteria (as colony forming units or CFU) in a sample, a dilution series was prepared and plated on LB agar (1.5% agar), after which the concentration was expressed as CFU/mL.

To obtain a H-19B phage lysate from lysogens, cells from 1 mL bacterial culture were pelleted by centrifugation (6000 × g, 15 min), after which 50 µL of CHCl₃ was added to sterilize the supernatant and to release progeny phage from infected but non-lysed cells. In case high phage titers were required, the cultures were grown in the presence of 2 µg/mL mitomycin C which induces the release of temperate phages. To obtain a T7 phage [44] lysate, T7 was propagated on MG1655, after which 1 mL of cells were pelleted by centrifugation (6000 × g, 15 min) and the resulting supernatant was sterilized by addition of 50 µL of CHCl₃. In order to determine the number of phages (as plaque forming units or PFU) in a sample of lysate or effluent, a dilution series was prepared and plated on *E. coli* MG1655 grown in LB soft agar (0.7% agar), after which the concentration was expressed as PFU/mL.

Where necessary, growth media were supplemented with tetracycline (20 µg/mL) and/or chloramphenicol (30 µg/mL) (Applichem, Darmstadt, Germany).

3.2. Analysis of H-19B::Tn10

After its construction as described in the text, the Tn10 insertions site in H-19B::Tn10 was mapped by subcloning of the Tn10 encoded chloramphenicol resistance marker in pUC18. Subsequently, genomic DNA flanking the Tn10 transposon was sequenced using a primer (5'-AAGCACCGCCGGACATC-3') reading outwards of the transposon. In addition, the isolated H-19B::Tn10 phage was also confirmed by PCR (directly on plaques or crude lysate) to still carry its stx1 locus, by using primers flanking this region (5'-CAGTGGATCCTGGCACGGAAACATGGGT-3' and 5'-TCAGTCTAGATTACGTCTTTGCAGT CGAGAAGTC-3').

3.3. Setup for Biofilm Formation

Biofilms were grown at 30 °C in three-channel flow chambers (Biocentrum DTU: Technical University of Denmark, Soltofts Plads 221 DK-2800 Kgs, Lyngby) with individual channel dimensions of $1 \times 4 \times 40$ mm^3 [45] that were covered with a microscope glass coverslip (st1; Knittel Gläser, Braunschweig, Germany). The setup makes use of a 16-channel peristaltic pump (Watson Marlow 205S, Zellik, Belgium) that feeds each channel with a flow of 3 mL/h (flow rate of 0.2 mm/s) of AB-trace medium [2 g/L $(NH_4)_2SO_4$; 6 g/L $Na_2HPO_4.2H_2O$; 3 g/L KH_2PO_4; 3 g/L NaCl; 9.5 mg/L $MgCl_2$; 11.1mg/L $CaCl_2$ and 0.1 mL/L of the following trace metal mixture (200 mg/L $CaSO_4.2H_2O$; 200 mg/L $FeSO_4.7H_2O$; 20 mg/L $MnSO_4.H_2O$; 20 mg/L $CuSO_4.5H_2O$; 20 mg/L $ZnSO_4.7H_2O$; 10 mg/L $CoSO_4.7H_2O$; 10 mg/L $NaMoO_4.H_2O$; 5 mg/L H_3BO_3)] supplemented with 0.3 mM glucose and 1 µg/mL thiamine dichloride. Bubble traps were placed in each channel before the flow cell to remove air bubbles. Before use, the flow system was sterilized by flushing with a solution of 0.5% sodium hypochlorite for 4 h, and rinsed with approximately 0.2 L of sterile water before the medium was pumped through.

Bacterial cultures for inoculation were prepared by diluting an overnight LB broth culture 1/100 in fresh LB medium and regrowing it for 4 h at 30 °C under shaking conditions. To inoculate the flow cells the medium flow was stopped, flow chambers were turned with the glass coverslip down and 250 µL of the diluted cell suspension was carefully injected through the silicon tubes into each flow channel with a small syringe. After 1 h, to allow adsorption of the cells to the coverslip surface, the flow channels were turned upright and the flow was resumed. For experiments with H-19B::Tn10 and T7, E. coli MG1655 lac::Tc and E. coli MG1655 were used, respectively. When macroscopically visible mature E. coli biofilms were formed after 3 days, they became exposed for 1 h to either sterile water, 50 PFU/mL of H-19B::Tn10 or ca. 10^8 PFU/mL of T7 via a separate input channel.

3.4. Analysis of Biofilms

Biofilms were analyzed by determining the number of viable cells (CFU) or phage particles (PFU) in either the effluent or in the biofilm itself. Effluent was collected using a small connector inserted in the tubing behind the flowcell, while the actual biofilm was harvested by carefully disconnecting the flowcells and vigorously pipetting up and down the channels with 250 µL potassium phosphate buffer (10 mM, pH 7.0).

After vortexing, dilution series of collected samples were plated on LB agar (in case E. coli MG1655 was used) or LB agar supplemented with tetracycline (in case E. coli MG1655 lac::Tc was used) to enumerate total cell counts, and on LB agar supplemented with chloramphenicol to enumerate H-19B::Tn10 lysogens. Phages were enumerated by first sterilizing the collected samples with $CHCl_3$, and subsequently plating them on E. coli MG1655.

4. Conclusions

Biofilms, already recalcitrant to therapy, can become an even larger burden through the acquisition of novel virulence determinants such as those conveyed by temperate phages. In Stx phages, the toxin genes are associated with the late gene region, and become highly expressed when the (pro)phage enters the lytic cycle which can be induced at initiation of the bacterial SOS response [19,20,46]. Of intestine isolated *E. coli* strains, 10%–30% were shown to be able to contribute to Shiga toxin production upon infection through transfer of the phage [19,47,48], while also actual *in vivo* transfer was demonstrated [35,49,50]. Certain antibiotics, through induction of the SOS response [51], even lead to the release of temperate phages [52] and thus contribute to the spread of phages (and genes) to the normal flora and elevated production of toxins.

In this study, we report the integration of a chloramphenicol resistance marker in the temperate H-19B Stx-phage without compromising either the lysogenic or the lytic cycle. This phage was shown to rapidly and massively establish itself as a prophage within a mature *E. coli* biofilm, despite the very low initial dose. During and after this establishment, the biofilm started to produce a high number of free H-19B virions, thus contributing to the further dissemination of the phage and its virulence determinants. Moreover, since the production and release of H-19B phage typically coincides with the production and release of its Shiga-toxin [19,20,53], such biofilms likely become a source of continuous toxin production. These findings underscore the ability of temperate phages to rapidly and stably establish themselves within a susceptible but matrix-embedded population, and demonstrate that biofilms can serve as a source of new phage particles and their possible toxins.

Upon challenging a biofilm with lytic phages, an initial reduction in cell numbers is often observed [54,55], but complete eradication by a single phage is never achieved and the establishment of equilibrium between virus and host with stable numbers of both organisms was reported earlier [56]. In addition, resistance is thought to quickly arise within the biofilm, but is only scarcely supported by quantitative data. Here, we report the application of a single pulse of lytic T7 phage to an existing *E. coli* biofilm and demonstrate a reduction of the biofilm even several days after phage application. Nevertheless, this lysis was far from complete and, although highly variable, up to 30% of phage resistant cells could be isolated. This development of resistance was earlier described as a mutual, escalating arms race in which phage arise that are capable of infecting resistant strains, from which in turn bacteria evolve that are resistant to those phages [57,58]. These results confirm the therapeutic potential of lytic phages mainly lies in the use of either well-characterized phage cocktails or by combining them with antibiotics. Synergistic phage-antibiotic combinations on *E. coli* biofilms were observed earlier [59,60] and show great promise for adequate treatment.

Acknowledgments

This work was supported by a doctoral fellowship of the Flemish Agency for Innovation by Science and Technology (IWT-Vlaanderen; to P.M.), and by the KU Leuven Research Fund (Grant CREA/09/017).

References

1. Flemming, H.C.; Wingender, J. The biofilm matrix. *Nat. Rev. Microbiol.* **2010**, *8*, 623–633.
2. Hall-Stoodley, L.; Costerton, J.W.; Stoodley, P. Bacterial biofilms: From the natural environment to infectious diseases. *Nat. Rev. Microbiol.* **2004**, *2*, 95–108.
3. Moons, P.; Michiels, C.W.; Aertsen, A. Bacterial interactions in biofilms. *Crit. Rev. Microbiol.* **2009**, *35*, 157–168.
4. Høiby, N.; Ciofu, O.; Johansen, H.K.; Song, Z.J.; Moser, C.; Jensen, P.; Molin, S.; Givskov, M.; Tolker-Nielsen, T.; Bjarnsholt, T. The clinical impact of bacterial biofilms. *Int. J. Oral Sci.* **2011**, *3*, 55–65.
5. Høiby, N.; Bjarnsholt, T.; Givskov, M.; Molin, S.; Ciofu, O. Antibiotic resistance of bacterial biofilms. *Int. J. Antimicrob. Agents* **2010**, *35*, 322–332.
6. Costerton, J.W.; Stewart, P.S.; Greenberg, E.P. Bacterial biofilms: a common cause of persistent infections. *Science* **1999**, *284*, 1318–1322.
7. Francolini, I.; Donelli, G. Prevention and control of biofilm-based medical-device-related infections. *FEMS Immunol. Med. Microbiol.* **2010**, *59*, 227–238.
8. Hall-Stoodley, L.; Stoodley, P.; Kathju, S.; Høiby, N.; Moser, C.; Costerton, J.W.; Moter, A.; Bjarnsholt, T. Towards diagnostic guidelines for biofilm-associated infections. *FEMS Immunol. Med. Microbiol.* **2012**, *65*, 127–145.
9. Beloin, C.; Roux, A.; Ghigo, J.M. *Escherichia coli* biofilms. *Curr. Top. Microbiol. Immunol.* **2008**, *322*, 249–289.
10. Newell, D.G.; Koopmans, M.; Verhoef, L.; Duizer, E.; Aidara-Kane, A.; Sprong, H.; Opsteegh, M.; Langelaar, M.; Threfall, J.; Scheutz, F.; *et al.* Food-borne diseases—The challenges of 20 years ago still persist while new ones continue to emerge. *Int. J. Food Microbiol.* **2010**, *139*, S3–S15.
11. Cantón, R.; Akóva, M.; Carmeli, Y.; Giske, C.G.; Glupczynski, Y.; Gniadkowski, M.; Livermore, D.M.; Miriagou, V.; Naas, T.; Rossolini, G.M.; *et al.* Rapid evolution and spread of carbapenemases among Enterobacteriaceae in Europe. *Clin. Microbiol. Infect.* **2012**, *18*, 413–431.
12. Jacobsen, S.M.; Stickler, D.J.; Mobley, H.L.; Shirtliff, M.E. Complicated catheter-associated urinary tract infections due to *Escherichia coli* and *Proteus mirabilis*. *Clin. Microbiol. Rev.* **2008**, *21*, 26–59.
13. Kaper, J.B.; Nataro, J.P.; Mobley, H.L. Pathogenic *Escherichia coli*. *Nat. Rev. Microbiol.* **2004**, *2*, 123–140.
14. Chaudhuri, R.R.; Henderson, I.R. The evolution of the *Escherichia coli* phylogeny. *Infect. Genet. Evol.* **2012**, *12*, 214–226.
15. Johnson, T.J.; Nolan, L.K. Pathogenomics of the virulence plasmids of *Escherichia coli*. *Microbiol. Mol. Biol. Rev.* **2009**, *73*, 750–774.
16. Farfan, M.J.; Torres, A.G. Molecular mechanisms that mediate colonization of Shiga toxin-producing *Escherichia coli* strains. *Infect. Immun.* **2012**, *80*, 903–913.
17. Werber, D.; Krause, G.; Frank, C.; Fruth, A.; Flieger, A.; Mielke, M.; Schaade, L.; Stark, K. Outbreaks of virulent diarrheagenic *Escherichia coli*—are we in control? *BMC Med.* **2012**, *10*, 11.

18. Melton-Celsa, A.; Mohawk, K.; Teel, L.; O'Brien, A. Pathogenesis of Shiga-toxin producing *Escherichia coli*. *Curr. Top. Microbiol. Immunol.* **2012**, *357*, 67–103.

19. Schmidt, H. Shiga-toxin-converting bacteriophages. *Res. Microbiol.* **2001**, *152*, 687–695.

20. Herold, S.; Karch, H.; Schmidt, H. Shiga toxin-encoding bacteriophages—genomes in motion. *Int. J. Med. Microbiol.* **2004**, *294*, 115–121.

21. Minot, S.; Sinha, R.; Chen, J.; Li, H.; Keilbaugh, S.A.; Wu, G.D.; Lewis, J.D.; Bushman, F.D. The human gut virome: Inter-individual variation and dynamic response to diet. *Genome Res.* **2011**, *21*, 1616–1625.

22. Donlan, R.M. Preventing biofilms of clinically relevant organisms using bacteriophage. *Trends Microbiol.* **2009**, *17*, 66–72.

23. Loc-Carrillo, C.; Abedon, S.T. Pros and cons of phage therapy. *Bacteriophage* **2011**, *1*, 111–114.

24. Ryan, E.M.; Gorman, S.P.; Donnelly, R.F.; Gilmore, B.F. Recent advances in bacteriophage therapy: how delivery routes, formulation, concentration and timing influence the success of phage therapy. *J. Pharm. Pharmacol.* **2011**, *63*, 1253–1264.

25. Azeredo, J.; Sutherland, I.W. The use of phages for the removal of infectious biofilms. *Curr. Pharm. Biotechnol.* **2008**, *9*, 261–266.

26. Rodríguez-Rubio, L.; Martínez, B.; Donovan, D.M.; Rodríguez, A.; García, P. Bacteriophage virion-associated peptidoglycan hydrolases: potential new enzybiotics. *Crit. Rev. Microbiol.* **2012**, doi:10.3109/1040841X.2012.723675.

27. Hughes, K.A.; Sutherland, I.W.; Jones, M.V. Biofilm susceptibility to bacteriophage attack: The role of phage-borne polysaccharide depolymerase. *Microbiology* **1998**, *144*, 3039–3047.

28. Domenech, M.; García, E.; Moscoso, M. *In vitro* destruction of *Streptococcus pneumoniae* biofilms with bacterial and phage peptidoglycan hydrolases. *Antimicrob. Agents Chemother.* **2011**, *55*, 4144–4148.

29. Rice, S.A.; Tan, C.H.; Mikkelsen, P.J.; Kung, V.; Woo, J.; Tay, M.; Hauser, A.; McDougald, D.; Webb, J.S.; Kjelleberg, S. The biofilm life cycle and virulence of *Pseudomonas aeruginosa* are dependent on a filamentous prophage. *ISME J.* **2009**, *3*, 271–282.

30. Meng, X.; Shi, Y.; Ji, W.; Zhang, J.; Wang, H.; Lu, C.; Sun, J.; Yan, Y. Application of a bacteriophage lysin to disrupt biofilms formed by the animal pathogen *Streptococcus suis*. *Appl. Environ. Microbiol.* **2011**, *77*, 8272–8279.

31. Siringan, P.; Connerton, P.L.; Payne, R.J.; Connerton, I.F. Bacteriophage-Mediated dispersal of *Campylobacter jejuni* biofilms. *Appl. Environ. Microbiol.* **2011**, *77*, 3320–3632.

32. Neely, M.N.; Friedman, D.I. Functional and genetic analysis of regulatory regions of coliphage H-19B: Location of shiga-like toxin and lysis genes suggest a role for phage functions in toxin release. *Mol. Microbiol.* **1998**, *28*, 1255–1267.

33. Kleckner, N.; Bender, J.; Gottesman, S. Uses of transposons with emphasis on Tn10. *Meth. Enzymol.* **1991**, *204*, 139–180.

34. Serra-Moreno, R.; Acosta, S.; Hernalsteens, J.P.; Jofre, J.; Muniesa, M. Use of the lambda Red recombinase system to produce recombinant prophages carrying antibiotic resistance genes. *BMC Mol. Biol.* **2006**, *7*, 31.

35. Acheson, D.W.; Reidl, J.; Zhang, X.; Keusch, G.T.; Mekalanos, J.J.; Waldor, M.K. *In vivo* transduction with shiga toxin 1-encoding phage. *Infect. Immun.* **1998**, *66*, 4496–4498.

36. Wu, B.; Skarina, T.; Yee, A.; Jobin, M.C.; Dileo, R.; Semesi, A.; Fares, C.; Lemak, A.; Coombes, B.K.; Arrowsmith, C.H.; *et al.* NleG Type 3 effectors from enterohaemorrhagic *Escherichia coli* are U-Box E3 ubiquitin ligases. *PLoS Pathog.* **2010**, *6*, e1000960.

37. Ogura, Y.; Ooka, T.; Iguchi, A.; Toh, H.; Asadulghani, M.; Oshima, K.; Kodama, T.; Abe, H.; Nakayama, K.; Kurokawa, K.; *et al.* Comparative genomics reveal the mechanism of the parallel evolution of O157 and non-O157 enterohemorrhagic *Escherichia coli. Proc. Natl. Acad. Sci. USA* **2009**, *106*, 17939–17944.

38. Smith, D.L.; Rooks, D.J.; Fogg, P.C.; Darby, A.C.; Thomson, N.R.; McCarthy, A.J.; Allison, H.E. Comparative genomics of Shiga toxin encoding bacteriophages. *BMC Genom.* **2012**, *13*, 311.

39. Radke, K.L.; Siegel, E.C. Mutation preventing capsular polysaccharide synthesis in *Escherichia coli* K-12 and its effect on bacteriophage resistance. *J. Bacteriol.* **1971**, *106*, 432–437.

40. Scholl, D.; Adhya, S.; Merril, C. Escherichia coli K1's capsule is a barrier to bacteriophage T7. *Appl. Environ. Microbiol.* **2005**, *71*, 4872–4874.

41. Guyer, M.S.; Reed, R.R.; Steitz, J.A.; Low, K.B. Identification of a sex-factor-affinity site in *E. coli* as gamma delta. *Cold Spring Harb. Symp. Quant. Biol.* **1981**, *45*, 135–140.

42. Moons, P.; van Houdt, R.; Aertsen, A.; Vanoirbeek, K.; Engelborghs, Y.; Michiels, C.W. Role of quorum sensing and antimicrobial component production by *Serratia. plymuthica* in formation of biofilms, including mixed biofilms with *Escherichia coli. Appl. Environ. Microbiol.* **2006**, *72*, 7294–7300.

43. Aertsen, A.; Faster, D.; Michiels, C.W. Induction of Shiga toxin-converting prophage in *Escherichia coli* by high hydrostatic pressure. *Appl. Environ. Microbiol.* **2005**, *71*, 1155–1162.

44. Dunn, J.J.; Studier, F.W. Complete nucleotide sequence of bacteriophage T7 DNA and the locations of T7 genetic elements. *J. Mol. Biol.* **1983**, *166*, 477–535.

45. Christensen, B.B.; Sternberg, C.; Andersen, J.B.; Palmer, R.J.; Nielsen, A.T.; Givskov, M.; Molin, S. Molecular tools for study of biofilm physiology. *Meth. Enzymol.* **1999**, *310*, 20–42.

46. Fogg, P.C.; Saunders, J.R.; McCarthy, A.J.; Allison, H.E. Cumulative effect of prophage burden on Shiga toxin production in *Escherichia coli. Microbiology* **2012**, *158*, 488–497.

47. James, C.E.; Stanley, K.N.; Allison, H.E.; Flint, H.J.; Stewart, C.S.; Sharp, R.J.; Saunders, J.R.; McCarthy, A.J. Lytic and lysogenic infection of diverse *Escherichia coli* and *Shigella.* strains with a verocytotoxigenic bacteriophage. *Appl. Environ. Microbiol.* **2001**, *67*, 4335–4337.

48. Gamage, S.D.; Strasser, J.E.; Chalk, C.L.; Weiss, A.A. Nonpathogenic *Escherichia coli* can contribute to the production of Shiga toxin. *Infect. Immun.* **2003**, *71*, 3107–3115.

49. Tóth, I.; Schmidt, H.; Dow, M.; Malik, A.; Oswald, E.; Nagy, B. Transduction of porcine enteropathogenic *Escherichia coli* with a derivative of a shiga toxin 2-encoding bacteriophage in a porcine ligated ileal loop system. *Appl. Environ. Microbiol.* **2003**, *69*, 7242–7247.

50. Cornick, N.A.; Helgerson, A.F.; Mai, V.; Ritchie, J.M.; Acheson, D.W. *In vivo* transduction of an Stx-encoding phage in ruminants. *Appl. Environ. Microbiol.* **2006**, *72*, 5086–5088.

51. Hastings, P.J.; Rosenberg, S.M.; Slack, A. Antibiotic-induced lateral transfer of antibiotic resistance. *Trends Microbiol.* **2004**, *12*, 401–404.

52. Zhang, X.; McDaniel, A.D.; Wolf, L.E.; Keusch, G.T.; Waldor, M.K.; Acheson, D.W. Quinolone antibiotics induce Shiga toxin-encoding bacteriophages, toxin production, and death in mice. *J. Infect. Dis.* **2000**, *181*, 664–670.

53. Aertsen, A.; van Houdt, R.; Michiels, C.W. Construction and use of an *stx1* transcriptional fusion to *gfp*. *FEMS Microbiol. Lett.* **2005**, *245*, 73–77.

54. Carson, L.; Gorman, S.P.; Gilmore, B.F. The use of lytic bacteriophages in the prevention and eradication of biofilms of *Proteus mirabilis* and *Escherichia coli*. *FEMS Immunol. Med. Microbiol.* **2010**, *59*, 447–455.

55. Chibeu, A.; Lingohr, E.J.; Masson, L.; Manges, A.; Harel, J.; Ackermann, H.W.; Kropinski, A.M.; Boerlin, P. Bacteriophages with the ability to degrade uropathogenic *Escherichia coli* biofilms. *Viruses* **2012**, *4*, 471–487.

56. Corbin, B.D.; McLean, R.J.; Aron, G.M. Bacteriophage T4 multiplication in a glucose-limited Escherichia coli biofilm. *Can. J. Microbiol.* **2001**, *47*, 680–684.

57. Buckling, A.; Rainey, P.B. Antagonistic coevolution between a bacterium and a bacteriophage. *Proc. Biol. Sci.* **2002**, *269*, 931–936.

58. Kashiwagi, A.; Yomo, T. Ongoing phenotypic and genomic changes in experimental coevolution of RNA bacteriophage Qβ and *Escherichia coli*. *PLoS Genet.* **2011**, *7*, e1002188.

59. Lu, T.K.; Collins, J.J. Engineered bacteriophage targeting gene networks as adjuvants for antibiotic therapy. *Proc. Natl. Acad. Sci. USA* **2009**, *106*, 4629–4634.

60. Ryan, E.M.; Alkawareek, M.Y.; Donnelly, R.F.; Gilmore, B.F. Synergistic phage-antibiotic combinations for the control of *Escherichia coli* biofilms *in vitro*. *FEMS Immunol. Med. Microbiol.* **2012**, *65*, 395–398.

12

Genomic Sequences of two Novel *Levivirus* Single-Stranded RNA Coliphages (Family *Leviviridae*): Evidence for Recombination in Environmental Strains

Stephanie D. Friedman [1,*], Wyatt C. Snellgrove [2] and Fred J. Genthner [1]

[1] US Environmental Protection Agency, Gulf Ecology Division, 1 Sabine Island Drive, Gulf Breeze, FL, 32561, USA; E-Mails: friedman.stephanie@epa.gov (S.D.F.); genthner.fred@epa.gov (F.J.G.)

[2] William Carey University College of Osteopathic Medicine, 498 Tuscan Avenue, Hattiesburg, MS 39401, USA; E-Mail: cliffsnellgrove@gmail.com

* Author to whom correspondence should be addressed; E-Mail: friedman.stephanie@epa.gov

Abstract: Bacteriophages are likely the most abundant entities in the aquatic environment, yet knowledge of their ecology is limited. During a fecal source-tracking study, two genetically novel *Leviviridae* strains were discovered. Although the novel strains were isolated from coastal waters 1130 km apart (North Carolina and Rhode Island, USA), these strains shared 97% nucleotide similarity and 97–100% amino acid similarity. When the novel strains were compared to nine *Levivirus* genogroup I strains, they shared 95–100% similarity among the maturation, capsid and lysis proteins, but only 84–85% in the RNA-dependent RNA polymerase gene. Further bioinformatic analyses suggested a recombination event occurred. To the best of our knowledge, this is the first description of viral recombinants in environmental *Leviviridae* ssRNA bacteriophages.

Keywords: male-specific coliphage; *Leviviridae*; viral recombinants; ssRNA virus; FRNA; bacteriophage

1. Introduction

Bacteriophages have played a major role contributing to our knowledge of molecular biology, not only in the role of model viruses but also as tools to investigate mRNA, genes, genetic codes and genomes. The first sequenced genomes were the RNA bacteriophage MS2 [1] and the DNA bacteriophage Φ-X174 [2]. As important as *Drosophila* was in shaping the field of genetics and Tobacco Mosaic Virus was in advancing the study of virology and biochemistry, the RNA phage MS2 (family *Leviviridae*) was fundamental in laying the foundation of molecular biology. Thus, it is important to continue adding to the basic understanding of phages. The observations presented in this study were rather serendipitous, in that the focus was not on searching for natural recombinant bacteriophages. Nonetheless, evidence of a recombination event was revealed during a ssRNA bacteriophage sequencing project [3].

Male-specific ssRNA (FRNA) coliphages belong to the family *Leviviridae*. They are classified into two genera (*Levivirus* and *Allolevivirus*) which are subdivided into four genogroups (genogroups I and II in *Levivirus* and genogroups III and IV in *Allolevivirus*). Investigating the genetic diversity of FRNA phages Vinjé *et al* [4] conducted a phylogenetic analysis of 32 *Levivirus* field strains using a 189 bp replicase gene fragment. This study revealed three main clusters: genogroup I, genogroup II and a potential novel group, designated JS, which clustered between genogroup I and genogroup II. The putative JS group, represented by phages, WWTP1_50 and 2GI13, had a >40% sequence diversity in the 189 bp replicase gene sequence when compared to strains from genogroups I and II. As these strains were isolated from widely separated geographical regions (Massachusetts and South Carolina) Vinjé *et al.*, [4] proposed that JS may form a stable lineage. This report suggested further genomic sequencing and serological data were needed to confirm whether these strains formed a novel genogroup or whether they were the result of recombination or rearrangement events [5].

In its simplest form, recombination occurs when two disparate DNA or RNA strands exchange or merge stretches of their sequences whereas mutation involves the substitution, deletion or insertion of a nucleotide resulting in the change of the nucleotide sequence of a gene or an amino acid sequence of a protein. In some RNA viruses, RNA recombination events can occur when two or more strains infect the same host. Proposed models for the formation of novel RNA sequences include (i) cleavage and ligation in RNA molecules or RNA secondary structures [6]; (ii) replicative template switching whereby the RNA-dependent RNA polymerase (replicase) switches from one template to another RNA template, also known as copy choice [5,7]; and (iii) RNA transesterification which occurs when the polymerase adds a separate RNA fragment to the 3' terminus of the original RNA template [5].

Historically, experiments with ssRNA coliphage mutants failed to provide evidence for recombination and the investigators concluded that RNA phages would not undergo recombination [8]. A potential flaw in the conclusion may have been that the study occurred at the time when FRNA phages were thought to possess only three genes, not four. In all likelihood, laboratory-applied selective pressure failed to detect or generate a specific recombinant. This failure may not necessarily reflect the lack of recombination or responsible mechanisms that could occur under actual environmental conditions encountered by ssRNA coliphages.

The first indication of RNA recombination in a male-specific FRNA phage was the report of small, non homologous, recombinant RNA molecules produced from a purified template-free Qβ replicase molecule [9]. The investigators noted similar RNA molecules were present in *E. coli* cells infected with phage Qβ. Chetverin *et al.*, [7] studied this phenomenon by observing the formation of novel sequences in RNA molecules which suggested that this recombination event occurred as a transesterification reaction catalyzed by a conformation acquired by Qβ replicase during RNA synthesis [5,7]. Nucleotide sequences of recombined RNA molecules non-homologous to the parent RNA were formed in the absence of DNA intermediates, demonstrating an RNA recombination mechanism in the presence of Qβ replicase [5]. Therefore, it was plausible to have recombination in environmental ssRNA male-specific coliphage (*Leviviridae*) isolates.

In the present study, two JS strains, DL52 and DL54, were isolated during an environmental genotyping study of *Leviviridae* FRNA phages [10,11]. As in the Vinjé study [4], strains DL52 and DL54 were isolated from separate coastal waters. These phages were placed into the putative JS subgroup using the genotyping methods of Vinjé *et al.* [4]. The objective of this study was to determine whether the existence of a novel JS-like subgroup representing a third *Levivirus* cluster as proposed by Vinjé *et al.*, [4] could be verified. The approach taken here was to compare sequences from the JS strains to nucleotide and amino acid sequence data from entire genomes of 10 levivirus genogroup I strains and 5 levivirus genogroup II strains [3]. Analysis of the novel JS strains provided evidence to determine whether these *Levivirus* strains clustered to genogroup I, II, a combination of groups I and II or a unique genogroup. To further understand the phylogeny of these JS strains, complete genomic sequencing, amino acid composition, phylogenetic, bioinformatic and statistical analyses were performed.

2. Results

2.1. Sequence Analyses and Open Reading Frames

Preliminary analysis of nucleotide sequences from a replicase 189 bp amplicon placed the two novel strains, DL52 and DL54, into a "JS-like" subgroup [4]. Reverse-line blot hybridization failed to genotype the two strains into genogroups I or II [4].

A total of 17 strains (Table 1) were used to examine the relationships among nucleotides and amino acids in the *Levivirus* genus. The first 9 strains in genogroup I, Table 1, *i.e.*, MS2, ST4, DL1, DL2, DL13, DL16, R17, M12 and J20, were referred to as "MS2-like."

Genogroup I MS2-like strains Open Reading Frame (ORF) start and stop codons were located at identical or very similar nucleotide positions as previously reported for strain MS2. The JS strains also had identical ORF start and stop codon positions as the MS-2 like strains (Table 2).

Nucleotide pairwise comparisons of full-length genomes were made between all strains within the *Levivirus* genome, including strains within genogroups I, JS and genogroups II. Within the nine strains of MS2-like genogroup I, full-length nucleotide sequence similarity was 91%–99% [3] whereas the two JS strains, DL52 and DL54, shared 96.73% sequence similarity to each other. In comparison, the JS nucleotide sequences were more similar to MS2-like genogroup I (80%–85%) than to the genogroup I strain fr (69%) or to genogroup II strains (52%–54%) (Table 3a).

Table 1. Male-specific ssRNA coliphages (FRNA), family *Leviviridae*, genus *Levivirus*, strain origins and identifications.

Strain	Genogroup	Source	Origin	Accession number
MS2	I	sewage	Berkeley, CA	NC_001417
M12	I	sewage	Germany	AF195778
DL1	I	river water	Tijuana River, CA	EF107159
DL2	I	bay water	Delaware Bay, DE	N/A
DL13	I	oyster	Whiskey Creek, NC	N/A
DL16	I	bay water	Great Bay, NH	EF108464
J20	I	chicken litter	South Carolina	EF204939
ST4	I	unknown	unknown	EF204940
R17	I	sewage	Philadelphia, PA	EF108465
fr	I	dung hill	Heidelberg, Germany	X15031
DL52	I-JS	bay water	Rachel Carson Reserve, NC	JQ966307
DL54	I-JS	bay water	Narragansett Bay, RI	JQ966308
GA	II	sewage	Ookayama, Japan	NC_001426
KU1	II	sewage	Kuwait	AF227250
DL10	II	mussel	Tijuana River, CA	FJ483837
DL20	II	clam	Narragansett Bay, RI	FJ483839
T72	II	bird	Talbert Marsh sandflats, CA	FJ483838

Table 2. Open Reading Frame positions and genome lengths of FRNA coliphage (family *Leviviridae*, genus *Levivirus*). Nucleotide positions are based on alignment. Number of amino acids for each gene is in parentheses [3].

			Open Reading Frame Locations (amino acids)			
Strain	Group	Full length	ORF1	ORF2	ORF3	ORF4
MS2 [a]	I	3569	130-1311(393)	1335-1727(130)	1678-1905(75)	1761-3398(545)
M12 [a,b]	I	3340[b]	130-1311(393)	1335-1727(130)	1678-1905(75)	ND
DL1	I	3570	130-1311(393)	1335-1727(130)	1678-1905(75)	1761-3398(545)
DL2	I	3491[c]	130-1311(393)	1335-1727(130)	1678-1905(75)	1761-3398(545)
DL13	I	3491[c]	130-1311(393)	1335-1727(130)	1678-1905(75)	1761-3398(545)
DL16	I	3569	130-1311(393)	1335-1727(130)	1678-1905(75)	1761-3398(545)
J20	I	3569	130-1311(393)	1335-1727(130)	1678-1905(75)	1761-3398(545)
ST4	I	3569	130-1311(393)	1335-1727(130)	1678-1905(75)	1761-3398(545)
R17	I	3569	130-1311(393)	1335-1727(130)	1678-1905(75)	1761-3398(545)
fr [a]	I	3575	129-1310(393)	1336-1728(130)	1691-1906(71)	1762-3399(545)
DL52	JS	3525	130-1311(393)	1335-1727(130)	1678-1905(75)	1761-3398[d] (545)
DL54	JS	3398[c]	130-1311(393)	1335-1727(130)	1678-1905(75)	1761-3398[d] (545)

[a] previously published in GenBank; [b] strain M12 was not fully sequenced in the GenBank submission; [c] nearly full-length genome; [d] contains numerous deletions and insertions in ORF4; ND = not determined.

Table 3. (a) Pairwise nucleotide full-length genome percent similarity. (i) *Levivirus* JS strains DL52 and DL54 compared to genogroup I; (ii) *Levivirus* JS strains DL52 and DL54 compared to genogroup II. Pairwise alignments were performed in BioEdit with DAYHOFF similarity parameters.

(i) Genogroup I and JS strains		
Strain	**DL52**	**DL54**
DL52	100	
DL54	96.73	100
DL1	81.48	81.87
DL16	85.41	84.72
ST4	80.30	80.11
R17	80.55	80.53
J20	82.00	82.01
MS2	80.12	80.01
fr	69.18	69.06
(ii) Genogroup II and JS strains		
Strain	**DL52**	**DL54**
DL52	100	
DL54	96.73	100
T72	53.96	53.53
DL10	54.07	53.89
DL20	52.87	52.65
GA	52.44	52.29
KU1	52.94	52.66

Despite their sequence similarities, genome lengths for JS strains (3525 nt) were shorter than all genogroup I strains (3569–3575 nt) (Table 2) but longer than genogroup II (3458–3486 nt) [3]. Numerous deletions in the 3' untranslated region and a portion of ORF4 (replicase) in JS strains accounted for the decreased genome length (data not shown) but did not alter the ORF positions when the genogroup I strains were aligned (Table 2).

Analysis of the replicase gene revealed a 2 nt insertion at the 1374 nucleotide region when counting ORF4 start site as nucleotide 1 (Figure 1). This insertion occurred upstream from the ORF4 stop codon. Beginning approximately 40 nt downstream from the replicase ORF4 stop codon and continuing to the 3' termini, 53 nt deletions were present in the JS strains when aligned to MS2-like genomes. Nucleotide alignment of the replicase and nontranslated regions (NTR) revealed numerous nt deletions in the JS strains when compared to genogroup I strains accounting for the change in amino acid composition. However, JS strains shared the 3' terminal "signature", ACCACCCA, present in *Levivirus* genogroups I and II [3].

Figure 1. Replicase recombinant region in two JS strains when compared to genogroup I strains (family *Leviviridae*). Alignment (BioEdit v7.0.1) of the replicase nucleotide sequences from *Levivirus* genogroup I strains DL1, DL2, DL13, DL16, ST4, R17, J20, MS2 with JS strains DL52 and DL54. For clarity, only a portion of the alignment is shown. Alignment of each genogroup is depicted in discontinuous blocks. The numbers along the top are the nucleotide positions within the replicase gene with the start position of ORF4 assigned as nucleotide 1. Genome sequences read 5'-3' direction. Dots indicate identity with the consensus sequence. Degenerate bases are noted in the standard IUB codes. The replicase start codon and two nucleotide insertions are highlighted in red. Dashes denote a nucleotide sequence deletion from the consensus sequence.

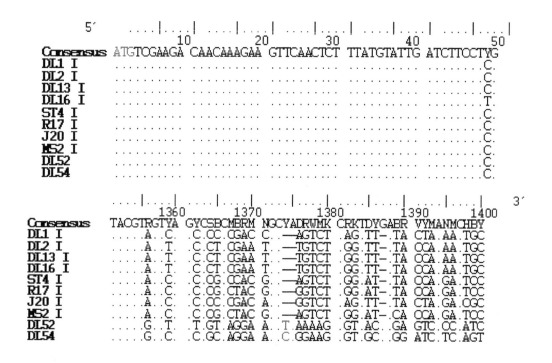

2.2. Amino Acid Analysis

Initially, nucleotide pairwise analyses of full-length genomes were made comparing all strains within the *Levivirus* genome, including genogroups I, JS and II; an 80%–85% nucleotide similarity between JS strains and the MS2-like strains was observed (Table 3a). In comparison, the amino acid sequences of the maturation, capsid and lysis proteins of the JS strains were very similar to those of the MS2-like genogroup I strains, sharing 97%–100%, 98%–100% and 95%–100% sequence similarities, respectively (Table 3b). Genogroup I strain fr, when compared to MS2-like and JS genogroup I strains, only shared an amino acid similarity to the maturation, capsid and lysis proteins ranging from 75.73%–91.85% (Table 3b). In contrast, the replicase protein sequences of the JS strains were quite dissimilar to the replicase protein sequences of the MS2-like genogroup I strains, displaying a similarity range of 79%–85% (Table 3c). However, a similarity of 97%–99% was observed among the highly conserved replicase genes for the MS2-like strains. Strain fr shared a 79% replicase similarity to JS strains and approximately 88%–89% similarity to MS2-like strains. Genogroup II replicase was approximately 52%–53% similar to JS strains, 50%–53% to MS2-like and fr strains and 92%–98% similar to other genogroup II strains (Table 3c).

Table 3. (b) Percent similarity in amino acid sequences between *Levivirus* JS strains and genogroup I maturation, capsid and lysis proteins. Amino acid pairwise computations were performed in Bionumerics.

Maturation Protein											
	ST4-I	MS2-I	R17-I	M12-I	DL1-I	J20-I	DL13-I	DL16-I	DL52-JS	DL54-JS	DL2-I	fr-I
ST4-I	100.0											
MS2-I	99.9	100.0										
R17-I	98.2	98.3	100.0									
M12-I	97.7	97.9	97.4	100.0								
DL1-I	97.7	97.8	97.2	97.2	100.0							
J20-I	97.9	98.0	97.1	97.1	98.3	100.0						
DL13-I	97.5	97.6	97.1	97.5	98.1	98.1	100.0					
DL16-I	97.5	97.6	97.1	97.5	98.1	98.1	100.0	100.0				
DL52-JS	97.5	97.6	97.1	97.5	98.1	98.1	100.0	100.0	100.0			
DL54-JS	97.5	97.6	97.1	97.5	98.1	98.1	100.0	100.0	100.0	100.0		
DL2-I	97.6	97.8	97.3	97.6	98.3	98.1	99.8	99.8	99.8	99.8	100.0	
fr-I	84.4	84.4	84.3	84.2	85.0	84.7	84.1	84.1	84.1	84.1	84.4	100.0

Capsid Protein											
	ST4-I	R17-I	MS2-I	M12-I	DL1-I	DL2-I	DL13-I	DL16-I	J20-I	DL52-JS	DL54-JS	fr-I
ST4-I	100.0											
R17-I	100.0	100.0										
MS2-I	100.0	100.0	100.0									
M12-I	98.7	98.7	98.7	100.0								
DL1-I	98.5	98.5	98.5	98.2	100.0							
DL2-I	98.5	98.5	98.5	98.2	100.0	100.0						
DL13-I	98.5	98.5	98.5	98.2	100.0	100.0	100.0					
DL16-I	98.5	98.5	98.5	98.2	100.0	100.0	100.0	100.0				
J20-I	98.5	98.5	98.5	98.2	100.0	100.0	100.0	100.0	100.0			
DL52-JS	98.5	98.5	98.5	98.2	100.0	100.0	100.0	100.0	100.0	100.0		
DL54-JS	98.5	98.5	98.5	98.2	100.0	100.0	100.0	100.0	100.0	100.0	100.0	
fr-I	91.2	91.2	91.2	90.9	91.9	91.9	91.9	91.9	91.9	91.9	91.9	100.0

Lysis Protein											
	ST4-I	MS2-I	R17-I	DL13-I	DL16-I	DL2-I	DL52-JS	DL54-JS	DL1-I	J20-I	M12-I	fr-I
ST4-I	100.0											
MS2-I	100.0	100.0										
R17-I	98.9	98.9	100.0									
DL13-I	95.0	95.0	95.0	100.0								
DL16-I	95.0	95.0	95.0	100.0	100.0							
DL2-I	95.0	95.0	95.0	100.0	100.0	100.0						
DL52-JS	95.0	95.0	95.0	100.0	100.0	100.0	100.0					
DL54-JS	95.0	95.0	95.0	100.0	100.0	100.0	100.0	100.0				
DL1-I	95.9	95.9	95.9	99.2	99.2	99.2	99.2	99.2	100.0			
J20-I	91.1	91.1	91.1	96.4	96.4	96.4	96.4	96.4	95.5	100.0		
M12-I	93.8	93.8	94.1	90.3	90.3	90.3	90.3	90.3	91.3	86.2	100.0	
fr-I	78.2	78.2	78.0	78.9	78.9	78.9	78.9	78.9	77.7	77.7	75.7	100.0

Table 3. (c) Amino acid percent similarity comparisons between *Levivirus* JS strains, DL52 and DL54, to *Levivirus* genogroup I and genogroup II RNA-dependent RNA polymerase (replicase) protein. Amino acid pairwise computations were performed in Bionumerics.

	ST4-I	R17-I	MS2-I	DL2-I	DL13-I	DL16-I	DL1-I	J20-I	fr-I	DL52-JS	DL54-JS	T72-II	KU1-II	DL10-II	DL20-II	GA-II
ST4-I	100.0															
R17-I	99.1	100.0														
MS2-I	99.0	98.6	100.0													
DL2-I	97.5	97.1	97.5	100.0												
DL13-I	97.4	97.0	97.4	99.9	100.0											
DL16-I	97.2	96.8	97.3	99.7	99.6	100.0										
DL1-I	98.0	97.6	97.8	99.4	99.3	99.1	100.0									
J20-I	97.7	97.2	97.5	99.0	98.9	98.7	99.1	100.0								
fr-I	88.5	88.4	88.2	89.0	88.9	88.7	88.9	88.8	100.0							
DL52-JS	84.4	83.9	84.0	85.4	85.4	85.2	85.3	84.8	79.7	100.0						
DL54-JS	84.2	84.1	83.9	85.1	85.0	84.8	85.0	84.8	79.8	97.2	100.0					
T72-II	51.9	51.5	51.1	52.6	52.6	52.3	52.6	52.7	52.8	52.9	53.1	100.0				
KU1-II	51.2	50.8	50.4	51.9	51.9	51.6	51.9	52.0	52.5	52.5	52.6	97.1	100.0			
DL10-II	52.7	52.3	51.9	53.4	53.4	53.1	53.4	53.7	52.7	53.7	53.8	93.3	93.0	100.0		
DL20-II	52.7	52.3	51.9	53.4	53.4	53.1	53.2	53.7	53.2	53.4	53.5	93.8	93.4	98.6	100.0	
GA-II	51.3	50.9	50.5	52.0	52.0	51.7	51.8	52.3	51.8	52.5	52.6	92.4	92.3	97.7	98.6	100.0

All genogroup I strains, including fr, and the two JS strains had a replicase protein length of 545 amino acids (Table 2) [3]. However, JS replicase differed from genogroup I replicase as it had one amino acid insertion at replicase position 467 and one amino acid deletion at the 3' termini of the stop codon, but maintained a total of 545 amino acids (data not shown). Identical to genogroup I strains, the replicase catalytic domain in the JS strains occurred between amino acid positions 243–373, thereby adding confidence to placing the grouping of JS into genogroup I [3]. Beginning at amino acid number 455 within the replicase gene, JS strains were unique in amino acid composition and diverged from the MS2-like strains.

2.3. Pfam and Protein Sequence Motifs

Individual proteins from JS strains DL52 and DL54 were grouped to protein families by Pfam analysis. The maturation protein generated "phage_mat-A" domain including all *Leviviridae* strains plus three additional bacteriophage, PRR1, PP7 and AP205. The capsid protein resulted in a "Levi-coat" domain including all *Leviviridae* strains plus bacteriophage PRR1. The lysis protein only generated results in a PfamB search matching the genus *Levivirus* strains from both genogroups I and II including KU1, JP34, M12, FP501, MS2, JP500, fr, TH1, SD, GA, BO1, TL2 and ZR. Replicase protein matched "RNA_replicase_B" domain within the *Leviviridae* family plus the additional bacteriophages PRR1, PP7 and AP205.

Common protein motifs such as casein kinase II phosphorylation, cAMP and cGMP-dependent protein kinase phosphorylation and protein kinase C phosphorylation occurred in DL52 and DL54 when compared to the *Levivirus* strains [3]. Interestingly, every amino acid motif position in all four genes was identical among these two JS strains.

2.4. Phylogenetic and Recombination Analyses

Cophenetic correlations showed the genogroup I strains, the JS subgroup strains, and the genogroup II strains all formed faithful clusters with correlations of 100, 90 and 98, respectively. The cluster cutoff method, however, showed only two relevant clusters being the genogroup I strains, which included fr and JS, and genogroup II strains (Figure 2).

Figure 2. Cophenetic cluster analysis of *Levivirus* (family *Leviviridae*) genogroups I and II strains generated from pairwise similarities of the replicase amino acid sequences. Horizontal bars at three of the branches show the standard deviations of the average similarities of the clusters. Numbers at each branch are the cophenetic correlations which represent the faithfulness of the clusters. Two relevant clusters, as determined by the cluster Cutoff method, are grouped as dictated by the dashed lines. Analysis performed in Bionumerics.

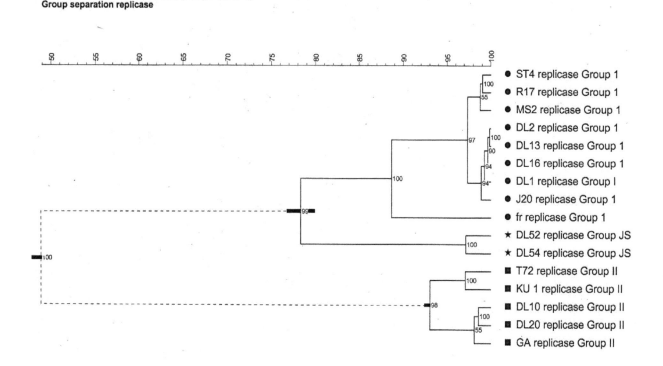

When referring to nucleotide or amino acid positions within the replicase gene, the numbering is in reference to the start codon as being position 1. In all analysis programs, the nucleotide or amino acid sequences were aligned to other strains and were therefore approximate positions on the replicase gene.

All recombination programs used, SimPlot, RAT, RDP3 and Recco, statistically predicted recombination in both JS strains, DL52 and DL54, when compared to genogroup I MS2-like strains. No recombination, however, was detected when DL52 and DL54 were compared to genogroup I strain fr and all genogroup II strains.

Figure 3. (**a**) The Simplot and bootscan analyses of the replicase nucleotides from JS strain DL52 queried to DL54, DL1, DL3, DL13, DL16, ST4, R17, J20 and MS2. The breakpoints are shown by the vertical red lines. The first recombination breakpoint occurred in the replicase gene in strain DL52 at nucleotide positions 787–818 where the χ^2 changed from 0.8 to 6.3 (sum χ^2 of 7.1). The second breakpoint occurred at nucleotide positions 979–1029 where the χ^2 changed from 0.6 to 7.0 (sum χ^2 of 7.6); (**b**) The Simplot and bootscan analyses of the replicase amino acids from JS strain DL52 queried to DL54, DL1, DL3, DL13, DL16, ST4, R17, J20 and MS2.

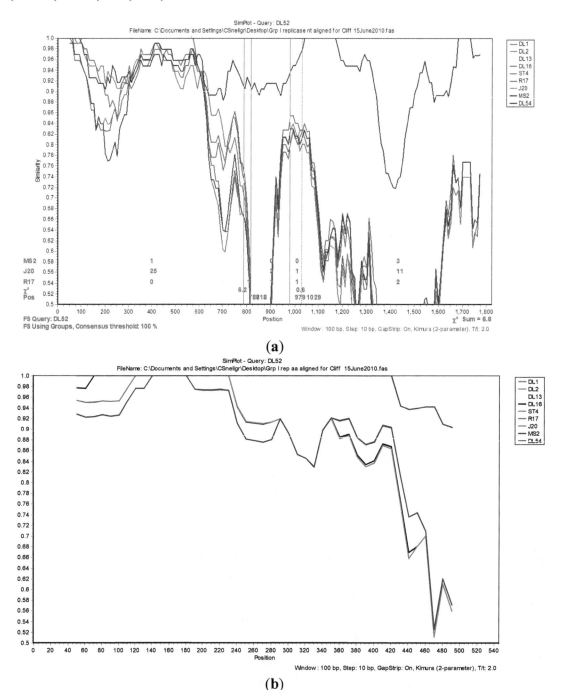

(**a**)

(**b**)

The Simplot and bootscan analyses of the replicase nucleotides from JS strains DL52 compared to *Levivirus* genogroup I strains DL54, DL1, DL3, DL13, DL16, ST4, R17, J20 and MS2 is shown in Figure 3A. Since the replicase nucleotide sequences in strain DL54 were 97% similar to strain DL52, DL52 was chosen as the query. The SimPlot analysis revealed the first recombination breakpoint occurred in the replicase from strain DL52 at nt positions 787–818 (approximate amino acid 262–273) where the χ^2 changes from 0.8 to 6.3 (sum χ^2 of 7.1). The second breakpoint occurred at nt positions 979–1029 (approximate amino acid 326–343) where the χ^2 changes from 0.6 to 7.0 (sum χ^2 of 7.6). However, Simplot amino acid analysis (Figure 3b) with strain DL52 showed a divergence at approximate amino acid position 460 region which is in agreement with the manual alignment (Figure 1).

Figure 4. Recombination analysis of the replicase nucleotide sequences from *Leviviridae* genogroup I strains DL13, DL16, ST4, R17, J20, MS2 and JS strains DL54, DL52 queried to DL1. Recombination Analysis Tool (RAT) was used to generate graphics with a window of 182 nt and step increments of 92 nt. The Y-axis represents the genetic distance and the X-axis is the sequence location along the genome. (**a**) The JS strains, depicted in green, diverged from the other genogroup I strains at approximate nucleotide (nt) position 660; (**b**) Recombination analysis of the replicase amino acid sequences from *Leviviridae* genogroup I strains DL13, DL16, ST4, R17, J20, MS2 and JS strains DL54, DL52 queried to DL1. Recombination Analysis Tool (RAT) was used to generate graphics with a window of 54 aa and step increments of 27 aa. The JS strains, DL52 and DL54, diverged from the other genogroup I strains at approximate amino acid 220 within the replicase gene.

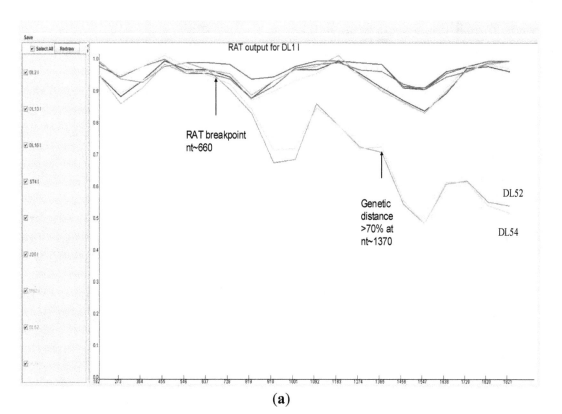

(**a**)

Figure 4. *Cont.*

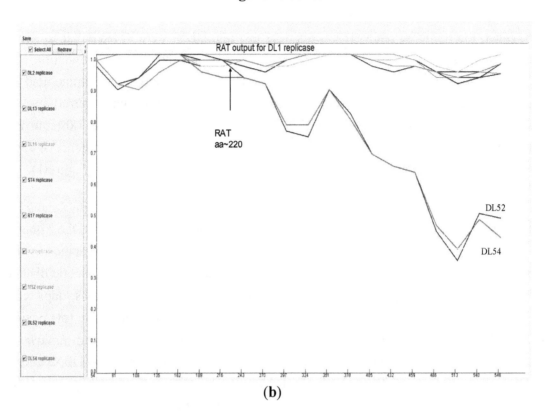

(b)

When analyzed with RAT, the nucleotide breakpoint (crossover) positions occurred at approximately nt 660 (Figure 4a) or amino acid 220 (Figure 4b) within the replicase gene. This crossover occurred when the two recombinant strains, DL52 and DL54, crossed the lines of the other MS-2 like strains and diverged by increasing genetic distances.

RDP3 predicted DL52 and DL54 as the recombinant strains using several detection methods and analysis algorithms (Table 4) and suggested DL16 as a minor parent strain. Breakpoint nucleotides for strains DL52 and DL54 (when aligned to genogroup I strains) occurred between nt 84–592 and 84–401, respectively (Figure 5a,b), corresponding to the approximate amino acid breakpoint positions of 133–197 within the replicase gene.

Table 4. Prediction of DL52 and DL54 as recombinant strains by analysis of *Levivirus* (family *Leviviridae*) genogroup I using Recombination Detection Program (RDP3).

Confirmation Table of Recombination Events		
Methods	**Events**	**Average p-value**
RDP	2	2.199×10^{-15}
GENECONV	1	3.031×10^{-27}
Bootscan	2	7.867×10^{-19}
MaxChi	2	1.445×10^{-10}
Chimaera	2	3.536×10^{-11}
SiScan	1	1.168×10^{-13}
3Seq	1	4.486×10^{-8}

Figure 5. (**a**) RDP3 analyses prediction of DL52 as a recombinant strain. Recombination area within the replicase gene is shown in pink beginning at nucleotide 84 and crossing over at 592, upstream from the catalytic domain. DL52 was queried to all *Levivirus* (family *Leviviridae*) genogroup I FRNA *Levivirus* strains. RDP3 suggested DL16 as the minor parental strain; (**b**) RDP3 analyses predicted DL54 as a recombinant strain. Recombination area within the replicase gene is shown in pink beginning at nucleotide 84 and crossing over at 401, upstream from the catalytic domain. RDP3 suggested DL16 as the minor parental strain.

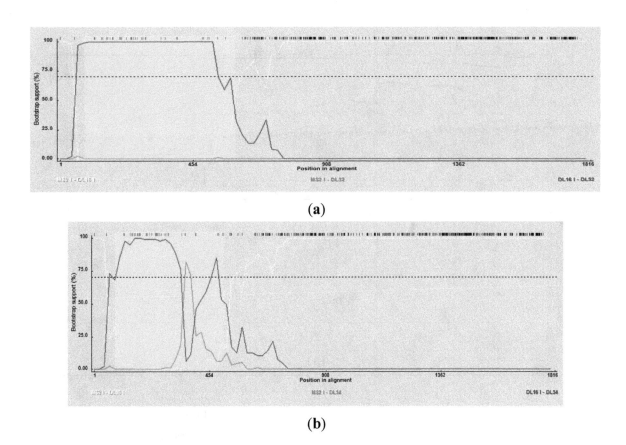

(a)

(b)

Manual alignment in BioEdit of the replicase nucleotides, counting the ATG start codon of the replicase gene as nt 1, showed an insertion of the nucleotides YA beginning at position 1374 (Figure 1) whereas the amino acid composition of the JS strains diverged from the other genogroup I strains slightly upstream from this insertion at amino acid position 455 (nucleotide 1366). Alignment also revealed numerous nt deletions as discussed in the "Sequence analyses and ORF" section.

The Recco p-value inspector predicted strain DL52 had recombined with strain DL1 (Figure 6a). In DL52, the recombinant region spanned from amino acids 181–212 whereas the DL1 region spanned from 396–457 with resulting sequence p-values of 0.000999 and 0.004995, respectively. Recco parametric cost curves predicted the highest preference for recombination in strains DL52 and DL54 (cost of 12.5–13) whereas the remaining genogroup I strains did not show a preference for recombination (cost of 0–3) (Figure 6b).

Figure 6. (**a**) Recco analysis of the RNA-dependent RNA polymerase (replicase) amino acid sequences in *Levivirus* genogroup I male-specific coliphages (FRNA). Recombination events are displayed by downward peaks in the graphics dataset. The upper graph represents the p-value for recombination at each position along the replicase gene. The lower graph is the breakpoint p-values for the entire set of *Levivirus* genogroup I and JS strains DL52 and DL54; (**b**) Recco parametric cost curve analysis of the RNA-dependent RNA polymerase (replicase) amino acid sequences for each FRNA strain in *Levivirus* genogroup I and JS strains DL52 and DL54. The y-axis corresponds to the cost curve and the x-axis represents α (0–1).

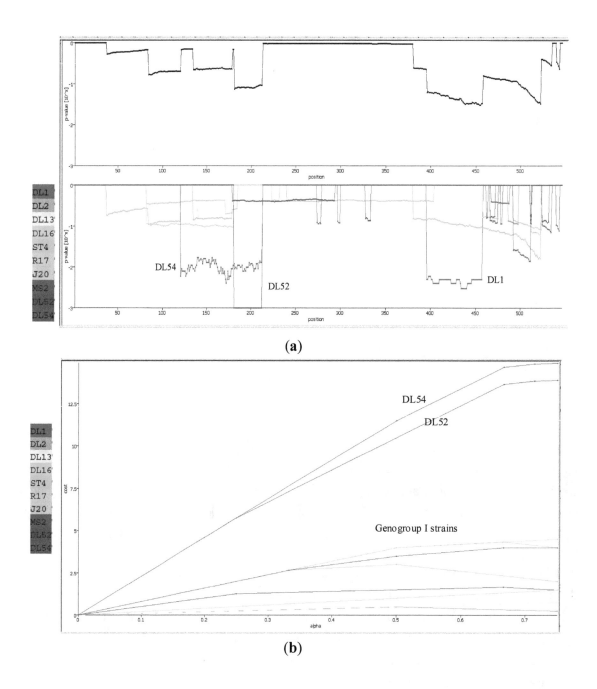

(a)

(b)

RAT, RDP3 and Recco all predicted recombination breakpoints ranging from amino acid positions 181–252 whereas Simplot agreed most closely with the manual alignment of 460 and 455, respectively. Also in agreement with the manual alignment was the crossover region between DL52 and DL1 occurring in the approximate amino acid region of 396–457 (Figure 6a). The predicted breakpoint regions occurred either upstream or downstream from the highly conserved catalytic domain amino acid positions 243–373 in *Levivirus* genogroup I [3].

3. Discussion

Reported here are whole genome sequence data and bioinformatic analyses supporting the hypothesis that two novel FRNA isolates, DL52 and DL54, were the result of natural recombination. Initially classified as a JS subgroup of genogroup I within the *Levivirus* genus (family *Leviviridae*), these strains were isolated from seawater approximately 1130 km apart in the Rachel Carson Reserve, Beaufort, NC, and Narragansett Bay, RI. Findings that JS strains were highly similar to three out of four genes (maturation, capsid and lysis) in genogroup I MS2-like strains, shared the catalytic site location in the RNA-dependent RNA polymerase (replicase) gene, had an identical 3' signature [3] and then greatly diverged along a stretch of the replicase gene all supported the occurrence of a recombination event.

In this study, two JS strains shared >95% amino acid identity in three (maturation, capsid and lysis) *Levivirus* genogroup I MS2-like genes but only an 84%–85% amino acid identity to the otherwise, highly conserved replicase protein. In comparison, genogroup I strain fr was uniformly different from all other genogroup I strains in all four proteins [3]. Cophenetic correlations and bootstrap analysis strengthen the possibility that JS strains were recombinants as the JS strains were only a subgroup of genogroup I and not a novel genogroup. Throughout the *Leviviridae* family, subgroups emerge within genogroups, however, subgroup strains differ in all four genes from the parent genogroup [3]. It is therefore plausible to propose natural recombination in these two novel JS-like FRNA coliphages as data presented here suggested a specific genetic rearrangement or recombination event in the replicase gene.

Interestingly, different *Leviviridae* subgrouped strains originating from across the globe display high amino acid similarity among subgrouped strains [3]. Recombination may explain why *Leviviridae* strains circulate as discrete subgroups independent of geographical location. Although the two unique JS-like strains were isolated from NC and RI, they shared 96.73% nucleotide similarity across the entire genome. Thus, either a single, natural recombination event occurred as a *de novo* mutation in each strain or identical, natural recombinations along a hot spot in these genomes formed these strains. In either case discovering that geographically-separated JS-like strains acquired the same recombination event is intriguing.

Largely responsible for the diversity of RNA viruses [12] RNA-RNA recombination was observed in several positive-sense, ssRNA human and animal viral taxa including caliciviruses, coronaviruses, hepatitis, dengue, enteroviruses and astroviruses [13–21]. For example, genetic exchange in ssRNA viruses was first demonstrated in polioviruses [22,23].

Recombination events frequently alter the RNA-dependent RNA polymerase region. Human Noroviruses, a positive sense ssRNA virus with a genome length of 7400–8300 nt, are considered to belong to a prototype strain if they share approximately 85% overall nucleotide sequence identity and a high amino acid sequence identity (>95%) in the polymerase gene [20]. A naturally occurring human Norovirus strain shared 95% amino acid sequence identity with the capsid sequences from a Mexico cluster and 95% amino acid identity to the polymerase in a Lordsdale virus cluster. Sequences from the natural strain were obtained from one viral isolate. The combination of sequences in the one strain being complementary to two distinct human Norovirus clusters led to the proposition that this strain was a naturally occurring recombinant [20].

Genetic recombination is known to occur in certain Enteroviruses, a positive ssRNA virus having an approximate 7500 nt genome. Poliovirus recombination occurs in vaccine-derived strains [24] in the human population as a single infected individual excretes a high proportion of recombinants [18]. To determine if other enteroviruses undergo natural recombination, isolates of echoviruses were collected from a meningitis outbreak. Nucleotide sequences were clustered based on a capsid protein (VP1) and RNA-dependent RNA polymerase (3D). Dendrogram relatedness of the echovirus strains grouped the VPI sequences to the prototype strains. However, the RNA polymerase sequences did not cluster to the prototype strains, suggesting genetic recombination among the outbreak strains [18].

Human astroviruses are positive sense, ssRNA with a genome length of approximately 6,800 nucleotides [19] and a polyadenylated 3' tail [21]. Two sets of strains were investigated for recombination; one set was identified from a child care center in Houston, TX, and the two other strains were found in stool samples from two children in Mexico City. The pool of strains shared >97% nucleotide sequence similarity in two out of three genomic regions. The novel strain clustered to one group based on the capsid region. When the RNA-dependent RNA polymerase was analyzed, the novel strain clustered to a separate human astrovirus group. The strains were identified as naturally occurring recombinants on the evidence of high sequence similarity to a few genes of one prototype and similarity to different genes in a second prototype. A total of 64 additional human astroviruses lacked these novel traits [19].

An enteric turkey astrovirus is a non-enveloped, positive sense ssRNA virus with a polyadenylated 3' tailed genome of approximately 7 kb. The most conserved gene in the avian and mammalian astrovirus is the RNA-dependent RNA polymerase or replicase. Genetic analysis of capsid and polymerase sequences from twenty-three turkey astrovirus strains resulted in 8 clusters for the capsid gene and two phylogenetic clusters for the RNA polymerase gene. Computer-generated analyses identified polymerase gene recombination in strains of turkey astrovirus [14].

In this study, four different recombination detection programs along with manual alignment predicted strains DL52 and DL54 as recombinants although the exact amino acid and/or nucleotide breakpoint varied somewhat along the RNA-dependent RNA polymerase gene. As expected the breakpoints did not occur within the catalytic-site domain. The sliding window approach as used with many recombination programs is based on an arbitrarily chosen window length, thus affecting the sensitivity and accuracy when pinpointing the precise breakpoint [25]. Overall, the use of a variety of recombination algorithms provided a stronger, more rigorous scientific case. When comparing the *Levivirus* strains, manual alignment provided a more accurate picture of where the recombination event occurred along the genome but it did not provide a statistical analysis. Therefore, statistical analysis in combination with manual alignment resulted in a more confident assessment of recombination. Evidence for recombination among positive ssRNA viruses exists within the RNA-dependent RNA polymerase (replicase) gene for numerous viruses as described here and supports the data that natural recombination can occur within the *Leviviridae* family.

4. Experimental Section

4.1. FRNA Coliphage Strains and RNA Extraction

FRNA phage strains CICEET 29 and CICEET 24 were isolated and placed into the putative JS subgroup [10,11] using the genotyping methods of Vinjé *et al.*, [4]. CICEET 29, renamed DL52, was isolated from estuarine waters in Rachel Carson W Reserve (Beaufort), NC, and CICEET 24, renamed DL54, and was isolated from Narragansett Bay, RI (Table 1).

Each strain was plaque purified and further enriched using *Escherichia coli* HS(pFamp)R as host [4]. Approximately 1–2 mL aliquots of the purified viral supernatant were frozen at &75 °C. Coliphage RNA was extracted from purified virus as described [26] using a QIAamp viral RNA mini kit (Qiagen, Valencia, CA, USA). Purified RNA was stored frozen at &20 °C.

4.2. Sequencing and Analysis

Full-length genome sequencing was performed by the "primer walking" approach as described [3]. Nucleotide and amino acid sequences from JS strains DL52 and DL54 were compared to nucleotide and amino acid sequences from 10 genogroup I strains (MS2, DL1, DL2, DL13, DL16, ST4, R17, J20, M12, fr) and 5 genogroup II strains (T72, DL10, DL20, GA, KUI) [3].

Nucleotide sequences from three individual clones were imported and aligned using BioEdit v7.0.1 [27] followed by Basic Local Alignment Search Tool (BLAST, National Center for Biotechnology Information) analyses for sequence and phylogenetic confirmation. Completed sequences from all strains were aligned with full-length prototype strains (GenBank) using BioEdit ClustalW application. For each strain, the Open Reading Frames (ORFs) were mapped using BioEdit.

4.3. Amino Acid Analysis

Deduced amino acid sequences for each of the four genes were determined using a computer-generated DNA-to-protein translation tool, ExPASY (http://ca.expasy.org/). Prediction of protein sequence motifs were identified by PROSITE (http://ca.expasy.org/) and protein families and domains were modeled in Pfam (http://pfam.janelia.org).

4.4. Phylogenetic, Statistical and Recombination Analyses

Sequence data were analyzed using BioNumerics Software v.3.5 (Applied Maths, Saint-Martens-Latem, Belgium). Phylogenetic trees were built by global cluster analysis performed on multiple aligned sequences and clustered by unweighted pair group method using arithmetic averages (UPGMA). A bootstrap analysis, based on 10,000 substitutions, was used to measure cluster significance. The reliability of each cluster was expressed on a percentage basis [3].

Nucleotide percent similarity and dendrograms were constructed using BioNumerics Software v.3.5 (Applied Maths, Saint-Martens-Latem, Belgium). Phylogenetic trees were built by global cluster analysis performed on multiple aligned sequences and clustered by UPGMA using the Jukes and Cantor correction [28]. Cophenetic correlations and cluster Cutoff method were employed to measure faithfulness and relevancy of the clusters (Applied Maths, Saint-Martens-Latem, Belgium). Average similarities with standard deviations were calculated for the relevant clusters.

Various approaches were used to examine recombination in *Levivirus* FRNA strains, using aligned nucleotides and aligned amino acids, as follows: (i) manual alignment using BioEdit; (ii) bootscan analysis in SimPlot v3.5.1 [29] (iii) Recombination Analysis Tool v1.0 [30], (iv) Recombination Detection Program v3.44 [31]; and (v) Recombination Analysis Using Cost Optimization (Recco) v0.93 [32] FRNA strains used in analyses were genogroup I strains DL1, DL2, DL13, DL16, ST4, R17, J20, MS2, fr; JS strains DL52 and DL54; and genogroup II strains GA, KU1, DL10, DL20 and T72.

SimPlot analyses was determined with a sliding window of 100 bp wide and a step size between plots of 10 bp when comparing reference strains to the queried sequences. Recombination events by SimPlot bootscan analysis occurred when the χ^2 value changes signifying a breakpoint position.

Aligned replicase amino acids and/or nucleotide sequences were analyzed in Recombination Analysis Tool (RAT), Recombination Detection Program (RDP3) and Recco. RAT uses a distance-based method of recombination in both DNA and protein multiple alignments [30]. Unless stated otherwise, default settings were used with each program. The RAT default settings were window size of 10% of the sequence length and an increment size being half of the window size. Both settings of "auto search" and "test sequence search" were used with RAT.

Recombination Detection Program v3 (RDP3) uses a number of recombination detection algorithms such as RDP, Bootscan, GENECONV, Maximum Chi Square, CHIMAERA, Sister Scanning (SISCAN) and 3SEQ [31]. The RDP3 program sorts the analyses from these various algorithms and statistical data to determine the unique recombination events. RDP3 used an alignment of all genogroup I and JS replicase nucleotide sequences and queried to DL16, MS2, DL52 and DL54.

The Recco p-value inspector was set at 3 and the permutation was set at 1000. Recco uses an algorithm that locates putative recombination points based on cost minimization. Recco compares the cost of mutation relative to recombination as represented by α.

4.5. Nucleotide Sequence Accession Numbers

The accession numbers for DL52 and DL54 are JQ966307 and JQ966308, respectively.

5. Conclusions

The results of this study provide genetic evidence, bioinformatic and statistical analyses suggesting a natural recombination event in the formation of a genogroup I subgroup JS-like *levivirus,* represented by two strains, DL52 and DL54. There was high nucleotide and amino acid identity in three genes, the maturation, capsid and lysis genes (≥95%) but a lack of similarity in the replicase gene (84%–85%) when JS strains were compared to genogroup I MS2-like strains. Four different recombination programs demonstrated one or two breakpoint regions in the replicase gene, signifying a recombination event. The recombination event occurred downstream of the replicase catalytic site thereby maintaining viral integrity and replication function. Thus, primers for oligonucleotide hybridization probes targeting the replicase beyond the catalytic site would not hybridize to JS strains. In contrast, molecular assays targeting the maturation, capsid or lysis sequences would presumptuously place JS strains as an MS-2 like genogroup.

Phylogenetic tree analysis produced a cophenetic correlation which showed (i) ten genogroup I strains, including strain fr; (ii) the JS subgroup strains; and (iii) the genogroup II strains all formed faithful clusters with correlations of 100, 90 and 98, respectively. The cluster cutoff method, however, revealed only two relevant clusters, (i) genogroup I strains, which included fr and JS; and (ii) genogroup II strains. Therefore, the novel JS strains are not a unique *Levivirus* genogroup. The proposed classification of JS strains is genogroup I subgroup "JS-like".

Although both JS strains were prepared for sequencing in the same laboratory, these strains were field-collected by different investigators and shipped to another location where they were plaque-purified and preliminarily classified. Therefore, the possibility that contamination resulted in false recombinants seems unlikely. Likewise, the possibility of cloning and/or PCR errors contributing to the nucleotide and amino acids changes would not have led to both JS strains being almost identical in the non recombinant regions as seven genogroup I strains (DL1, DL2, DL13, DL16, R17, J20 and ST4), three genogroup II strains (DL10, DL20, T72) and two JS strains (DL52 and DL54) were sequenced in this lab in no certain strain or fragment order using the same methods and sequencing company [3]. Finally, to the best of our knowledge, this is the first description of recombinant viruses from natural isolates in ssRNA *Leviviridae* bacteriophages.

Acknowledgments

This research was funded, in part, through EPA's New England Regional Applied Research Effort (RARE). We gratefully acknowledge the assistance of Jack Paar, III, U.S. EPA New England Regional Laboratory for initiating and sponsoring this program.

We wish to thank Jan Vinjé, Centers for Disease Control and Prevention, Atlanta, GA, for his intellectual contribution. An acknowledgement is extended to Syed Muaz Khalil for providing a portion of the sequence data. We thank Greg Lovelace and David Love (Johns Hopkins University) for isolating and providing some of the strains used in this study and Emilie Cooper (CDC) for reviewing the manuscript.

The information in this document has been funded wholly (or in part) by the U.S. Environmental Protection Agency. It has been subjected to review by the National Health and Environmental Effects Research Laboratory and approved for publication. Approval does not signify that the contents reflect the views of the Agency, nor does mention of trade names or commercial products constitute endorsement or recommendation for use.

References

1. Fiers, W.; Contreras, R.; Duerinck, F.; Haegeman, G.; Iserentant D.; Merregaert, J.; Min Jou, W.; Molemans, F.; Raeymaekers, A.; van den Berghe, A.; *et al.* Complete nucleotide sequence of bacteriophage MS2 RNA: Primary and secondary structure of the replicase gene. *Nature* **1976**, *260*, 500–507.

2. Sanger, F.; Air, G.M.; Barrell, B.G.; Brown, N.L.; Coulson, A.R.; Fiddes, J.C.; Hutchinson, C.A. III; Slocombe, P.M.; Smith M. Nucleotide sequence of bacteriophage Φ-X174 DNA. *Nature* **1977**, *265*, 687–695.

3. Friedman, S.D.; Genthner, F.J.; Gentry, J.; Sobsey, M.D; Vinjé, J. Gene mapping and phylogenetic analysis of the complete genome from 30 single-stranded RNA male-specific coliphages (family *Leviviridae*). *J. Virol.* **2009a**, *83*, 11233–11243.

4. Vinjé, J.; Oudejans, S.J.G.; Stewart, J.R.; Sobsey, M.D.; Long, S.C. Molecular detection and genotyping of male-specific coliphages by reverse transcription-PCR and reverse line blot hybridization. *Appl. Environ. Microbiol.* **2004**, *70*, 5996–6004.

5. Chetverin, A.B. The puzzle of RNA recombination. *FEBS Lett.* **1999**, *460*, 1–5.

6. Lutay, A.V.; Zenkova, M.A.; Vlassov, V.V. Nonenzymatic recombination of RNA: Possible mechanisms for the formation of novel sequences. *Chem. Biodiv.* **2007**, *4*, 762–767.

7. Chetverin, A.B.; Kopein, D.S.; Chetverina, H.V.; Demidenko A.A.; Ugarov V.I. Viral RNA-directed RNA polymerases use diverse mechanisms to promote recombination between RNA molecules. *J. Biol. Chem.* **2005**, *280*, 8748–8755.

8. Horiuchi, K. Genetic Studies of RNA Phages. In *RNA Phages*; Zinder, N.D., Ed.; Cold Spring Harbor Laboratory: Cold Spring Harbor, NY, USA, 1975; pp. 29–50.

9. Munishkin, A.V.; Voronin, L.A.; Chetverin, A.B. An *in vivo* recombinant RNA capable of autocatalytic synthesis by Qβ replicase. *Nature* **1988**, *333*, 473–475.

10. Sobsey, M.D.; Love, D.C.; Lovelace, G.L. F+RNA coliphages as source tracking viral indicators of fecal pollution. A final report submitted to the NOAA/UNH Cooperative Institute for Coastal and Estuarine Environmental Technology (CICEET). **2006**.

11. Love, D.C.; Vinjé, J.; Khalil, S.M.; Murphy, J.; Lovelace, G.L.; Sobsey, M.D. Evaluation of RT-PCR and reverse line blot hybridization for detection and genotyping F+ RNA coliphages from estuarine waters and molluscan shellfish. *J. Appl. Microbiol.* **2008**, *104*, 1203–1212.

12. Lai, M.M.C. RNA recombination in animal and plant viruses. *Microbiol. Rev.* **1992**, *56*, 61–79.

13. Cristina, J.; Colina, R. Evidence of structural genomic region recombination in Hepatitis C virus. *Virol. J.* **2006**, *3*, 53–60.

14. Pantin-Jackwood, M.J.; Spackman, E.; Woolcock, P.R. Phylogenetic analysis of turkey astroviruses reveals evidence of recombination. *Virus Genes* **2006**, *32*, 187–192.

15. Holmes, E.C.; Worobey, M.; Rambaut, A. Phylogenetic evidence for recombination in Dengue virus. *Mol. Biol. Evol.* **1999**, *16*, 405–409.

16. Oberste, M.S.; Maher, K.; Pallansch, M.A. Evidence for frequent recombination within species *Human Enterovirus B* based on complete genomic sequences of all thirty-seven serotypes. *J. Virol.* **2004**, *78*, 855–867.

17. Banner, L.R.; Lai, M.M.C. Random nature of coronavirus RNA recombination in the absence of selection pressure. *Virolgy* **1991**, *185*, 441–445.

18. Oprisan, G.; Combiescu, M.; Guillot, S.; Caro, V.; Combiescu, A.; Delpeyroux, F.; Crainic, R. Natural genetic recombination between co-circulating heterotypic enteroviruses. *J. Gen. Virol.* **2002**, *83*, 2193–2200.

19. Walter, J.E.; Briggs, J.; Guerrero, M.L.; Matson, D.O.; Pickering, L.K.; Ruiz-Palacios, G.; Berke, T.; Mitchell, D.K. Molecular characterization of a novel recombinant strain of human astrovirus associated with gastroenteritis in children. *Arch. Virol.* **2001**, *146*, 2357–2367.

20. Jiang, X.; Espul, N.; Zhong, W.M.; Cuello, H.; Matson, D.O. Characterization of a novel human calicivirus that may be a naturally occurring recombinant. *Arch. Virol.* **1999**, *144*, 2477–2387.

21. Belliot, G.; Laveran, H.; Monroe, S.S. Detection and genetic differentiation of human astroviruses: Phylogenetic grouping varies by coding region. *Arch. Virol.* **1997**, *142*, 1323–1334.

22. Hirst, G.K. Genetic recombination with Newcastle disease virus, polioviruses, and influenza. *Cold Spring Harb. Symp. Quant. Biol.* **1962**, *27*, 303–309.

23. Ledinko, N. Genetic recombination with poliovirus type 1 studies of crosses between a normal horse serum-resistant mutant and several guanidine-resistant mutants of the same strain. *Virology* **1963**, *20*, 107–119.

24. Kew, O.; Morris-Glasgow V.; Landaverde, M.; Burns C.; Shaw J.; Garib, Z.; André J.; Blackman, E.; Freeman C.J.; Jorba, J.; *et al.* Outbreak of poliomyelitis in Hispaniola associated with circulating type 1 vaccine-derived poliovirus. *Science* **2002**, *296*, 356–359.

25. Lee, W.H.; Sung, W.K. RB-finder: An improved distance-based sliding window method to detect recombination breakpoints. *J. Comput. Biol.* **2008**, *15*, 881–898.

26. Stewart, J.R.; Vinjé, J.; Oudejans, S.J.G.; Scott, G.I. Sequence variation among group III F-specific RNA coliphages from water samples and swine lagoons. *Appl. Environ. Microbiol.* **2006**, *72*, 1226–1230.

27. Hall, T.A. BioEdit: A user-friendly biological sequence alignment editor and analysis program for Windows 95/98/NT. *Nucleic Acids Symp. Ser.* **1999**, *41*, 95–98.

28. Jukes, T.H.; Cantor, C.R. Evolution of Protein Molecules. In *Mammalian Protein Metabolism*; Munro, H.N., Ed.; Academic Press: New York, NY, USA, 1969; pp. 21–132.

29. Lole, K.S.; Bollinger, R.C.; Paranjape, R.S.; Gadkari, D.; Kulkarni, S.S.; Novak, N.G.; Ingersoll, R.; Sheppard, H.W.; Ray S.C. Full-length human immunodeficiency virus type 1 genomes from subtype c-infected seroconverters in India, with evidence of intersubtype recombination. *J. Virol.* **1999**, *73*, 152–160.

30. Etherington, G.J.; Dicks, J.; Roberts, I.N. Recombination analysis tool (RAT): A program for the high-throughput detection of recombination. *Bioinformatics* **2005**, *21*, 278–281.

13

Function and Regulation of Clustered Regularly Interspaced Short Palindromic Repeats (CRISPR)/CRISPR Associated (Cas) Systems

Corinna Richter, James T. Chang and Peter C. Fineran *

Department of Microbiology and Immunology, University of Otago, PO Box 56, Dunedin 9054, New Zealand; E-Mails: ricco896@student.otago.ac.nz (C.R.); james.chang@otago.ac.nz (J.T.C.)

* Author to whom correspondence should be addressed; E-Mail: peter.fineran@otago.ac.nz

Abstract: Phages are the most abundant biological entities on earth and pose a constant challenge to their bacterial hosts. Thus, bacteria have evolved numerous 'innate' mechanisms of defense against phage, such as abortive infection or restriction/modification systems. In contrast, the clustered regularly interspaced short palindromic repeats (CRISPR) systems provide acquired, yet heritable, sequence-specific 'adaptive' immunity against phage and other horizontally-acquired elements, such as plasmids. Resistance is acquired following viral infection or plasmid uptake when a short sequence of the foreign genome is added to the CRISPR array. CRISPRs are then transcribed and processed, generally by CRISPR associated (Cas) proteins, into short interfering RNAs (crRNAs), which form part of a ribonucleoprotein complex. This complex guides the crRNA to the complementary invading nucleic acid and targets this for degradation. Recently, there have been rapid advances in our understanding of CRISPR/Cas systems. In this review, we will present the current model(s) of the molecular events involved in both the acquisition of immunity and interference stages and will also address recent progress in our knowledge of the regulation of CRISPR/Cas systems.

Keywords: phages; plasmids; horizontal gene transfer; CRISPR; Cas; cascade; PAM; crRNA; resistance

1. Introduction

Bacteria and Archaea are frequently exposed to stresses such as infection from bacteriophages (phage) and other genetic elements. These events can result in horizontal gene transfer (HGT), which is mediated by transduction, transformation or by conjugation of mobile elements such as plasmids [1,2]. HGT can be beneficial for adaptation and survival by rapidly promoting the transfer of genes encoding antibiotic resistance, virulence factors, or the ability to degrade toxic compounds [1]. However, in the absence of any selective advantage, foreign genetic elements can have a fitness cost. This is most evident when prokaryotes are infected by virulent phages. The sheer scale of viral abundance ($\sim 10^{31}$), diversity and rates of infection ($\sim 10^{25}$/s) [3,4] has resulted in the evolution of prokaryotic defenses to maintain the balance in this arms race [5].

Bacteria possess multiple methods to regulate genetic flux and resist phage infection. These mechanisms include the mutation or masking of cell surface receptors, restriction-modification, abortive infection and the clustered regularly interspaced short palindromic repeats (CRISPR) systems [5,6]. CRISPR systems are a widespread mechanism that equips bacteria with a sequence-specific heritable 'adaptive immune system' that has a genetic memory of past genetic incursions (for recent reviews see [7–9]).

CRISPRs use small non-coding RNAs for defense and function in conjunction with CRISPR associated (Cas) proteins. The mechanism of CRISPR/Cas interference involves three phases (Figure 1). Firstly, resistance is acquired via the integration of short sequences from foreign genetic elements (termed spacers) into repetitive genetic elements known as CRISPR arrays. Secondly, CRISPR arrays are then transcribed and processed into small RNAs (crRNAs) by Cas proteins. In the third and final step, targeting of the invading phage or plasmid is mediated by a Cas protein complex that contains crRNAs. During this stage, the crRNA-Cas protein complex then interferes, in a sequence-specific manner, with the foreign nucleic acids.

Despite sharing mechanistic similarities, there is significant diversity amongst CRISPR/Cas systems. There are three major CRISPR/Cas system types (I-III), which are characterized by a signature protein [10]. The main types are further divided into subtypes (e.g., I-A to I-F) based mainly on the presence of a subtype-specific set of Cas proteins and for some types the repeat sequence of the associated CRISPR array is also taken into account. The last five years have seen rapid advances in our mechanistic/molecular understanding of CRISPR/Cas systems. In this review, we provide an historical context, highlight the current models for the adaptation, processing and interference stages and finally, discuss the regulation of the CRISPR/Cas systems.

2. Discovery of CRISPR/Cas Systems

The first report of interspaced palindromic repeat sequences was made by Ishino *et al*. who detected five 29 bp repeats with 32 bp spacers near the *iap* gene of *Escherichia coli* [11]. Further reports were made for *Mycobacterium tuberculosis* [12], *Haloferax* spp. [13] and *Archaeoglobus fulgidus* [14,15]; however, they were not detected in eukaryotic or virus sequences [15]. Mojica and co-workers performed a comparative *in silico* study of those repetitive elements to determine structure and sequence similarity, as well as their phylogenetic distribution [16] showing that CRISPRs display a high degree of homology between phylogenetically distant species and a wide distribution in bacteria and archaea. To date CRISPR/Cas systems have been found in almost 50% of bacterial and 85% of archaeal genome sequences available [17].

Figure 1. Overview of clustered regularly interspaced short palindromic repeats (CRISPR)/CRISPR associated (Cas) adaptive immunity. (**A**) Adaptation. The CRISPR arrays are composed of short repeats and intervening sequences derived from foreign invaders. Upon infection with a foreign element (e.g., phages or plasmids), part of the genome is typically incorporated into the leader end of the CRISPR array and the repeat is duplicated. The CRISPR arrays are located adjacent to a cluster of *cas* genes; (**B**) crRNA generation. The CRISPRs are transcribed into pre-crRNAs that are then processed into mature crRNAs; (**C**) Interference. The crRNA, in a complex with Cas proteins, binds and degrades the target nucleic acid of the invading element.

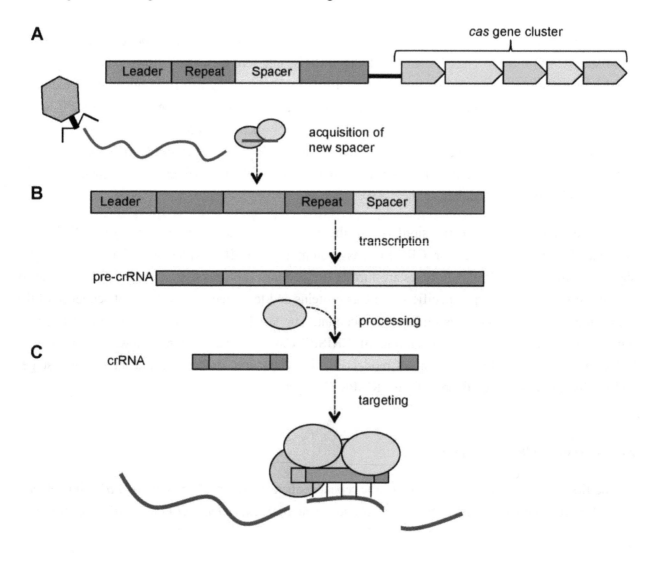

Terminology for these repeated elements was inconsistent and included interspaced short sequence repeats (SSR) [18], spacer interspersed direct repeats (SPIDR) [15] or short regularly spaced repeats (SRSRs) [16]. For simplicity, the name CRISPR (clustered regularly interspaced short palindromic repeats) was accepted since it reflects the most important characteristics [19].

An important step towards the understanding of the role of CRISPRs was the identification of four genes closely located to the CRISPR arrays [19], which were termed CRISPR-associated (*cas*) genes 1-4. Cas1-4 were only present in genomes containing CRISPR arrays and their predicted functions included helicase and nuclease activities, which led to the hypothesis that CRISPR/Cas systems might be involved in DNA repair [20].

A critical discovery came in 2005 with three independent studies showing that spacers matched sequences of extrachromosomal origin, including phages, prophages and plasmids [21–23]. Furthermore, there was a positive correlation between the presence of spacers matching a particular phage and phage resistance [22,23]. In combination with detailed bioinformatics analyses of the Cas proteins, it was hypothesized that CRISPR/Cas systems provide an RNAi-like mechanism of resistance against invading genetic elements [21–25].

The first direct evidence that CRISPR/Cas could protect bacteria from phages or plasmids was provided by two key studies. Firstly, Barrangou *et al.* obtained phage resistant mutants following phage challenge of *Streptococcus thermophilus*. The phage resistant strains had incorporated new spacers matching to the viral genome while removal or addition of spacers matching to the phages resulted in phage sensitivity or resistance, respectively [26]. Subsequently, Marraffini and Sontheimer showed that CRISPR/Cas systems can also prevent both conjugation and transformation of plasmids in *Staphylococcus epidermidis* [27]. Recently, it was shown that the *Streptococcus pneumoniae* CRISPR/Cas can prevent natural transformation. Non-capsulated *S. pneumoniae* with CRISPRs programmed to target capsule genes were unable to be transformed into capsulated strains during infection in mice [28].

3. Components of CRISPR/Cas Systems

CRISPR/Cas systems comprise the following critical elements: a CRISPR array, an upstream leader sequence and the *cas* genes (Figure 1A) that are discussed below [19,29].

3.1. CRISPR Arrays: Repeats, Spacers and Leader Sequence

In the CRISPR array, repeats alternate with spacer sequences [11,30–32]. Within arrays, the repeats are typically identical in terms of length and sequence, but small differences do occur [16,29]. In particular, the repeat at the end of an array is often truncated or deviates more from the consensus sequence [17,33]. Repeats are usually between 23 and 47 bp in length and those of the same subtype share a consensus sequence in type I-C, I-E, I-F, and II [10,34,35]. Many repeat sequences show palindromes or short inverted repeats [16] and a predicted stable secondary structure in the form of a stem-loop [34]. However, there are repeats, which are not palindromic and are predicted to be unstructured. This difference in the ability to form secondary structures also has implications for the mechanism of pre-crRNA processing, which will be discussed later [10,34].

In contrast, the spacer sequences are mostly unique within a genome. Many spacers have been matched to sequences originating from extra-chromosomal sources such as phage or plasmids and other transferable elements [21–23,36]. Indeed, it is the spacers that confer the sequence-specific immunity against those extra-chromosomal agents [26,33,37]. The sequences in the foreign genome from which spacers are derived are termed protospacers. It is likely that the majority of spacers are phage- or plasmid-derived, but the limited depth of sequence data on these abundant and diverse elements leads to an underestimation. Interestingly, some CRISPR/Cas systems also possess spacers that match sequences elsewhere within their own genome. Their function, if any, is not resolved, but it has been proposed that they have been accidentally incorporated [38]. Spacer length can vary slightly throughout an array (typically by 1–2 nt) and spacer lengths up to 72 bp have been reported, but usually the size is similar to that of the repeats in the same array [17].

The third component of the CRISPR array is the leader sequence, which is located upstream of the first repeat [15]. This AT-rich sequence is about 200–500 bp long and includes the promoter necessary for transcription of the array (see below) [39–42]. The leader region is also important for the acquisition of new spacers [43].

A single genome can harbor more than one CRISPR array. Those can vary considerably in size with the largest identified to date, in *Haliangium ochraceum* DSM 14365, containing 587 repeats [35]. The length of CRISPR arrays appears to correlate with the degree of activity in spacer acquisition, with longer arrays being more active than short arrays and those with degenerate repeat sequences [33].

3.2. Cas Proteins

In close proximity to the CRISPR array are genes that encode the **CRISPR as**sociated (Cas) proteins (Figure 1A) [19,20,24,25,44]. Cas proteins provide the enzymatic machinery required for the acquisition of new spacers from, and targeting, invading elements.

CRISPR/Cas systems are currently classified into type I, II and III, based on the phylogeny and presence of particular Cas proteins [10]. There is further division within each type into subtypes (e.g., type I is composed of type I-A to I-F). The Cas proteins are important for the differentiation between both the major CRISPR/Cas types and the subtypes. Two main groups of Cas proteins can be distinguished. The first group, which includes the core Cas proteins, is found across multiple types or subtypes. Cas1 and Cas2 (Cas2 is sometimes fused to another Cas protein) are found in every Cas operon across the three main types, and Cas1 is considered the universal marker of CRISPR/Cas systems [10,24,25,44]. Cas3, Cas9, and Cas10 are each specific for one major type, serving as signature proteins for type I, II, and III, respectively [10].

The second major group consists of proteins only found within the gene clusters of one particular subtype [10,25]. For example, the Cse and Csy proteins are subtype-specific for type I-E and type I-F systems, respectively. The subtype specific proteins of some systems form a complex involved in targeting and interference, which in the type I-E system is referred to as Cascade (CRISPR-associated complex for antiviral defense) [45–49]. These complexes will be discussed in detail below. The systems can be complex since some bacteria contain multiple CRISPR/Cas subtypes, each of which can have multiple CRISPR arrays that function with the appropriate Cas cluster [18]. In summary, the *cas* cluster for each CRISPR/Cas subtype contains genes encoding Cas1 and Cas2, the signature protein defining the major type and a set of subtype specific proteins. Interestingly, the total number of proteins differs between the types of system, even though the functional principles are similar in the cases studied thus far. It is possible that unexpected functions may reside within Cas proteins that are yet to be characterized.

4. CRISPR/Cas Mechanism

As highlighted in Figure 1, the mechanism of CRISPR/Cas defense involves three stages. Resistance must first be acquired by integrating spacers into the CRISPR arrays. These arrays are then transcribed and processed into short crRNAs. Finally, a crRNA-Cas ribonucleoprotein complex targets the invading nucleic acid for degradation. In the following sections we outline the processes of spacer acquisition, crRNA biogenesis and interference.

4.1. Adaptation via Spacer Acquisition

The least characterised stage of CRISPR/Cas defense is adaptation, whereby new spacers are acquired from phages and plasmids and, with duplication of the repeat, are added to the leader proximal end of the CRISPR array (Figure 2). A number of metagenomic studies of phage-bacterial dynamics in environmental [50,51] and human [52] niches have provided clear evidence that CRISPR/Cas systems actively acquire new spacers from mobile genetic elements over short time-scales. The first demonstration of spacer incorporation from phages and plasmids in the laboratory was in the type II-A system of *S. thermophilus* [26,53]. Very recently, studies have observed spacer acquisition in the *E. coli* type I-E [43,54,55], *P. aeruginosa* type I-F [56], *Streptococcus agalactiae* type II-A [57] and *Sulfolobus solfataricus* type I and III-B [58] systems. The most well characterized system is the *E. coli* type I-E, so this will be the focus of the following discussion.

4.1.1. Cas Proteins Required for Adaptation

In the *E. coli* type I-E system, Cas1 and Cas2 are required for spacer incorporation from plasmids and phages [43,55]. These recent results are consistent with the dispensable role of Cas1 and Cas2 in crRNA biogenesis / targeting in type I-E [45], type II-A [59] and type III-A [60] systems. The type I-E data is of broad relevance because Cas1 and Cas2 are conserved across all CRISPR/Cas types [10,24]. It is interesting that an insertion mutation of *csn2* inhibited spacer acquisition in the *S. thermophilus* type II-A system, suggesting that, in addition to Cas1 and Cas2, other Cas proteins might play a role [26].

Figure 2. Model of CRISPR/Cas adaptation (based on type I-E systems). (**A**) Naïve acquisition. Cas1 and Cas2 are required for acquisition of new spacers. The first repeat at the leader end of the CRISPR array is duplicated and incorporates a new spacer sequence from a protospacer. The final nucleotide of the repeat is not duplicated but is provided by the PAM nucleotide immediately adjacent to the protospacer; (**B**) Priming acquisition. Expression of the CRISPR array and generation of the crRNAs against the foreign DNA results in binding/targeting, which is hypothesized to aid in the generation of precursors for integration (possibly single- or double-stranded). Productive interference is not required for priming. Priming acquisition requires Cascade-crRNA, Cas3 and Cas1 and Cas2 and results in new spacers derived from the same strand as the initial spacer.

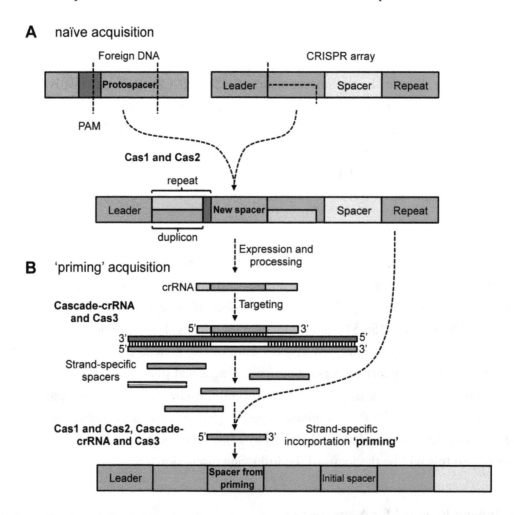

Exactly how Cas1 and Cas2 function is unclear, but biochemical and structural studies offer some clues [61–63]. Cas1 is a metal-dependent endonuclease and the *Pseudomonas* Cas1 cleaves dsDNA, generating ~80 bp fragments [61]. The 80 bp dsDNA fragments generated from Cas1 are larger than expected for integration (~32–33 nt), indicating that other factors are involved. Cas1 dimers contain a stirrup-like structure and a positively charged surface that together might be involved in binding dsDNA [9]. Cas2 from *S. solfataricus* cleaves ssRNA at U-rich regions *in vitro* [64], but ssRNA or ssDNA binding or cleavage was not detected for *Desulfovibrio vulgaris* Cas2 [65]. Recently, the *Bacillus halodurans* Cas2 dimer was shown to contain Mg^{2+}-dependent endonuclease activity against dsDNA and generated 120 bp products [66]. Despite these studies the exact mechanistic roles of Cas1 and Cas2 in acquisition are unknown.

There is increasing evidence that some Cas1 and Cas2 proteins may function together directly (e.g., via protein-protein interactions). For example, the type I-A system from *Thermoproteus tenax* contains a Cas1-Cas2 fusion protein, which interacts with Csa4 and Csa1, in what the authors name Cascis (CRISPR associated complex for the integration of spacers) [67]. It is noteworthy that the type I-F Cas3 is a Cas2-Cas3 hybrid, containing an N-terminal Cas2-like domain fused to Cas3 [10,24].

4.1.2. Protospacer Selection and Incorporation into CRISPR Arrays

The minimal CRISPR requirements for spacer acquisition in the type I-E system are a single 'repeat' and 60 bp of upstream sequence [43]. As the CRISPR promoter is not contained within this 60 bp, this suggests that CRISPR transcription is not required for incorporation, but indicates this region might be recognized by Cas1 and/or Cas2 to enable the directionality of integration at the leader end of the CRISPR array.

In type I-E arrays with multiple repeats, the leader proximal repeat is the one that is duplicated (Figure 2A) [43]. In addition, the protospacer added to the CRISPR array contains the last nt of the PAM motif, which becomes the final nt of the 5' repeat (Figure 2A) [54,55,68]. Therefore, only the first 28 nt of the repeat are duplicated and thus, this unit has been called the 'duplicon' [68] and might provide fidelity to the direction of incorporation [69]. This duplicon mechanism cannot apply to all CRISPR/Cas types since the last nucleotide of some PAMs (e.g., CCN in *S. solfataricus*) is not conserved, so could not contribute to the conserved repeats [58]. Interestingly, internal spacer acquisition was recently observed for one CRISPR array in *S. solfataricus* [58], which challenges the dogma that spacer acquisition is always leader proximal [69].

One critical question is how do the CRISPR/Cas systems acquire foreign DNA and not DNA from their own chromosome? In *S. thermophilus*, examination of the protospacers in the phage genomes enabled the identification of a short 3' protospacer adjacent motif (PAM) [37]. PAM sequences have since been identified via bioinformatics for other CRISPR/Cas systems [70]. In *E. coli*, spacer selection was shown to require the presence of a PAM [43], but the short length of PAMs cannot provide the level of discrimination to avoid sampling chromosomal DNA. A self / non-self mechanism does occur, since the incorporation of spacers from the chromosome is rare [43]. It is possible that DNA modification, such as by a restriction-modification-like mechanism, might play a role.

4.1.3. Priming Acquisition

In the type I-E system there is now evidence that acquisition is divided into two stages; (1) naïve- and (2) priming-based acquisition [54,55]. As discussed above naïve acquisition is the initial spacer incorporation from a new foreign element. During priming, the initial spacer 'primes' the array for the addition of multiple invader-derived spacers from the same DNA strand as the initial spacer (Figure 2B) [54,55]. Cas1, Cas2, Cascade, Cas3 and a crRNA that matches the invading element are required for priming [55]. The nuclease activity of Cas3 was proposed to be required for the feedback loop, generating precursors for incorporation following the initial targeting event [54]. However, spacers that match the foreign DNA but do not support targeting, due to single nucleotide mutations, still promote priming [55]. Priming is proposed to allow adaptation to phage and plasmids that mutate and escape the initial spacers [55] and multiple spacers would increase resistance and decrease the frequency of escape.

In summary, details of CRISPR adaptation are beginning to come to light, but there are still many questions regarding the exact mechanisms of protospacer recognition, spacer generation, repeat duplication, integration of the new spacer and the roles of the Cas proteins.

4.2. Expression, crRNA Generation and Interference

Following the acquisition of spacers, the CRISPR array is expressed as a long pre-crRNA that is processed into crRNAs. The mature crRNAs form part of a ribonucleoprotein complex, which targets and degrades the foreign genetic material. As will be outlined below, these processes are similar, yet differ between type I, II, and III systems (Figure 3).

4.2.1. Type I Systems

In type I CRISPR/Cas systems, regions within each repeat form a stem-loop, which is bound by the processing endoribonuclease. The endoribonucleases in type I systems belong to the group of Cas6 proteins. Characterized members of this group include Cas6b in type I-B [71], Cas5d in type I-C [49], Cas6e (CasE, Cse3) in type I-E [45,72,73] and Cas6f (Csy4) in type I-F [42,74]. The pre-crRNA is cleaved in a sequence-specific manner at the downstream base of the stem-loop, producing the mature crRNA consisting of a short 5' repeat handle followed by the spacer sequence and the stem-loop of the next repeat (Figure 3B) [49,71,74,75]. The crRNA and the endoribonuclease stay associated, and might serve as nucleation point [48,49,76] for the formation of Cascade (Figure 3C). Cascade is a ribonucleoprotein complex formed by the subtype-specific Cas proteins and the crRNA, initially identified in type I-E, but later also found in type I-C and I-F [9,45,46,48,49].

Figure 3. crRNA generation and target interference in type I, II and III CRISPR/Cas systems. (**A**) Transcription of the CRISPR array into a pre-crRNA; (**B**) Processing of the pre-crRNA into mature short crRNAs. In type I, RNA cleavage is performed by Cas6-homologues, which bind the repeat stem-loop and stay associated for Cascade formation. In type II, tracrRNA is required for binding and processing of the pre-crRNA by Cas9 and RNaseIII. In type III, Cas6 binds to non-structured repeats and processes the pre-crRNA into crRNA and then dissociates; (**C**) Target binding and cleavage. Type I Cascade binds the DNA target before recruiting Cas3 for degradation. In type II, Cas9 stays associated with the tracrRNA:crRNA complex after processing and subsequently binds and cleaves target DNA. The type III-B CMR-complex binds spacer sequence and targets RNA. It is hypothesized that a type III-A Csm complex forms and this system targets DNA.

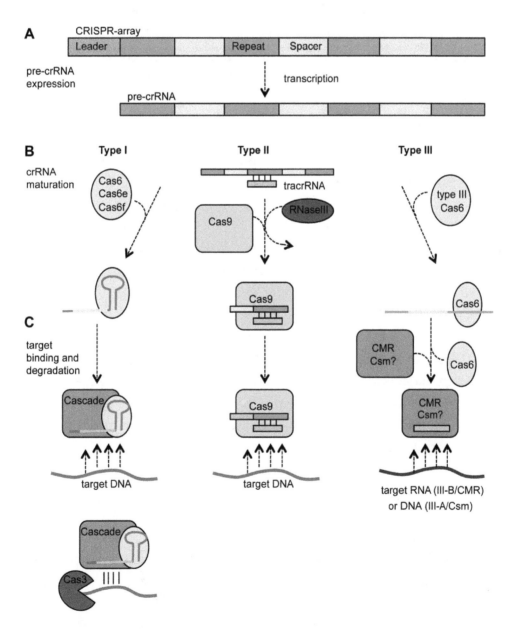

The type I-E Cascade has a shape referred to as seahorse-like and has a hexameric backbone formed by Cas7 (CasC, Cse4). The stem-loop of the crRNA is bound to the endoribonuclease Cas6e, which forms the "head" of the seahorse. Cse1 (CasA) forms the "tail" and binds the 3' end of the crRNA at the other end. Cse1 is bridged to Cas6e by a Cse2 (CasB) dimer, and linked by Cas5 (CasD) to the sixth Cas7 protein of the backbone [46,47]. The I-F and I-C complexes share common features with the I-E Cascade in terms of their shape, but also show variation, which is mainly due to different proteins involved. Different numbers of subtype-specific proteins constitute Cascade, three in the type I-C, four in I-F systems, while five distinct proteins are present in the type I-E Cascade [10]. Additional variation occurs at the level of the stoichiometry of individual protein subunits in different subtype Cascade complexes. When three or four proteins are used, the endoribonuclease, with the crRNA bound via the stem-loop, is located at one end of the complex. A homo-multimer of six proteins (Csy3 and Csd2 for I-F, and I-C, respectively) forms the neck or backbone of the complex, which holds the spacer and ensures it is extended for optimal presentation to invading nucleic acids. The end of the complex is formed by a hetero-dimer [48,49]. A type I-A complex from *S. solfataricus* (termed aCASCADE by the authors) has also been partially characterized and contained the backbone protein, Cas7 (Csa2), Cas5a, Csa5, Cas6 and crRNAs [77]. Although not directly shown, based on the data from other type I systems, the type I-A Cascade is likely to consist of the Cas6 endoribonuclease bound to the crRNA, a backbone of six Cas7 proteins and a Cas5a/Csa5 heterodimer.

During the final step of interference, the type I Cascade containing a specific crRNA recognizes and binds the appropriate target DNA (Figure 3C). In type I-E systems, a short loop within CasA recognizes potential targets via the PAM sequence and it is proposed to bind them before the target DNA is incorporated into Cascade [78]. In addition to the PAM requirement in the target, a non-contiguous 7 nt sequence (nucleotides in position 1–5 and 7–8) within the first 8 nt of the spacer, relative to the 5' end of the spacer within the crRNA, are essential for initial target recognition and hence are termed the 'seed sequence' [48,79]. Cascade with bound crRNA specifically binds to negatively supercoiled target DNA [80]. The negative supercoiled topology serves as energy source for formation of an R-loop. Furthermore, Cascade induces DNA bending before Cas3 is recruited for subsequent DNA degradation (Figure 3C) [80]. Cas3 is the signature protein of type I systems and possesses ATP-dependent helicase activity as well as the ability to cleave ssDNA via an HD nuclease domain [81–83]. Even though the exact mechanism is not yet fully understood, it is hypothesized that once recruited, Cas3 further unwinds and cleaves the target DNA in a 5' to 3' direction, while Cascade dissociates from the target and is recycled [80,83].

4.2.2. Type II Systems

A very different mechanism of CRISPR expression and crRNA maturation was discovered in the type II-A system of *Streptococcus pyogenes* [84]. An abundant short *trans*-encoded transcript is essential for processing of pre-crRNA into mature crRNA. Therefore, these RNAs are termed *trans*-activating crRNAs (tracrRNA). These tracrRNAs contain a 25 bp stretch with almost perfect complementarity to the type II CRISPR repeats within the pre-crRNA (Figure 3B). The type II repeats do not form stem-loops and the authors suggest this deficiency is overcome by pairing with the tracrRNA [84].

In addition to the tracrRNA, Cas9 (Csn1) and host RNase III are needed for the processing of pre-crRNA into mature crRNA. The requirement of RNase III is the first discovery of a non-Cas protein being an essential part of CRISPR/Cas defense machinery (Figure 3B) [84]. Interestingly, the *S. thermophilus* type II-A system provides phage resistance in *E. coli* when recombinantly expressed [59], which suggests diverse RNase III enzymes are compatible for crRNA generation. Cas9 is not only involved in crRNA processing but, along with the tracrRNA:crRNA hybrid, recognizes and degrades the target (Figure 3C) [85]. In complex with the tracrRNA:crRNA structure, Cas9 binds the target dsDNA and creates a double strand break by cleaving the complementary and non-complementary strand with its HNH- and RuvC-like domains, respectively [85]. *In vivo*, blunt cleavage of phage and plasmid dsDNA was observed at a position 3 bp from the PAM within the protospacer [53]. The PAM is important for the affinity towards the target and may have implications on the unwinding of the duplex, as cleavage of targets with mutated PAMs is only impaired in dsDNA but not ssDNA. Indeed, phages with PAM mutations also have the ability to avoid targeting by type II systems [37]. Furthermore, similar to the seed sequence in type I systems, complementarity between crRNA and target over a 13 bp stretch proximal to the PAM is required for interference [85] and hence, phages with mutations in this region of the protospacer region can evade interference [37].

4.2.3. Type III Systems

In type III CRISPR/Cas systems, expression and interference are predominantly similar to those described for type I systems. However, the significant difference is that the endoribonuclease Cas6 does not bind to a repeat stem-loop but to the first bases of the repeat and that further crRNA maturation steps occur (Figure 3B) [86,87]. Indeed, not all repeats in those systems contain a strong palindrome and those associated with type III-A are predicted to be mostly non-structured [34]. In *Pyrococcus furiosus*, the binding site of Cas6 is located within nt 2–8 of the repeat and cleaves at a distal site between bases 22 and 23 resulting in crRNAs with an 8 nt repeat handle on the 5' end, the spacer sequence and 22 nt of the following repeat at the 3' end [86–88]. Interestingly, this RNA is not the final mature crRNA product. Cas6, unlike the type I endoribonucleases, only transfers the crRNA to the targeting complex termed CMR for type III-B where they are further processed (Figure 3C) [10,89,90]. In the type III-A system of *S. epidermidis*, a similar mechanism for processing by Cas6 was observed and the *csm2*, *csm3* and *csm5* genes were required for further 3' processing and crRNA maturation [60].

The *P. furiosus* CMR complex includes the full subset of Cmr proteins (Cmr1–6) and can contain two different species of crRNA [90], both with the 8 nt 5' end derived from the repeat and either 31 nt or 37 nt of the spacer [91]. The type III-B CMR complex from *S. solfataricus* contains an additional protein, Cmr7 and a stoichiometry of one protein of each of Cmr1–6 and 6 Cmr7 proteins [92]. Interestingly, the *P. furiosus* CMR complex has been shown to target RNA [89,90,92]. Basepairing between the last 14 nt of the crRNA and the RNA target seem to be essential for cleavage [90]. Target RNA cleavage occurs 14 nt upstream from the 3' end of the two mature crRNA species, generating two cleavage sites [90]. In contrast to type III-B, DNA is the target of type III-A systems [27], but there is not yet data about complex formation for the Csm proteins.

The ability of CRISPR/Cas systems to target DNA raised the question of how they avoid targeting their own CRISPR arrays, which have perfect complementarity to the crRNAs they produce. For the type III-A system, this is avoided by requiring spacer:protospacer complementarity and an absence of base-pairing to the 5' handle in the crRNA [93]. This ensures that only non-self targets are licensed for degradation, but whether the same principle applies to other types remains unknown.

5. Regulation of CRISPR/Cas Systems

As CRISPR/Cas systems function as a defense mechanism their expression might be expected to respond to invasion by extrachromosomal elements. This was recently proven in a shotgun proteomics approach in *S. thermophilus*, which showed an increase in Cas protein expression following phage infection [94]. Likewise, CRISPR and Cas expression were increased by phage infection in *Thermus thermophilus* [39]. Furthermore, CRISPR/Cas systems can be modulated in response to UV and play a role in sensitivity to DNA damage, hinting towards other possible roles besides the neutralization of invading elements [63,67]. However, to date little is known about how CRISPR/Cas systems are regulated in response to those external stimuli or during periods in which the system is not required [63,67]. Under certain conditions it could be viewed as favorable to down-regulate CRISPR/Cas activity when beneficial elements are available. However, most evidence demonstrates up-regulation of CRISPR/Cas upon exposure to phage. Since CRISPR/Cas systems are not 100% effective, some beneficial elements will be acquired and when they provide a selective advantage they are likely to be maintained. An overview of the regulatory inputs is shown in Figure 4.

Figure 4. Regulation of CRISPR/Cas activity. Network model summarizing the regulation of CRISPR/Cas activity. Triangular and flat arrowheads indicate positive and negative effects on CRISPR/Cas activity, respectively for *E. coli* (red), *Salmonella* (cyan), *M. xanthus* (green), *Sulfolobus* (black), *T. thermophilus* (purple) and *S. thermophilus* (blue). For details see text.

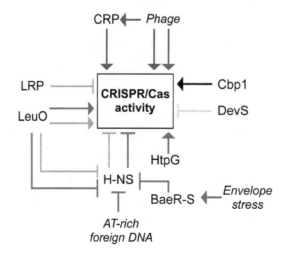

Most information about CRISPR/Cas regulation is available for the type I-E systems in *E. coli* and *Salmonella enterica* serovar Typhi backgrounds. The Histone-like nucleoid structuring protein (H-NS) is a global regulator involved in compacting the bacterial chromosome, via high affinity binding to AT-rich, curved DNA (reviewed by Dorman *et al.* [95]). Many promoters are located in close proximity to curved DNA. Binding of H-NS to these promoters prevents RNA polymerase (RNAP) binding, which results in gene-silencing [95]. In *E. coli*, H-NS negatively regulates *cas* gene expression, due to H-NS binding sites near the *cas* operon promoters [41,96]. It is hypothesized that H-NS will bind to invading nucleic acids when they enter the cell, due to the higher AT-content [97,98]. Sequestration of H-NS is predicted to free the *cas* and/or LeuO promoter (see below) for RNAP recognition and thus activate expression, allowing effective CRISPR/Cas-mediated defense [41,96,99].

LRP (lysine-responsive regulatory protein) is a negative regulator for *cas* expression in *S. typhi* [100]. LRP functions in a similar fashion to H-NS by binding to the *cas* promoter and competitively excluding binding of RNAP. This appears to be independent of H-NS. However, LRP is able to bind simultaneously alongside H-NS, indicating that the two proteins interact to generate a nucleosome structure for *cas* gene repression [100]. Interestingly, LRP does not regulate the *cas* operon in *E. coli*, suggesting that the type I-E CRISPR/Cas systems in these two closely-related organisms function differently at the regulatory level [41,100].

LeuO, a LysR-type transcriptional regulator that also responds to amino acid starvation, affects *cas* gene expression via competition with H-NS in *S.* Typhi and *E. coli* [96,100,101]. LeuO binds to a region flanking the *cas* promoter and the H-NS binding site. The binding of LeuO to the DNA competes with H-NS binding, and relieves the inhibition of *cas* expression by enabling promoter recognition by RNAP [96,100,101]. During amino acid starvation expression of LeuO is increased in response to accumulation of the small molecules guanosine 3'-diphosphate 5'-triphosphate and guanosine 3',5'-bis(diphosphate), collectively called (p)ppGpp [102]. Interestingly, accumulation of (p)ppGpp does not occur during phage lambda infection in *E. coli* [103]. In theory, it is possible that infection with other phages triggers amino acid starvation, leading to LeuO-dependent activation of the CRISPR/Cas system.

The BaeR-S two-component regulatory system is activated in response to envelope stress (reviewed by MacRitchie *et al.* [104]). Phage infection potentially causes envelope stress by accumulation of viral proteins in the membrane. Upon detection of membrane stress, the histidine sensor kinase BaeS is activated by phosphorylation and in turn activates the BaeR protein. Activated BaeR gains the ability to bind DNA, modulating gene expression [104]. In *E. coli*, the BaeR binding site is located at the H-NS binding site near the *cse1* (*casA*) promoter. Therefore, bound BaeR serves as an antagonist for H-NS binding, predicted to result in the release of the *cas* promoters for RNA polymerase recognition and *cas* expression [105,106].

The high-temperature protein G (HtpG) chaperone affects the CRISPR/Cas system in *E. coli* by stabilizing Cas3. If HtpG is absent, the CRISPR/Cas system is no longer able to efficiently prevent phage infection [107]. However, HtpG is not present in all bacteria containing CRISPR/Cas systems. Indeed, most archaeal strains that contain Cas3 homologues do not carry an HtpG homolog, suggesting that it is not universally required. Therefore, other bacteria may require other chaperones in order for CRISPR/Cas interference to occur [108].

Other systems that affect CRISPR/Cas regulation were identified in *Myxococcus xanthus*, *T. thermophilus* and Sulfolobales. In *Myxococcus*, the *dev* operon is co-transcribed with *cas* genes, and this operon is negatively auto-regulated by DevS (a Cas5 protein) [109]. *T. thermophilus* contains 12 CRISPRs, and 3 different *cas* operons, type I-E, type III-A and type III-B [10,39]. During phage infection, the *cas* genes for the type I-E and type III-A systems were up-regulated via CRP (cAMP receptor protein) [39,110]. When CRP is bound to cAMP, it recognizes promoters and activates gene expression. Interestingly, inactivation of *crp* did not abolish the *cas* activation, suggesting both *crp*-dependent and -independent pathways. In addition, some *cas* genes and CRISPR arrays were regulated by phage infection in a CRP-independent manner [39]. In *Sulfolobus islandicus* and *S. solfataricus*, the expression of pre-crRNA is regulated by Cbp1 (CRISPR DNA repeat binding protein), which directly binds the DNA repeats and might influence the activity of promoters and terminators present within some spacers or repeats [111,112]. In summary, CRISPR/Cas systems can be regulated at the level of transcription of the *cas* genes and CRISPR arrays and post-translational level on the Cas proteins (Figure 4). The picture of how these systems are regulated is far from complete and diverse and complex regulatory strategies exist.

6. Conclusions

CRISPR/Cas systems are widespread and versatile prokaryotic defense mechanisms providing adaptive and heritable immunity to extrachromosomal elements. In recent years research performed on these systems has led to a deeper understanding of the underlying mechanistic principles. Despite this rapid increase in knowledge, many questions remain to be answered. For example, challenges in the future include elucidating why some prokaryotes carry multiple CRISPR/Cas types and how their activity is coordinated. Furthermore, the adaptation step is still poorly understood, especially with respect to the specific roles of the Cas proteins involved and how foreign DNA is selected for integration. Major progress has been made regarding structure, function and interactions of some Cas proteins and their complexes. However, for many subtype-specific proteins, functional and structural information is still lacking and their precise role in the CRISPR/Cas mechanism remains unknown. Finally, the question, how CRISPR/Cas activity is regulated and to what extend it plays a role in areas not related to defense, e.g., DNA repair, needs to be addressed. Future research on CRISPR/Cas systems will also open up vast opportunities to utilize them, such as genome engineering for the creation of more robust industrial starter strains or as a gene targeting or gene silencing mechanism similar to that used in eukaryotic cells.

Acknowledgments

We thank Rita Przybilski for critically reading the manuscript and the referees for their constructive input. This work was supported by a University of Otago Research Grant and a Rutherford Discovery

Fellowship to PCF from the Royal Society of New Zealand. CR was supported by a University of Otago Postgraduate Scholarship and a DAAD Doktorandenstipendium.

References

1. Frost, L.S.; Leplae, R.; Summers, A.O.; Toussaint, A. Mobile genetic elements: The agents of open source evolution. *Nat. Rev. Microbiol.* **2005**, *3*, 722–732.

2. Koonin, E.V.; Wolf, Y.I. Genomics of bacteria and archaea: The emerging dynamic view of the prokaryotic world. *Nucleic Acids Res.* **2008**, *36*, 6688–6719.

3. Fuhrman, J.A. Marine viruses and their biogeochemical and ecological effects. *Nature* **1999**, *399*, 541–548.

4. Hendrix, R.W. Bacteriophage genomics. *Curr. Opin. Microbiol.* **2003**, *6*, 506–511.

5. Labrie, S.J.; Samson, J.E.; Moineau, S. Bacteriophage resistance mechanisms. *Nat. Rev. Microbiol.* **2010**, *8*, 317–327.

6. Petty, N.K.; Evans, T.J.; Fineran, P.C.; Salmond, G.P. Biotechnological exploitation of bacteriophage research. *Trends Biotechnol.* **2007**, *25*, 7–15.

7. Marraffini, L.A.; Sontheimer, E.J. CRISPR interference: RNA-directed adaptive immunity in bacteria and archaea. *Nat. Rev. Genet.* **2010**, *11*, 181–190.

8. Terns, M.P.; Terns, R.M. CRISPR-based adaptive immune systems. *Curr. Opin. Microbiol.* **2011**, *14*, 321–327.

9. Wiedenheft, B.; Sternberg, S.H.; Doudna, J.A. RNA-guided genetic silencing systems in bacteria and archaea. *Nature* **2012**, *482*, 331–338.

10. Makarova, K.S.; Haft, D.H.; Barrangou, R.; Brouns, S.J.; Charpentier, E.; Horvath, P.; Moineau, S.; Mojica, F.J.; Wolf, Y.I.; Yakunin, A.F.; *et al.* Evolution and classification of the CRISPR-Cas systems. *Nat. Rev. Microbiol.* **2011**, *9*, 467–477.

11. Ishino, Y.; Shinagawa, H.; Makino, K.; Amemura, M.; Nakata, A. Nucleotide sequence of the *iap* gene, responsible for alkaline phosphatase isozyme conversion in *Escherichia coli*, and identification of the gene product. *J. Bacteriol.* **1987**, *169*, 5429–5433.

12. Groenen, P.M.; Bunschoten, A.E.; van Soolingen, D.; van Embden, J.D. Nature of DNA polymorphism in the direct repeat cluster of *Mycobacterium tuberculosis*; application for strain differentiation by a novel typing method. *Mol. Microbiol.* **1993**, *10*, 1057–1065.

13. Mojica, F.J.; Ferrer, C.; Juez, G.; Rodríguez-Valera, F. Long stretches of short tandem repeats are present in the largest replicons of the Archaea *Haloferax mediterranei* and *Haloferax volcanii* and could be involved in replicon partitioning. *Mol. Microbiol.* **1995**, *17*, 85–93.

14. Klenk, H.P.; Clayton, R.A.; Tomb, J.F.; White, O.; Nelson, K.E.; Ketchum, K.A.; Dodson, R.J.; Gwinn, M.; Hickey, E.K.; Peterson, J.D.; *et al.* The complete genome sequence of the hyperthermophilic, sulphate-reducing archaeon *Archaeoglobus fulgidus*. *Nature* **1997**, *390*, 364–370.

15. Jansen, R.; van Embden, J.D.; Gaastra, W.; Schouls, L.M. Identification of a novel family of sequence repeats among prokaryotes. *OMICS* **2002**, *6*, 23–33.

16. Mojica, F.J.; Díez-Villaseñor, C.; Soria, E.; Juez, G. Biological significance of a family of regularly spaced repeats in the genomes of Archaea, Bacteria and mitochondria. *Mol. Microbiol.* **2000**, *36*, 244–246.

17. Grissa, I.; Vergnaud, G.; Pourcel, C. The CRISPRdb database and tools to display CRISPRs and to generate dictionaries of spacers and repeats. *BMC Bioinformatics* **2007**, *8*, 172.

18. van Belkum, A.; Scherer, S.; van Alphen, L.; Verbrugh, H. Short-sequence DNA repeats in prokaryotic genomes. *Microbiol. Mol. Biol. Rev.* **1998**, *62*, 275–293.

19. Jansen, R.; Embden, J.D.; Gaastra, W.; Schouls, L.M. Identification of genes that are associated with DNA repeats in prokaryotes. *Mol. Microbiol.* **2002**, *43*, 1565–1575.

20. Makarova, K.S.; Aravind, L.; Grishin, N.V.; Rogozin, I.B.; Koonin, E.V. A DNA repair system specific for thermophilic Archaea and bacteria predicted by genomic context analysis. *Nucleic Acids Res.* **2002**, *30*, 482–496.

21. Mojica, F.J.; Díez-Villaseñor, C.; García-Martínez, J.; Soria, E. Intervening sequences of regularly spaced prokaryotic repeats derive from foreign genetic elements. *J. Mol. Evol.* **2005**, *60*, 174–182.

22. Bolotin, A.; Quinquis, B.; Sorokin, A.; Ehrlich, S.D. Clustered regularly interspaced short palindrome repeats (CRISPRs) have spacers of extrachromosomal origin. *Microbiology* **2005**, *151*, 2551–2561.

23. Pourcel, C.; Salvignol, G.; Vergnaud, G. CRISPR elements in *Yersinia pestis* acquire new repeats by preferential uptake of bacteriophage DNA, and provide additional tools for evolutionary studies. *Microbiology* **2005**, *151*, 653–663.

24. Makarova, K.S.; Grishin, N.V.; Shabalina, S.A.; Wolf, Y.I.; Koonin, E.V. A putative RNA-interference-based immune system in prokaryotes: Computational analysis of the predicted enzymatic machinery, functional analogies with eukaryotic RNAi, and hypothetical mechanisms of action. *Biol. Direct* **2006**, *1*, doi:10.1186/1745-6150-1-7.

25. Haft, D.H.; Selengut, J.; Mongodin, E.F.; Nelson, K.E. A guild of 45 CRISPR-associated (Cas) protein families and multiple CRISPR/Cas subtypes exist in prokaryotic genomes. *PLoS Comput. Biol.* **2005**, *1*, e60.

26. Barrangou, R.; Fremaux, C.; Deveau, H.; Richards, M.; Boyaval, P.; Moineau, S.; Romero, D.A.; Horvath, P. CRISPR provides acquired resistance against viruses in prokaryotes. *Science* **2007**, *315*, 1709–1712.

27. Marraffini, L.A.; Sontheimer, E.J. CRISPR interference limits horizontal gene transfer in *Staphylococci* by targeting DNA. *Science* **2008**, *322*, 1843–1845.

28. Bikard, D.; Hatoum-Aslan, A.; Mucida, D.; Marraffini, L.A. CRISPR Interference Can Prevent Natural Transformation and Virulence Acquisition during *In Vivo* Bacterial Infection. *Cell Host Microbe* **2012**, *12*, 177–186.

29. Lillestøl, R.K.; Redder, P.; Garrett, R.A.; Brugger, K. A putative viral defence mechanism in archaeal cells. *Archaea* **2006**, *2*, 59–72.

30. Nakata, A.; Amemura, M.; Makino, K. Unusual nucleotide arrangement with repeated sequences in the *Escherichia coli* K-12 chromosome. *J. Bacteriol.* **1989**, *171*, 3553–3556.

31. She, Q.; Singh, R.K.; Confalonieri, F.; Zivanovic, Y.; Allard, G.; Awayez, M.J.; Chan-Weiher, C.C.; Clausen, I.G.; Curtis, B.A.; De Moors, A.; *et al.* The complete genome of the crenarchaeon *Sulfolobus solfataricus* P2. *Proc. Natl. Acad. Sci. USA* **2001**, *98*, 7835–7840.

32. Van Embden, J.D.; van Gorkom, T.; Kremer, K.; Jansen, R.; van der Zeijst, B.A.; Schouls, L.M. Genetic variation and evolutionary origin of the direct repeat locus of *Mycobacterium tuberculosis* complex bacteria. *J. Bacteriol.* **2000**, *182*, 2393–2401.

33. Horvath, P.; Romero, D.A.; Côuté-Monvoisin, A.C.; Richards, M.; Deveau, H.; Moineau, S.; Boyaval, P.; Fremaux, C.; Barrangou, R. Diversity, activity, and evolution of CRISPR loci in *Streptococcus thermophilus*. *J. Bacteriol.* **2008**, *190*, 1401–1412.

34. Kunin, V.; Sorek, R.; Hugenholtz, P. Evolutionary conservation of sequence and secondary structures in CRISPR repeats. *Genome Biol.* **2007**, *8*, doi:10.1186/gb-2007-8-4-r61.

35. Grissa, I.; Vergnaud, G.; Pourcel, C. CRISPRFinder: A web tool to identify clustered regularly interspaced short palindromic repeats. *Nucleic Acids Res.* **2007**, *35*, W52–W57.

36. Brodt, A.; Lurie-Weinberger, M.N.; Gophna, U. CRISPR loci reveal networks of gene exchange in archaea. *Biol. Direct* **2011**, *6*, doi:10.1186/1745-6150-6-65.

37. Deveau, H.; Barrangou, R.; Garneau, J.E.; Labonté, J.; Fremaux, C.; Boyaval, P.; Romero, D.A.; Horvath, P.; Moineau, S. Phage response to CRISPR-encoded resistance in *Streptococcus thermophilus*. *J. Bacteriol.* **2008**, *190*, 1390–1400.

38. Stern, A.; Keren, L.; Wurtzel, O.; Amitai, G.; Sorek, R. Self-targeting by CRISPR: Gene regulation or autoimmunity? *Trends Genet.* **2010**, *26*, 335–340.

39. Agari, Y.; Sakamoto, K.; Tamakoshi, M.; Oshima, T.; Kuramitsu, S.; Shinkai, A. Transcription profile of *Thermus thermophilus* CRISPR systems after phage infection. *J. Mol. Biol.* **2009**, *395*, 270–281.

40. Pougach, K.; Semenova, E.; Bogdanova, E.; Datsenko, K.A.; Djordjevic, M.; Wanner, B.L.; Severinov, K. Transcription, processing and function of CRISPR cassettes in *Escherichia coli*. *Mol. Microbiol.* **2010**, *77*, 1367–1379.

41. Pul, Ü.; Wurm, R.; Arslan, Z.; Geissen, R.; Hofmann, N.; Wagner, R. Identification and characterization of *E. coli* CRISPR-cas promoters and their silencing by H-NS. *Mol. Microbiol.* **2010**, *75*, 1495–1512.

42. Przybilski, R.; Richter, C.; Gristwood, T.; Clulow, J.S.; Vercoe, R.B.; Fineran, P.C. Csy4 is responsible for CRISPR RNA processing in *Pectobacterium atrosepticum*. *RNA Biol.* **2011**, *8*, 517–528.

43. Yosef, I.; Goren, M.G.; Qimron, U. Proteins and DNA elements essential for the CRISPR adaptation process in *Escherichia coli*. *Nucleic Acids Res.* **2012**, *40*, 5569–5576.

44. Godde, J.S.; Bickerton, A. The repetitive DNA elements called CRISPRs and their associated genes: Evidence of horizontal transfer among prokaryotes. *J. Mol. Evol.* **2006**, *62*, 718–729.

45. Brouns, S.J.; Jore, M.M.; Lundgren, M.; Westra, E.R.; Slijkhuis, R.J.; Snijders, A.P.; Dickman, M.J.; Makarova, K.S.; Koonin, E.V.; van der Oost, J. Small CRISPR RNAs guide antiviral defense in prokaryotes. *Science* **2008**, *321*, 960–964.

46. Jore, M.M.; Lundgren, M.; van Duijn, E.; Bultema, J.B.; Westra, E.R.; Waghmare, S.P.; Wiedenheft, B.; Pul, Ü.; Wurm, R.; Wagner, R.; *et al.* Structural basis for CRISPR RNA-guided DNA recognition by Cascade. *Nat. Struct. Mol. Biol.* **2011**, *18*, 529–536.

47. Wiedenheft, B.; Lander, G.C.; Zhou, K.; Jore, M.M.; Brouns, S.J.; van der Oost, J.; Doudna, J.A.; Nogales, E. Structures of the RNA-guided surveillance complex from a bacterial immune system. *Nature* **2011**, *477*, 486–489.

48. Wiedenheft, B.; van Duijn, E.; Bultema, J.B.; Waghmare, S.P.; Zhou, K.; Barendregt, A.; Westphal, W.; Heck, A.J.; Boekema, E.J.; Dickman, M.J.; *et al*. RNA-guided complex from a bacterial immune system enhances target recognition through seed sequence interactions. *Proc. Natl. Acad. Sci. USA* **2011**, *108*, 10092–10097.

49. Nam, K.H.; Haitjema, C.; Liu, X.; Ding, F.; Wang, H.; DeLisa, M.P.; Ke, A. Cas5d Protein Processes Pre-crRNA and Assembles into a Cascade-like Interference Complex in Subtype I-C/Dvulg CRISPR-Cas System. *Structure* **2012**, *20*, 1574–1584.

50. Andersson, A.F.; Banfield, J.F. Virus population dynamics and acquired virus resistance in natural microbial communities. *Science* **2008**, *320*, 1047–1050.

51. Tyson, G.W.; Banfield, J.F. Rapidly evolving CRISPRs implicated in acquired resistance of microorganisms to viruses. *Environ. Microbiol.* **2008**, *10*, 200–207.

52. Pride, D.T.; Salzman, J.; Haynes, M.; Rohwer, F.; Davis-Long, C.; White, R.A., 3rd.; Loomer, P.; Armitage, G.C.; Relman, D.A. Evidence of a robust resident bacteriophage population revealed through analysis of the human salivary virome. *ISME J.* **2012**, *6*, 915–926.

53. Garneau, J.E.; Dupuis, M.E.; Villion, M.; Romero, D.A.; Barrangou, R.; Boyaval, P.; Fremaux, C.; Horvath, P.; Magadan, A.H.; Moineau, S. The CRISPR/Cas bacterial immune system cleaves bacteriophage and plasmid DNA. *Nature* **2010**, *468*, 67–71.

54. Swarts, D.C.; Mosterd, C.; van Passel, M.W.; Brouns, S.J. CRISPR interference directs strand specific spacer acquisition. *PLoS One* **2012**, *7*, e35888.

55. Datsenko, K.A.; Pougach, K.; Tikhonov, A.; Wanner, B.L.; Severinov, K.; Semenova, E. Molecular memory of prior infections activates the CRISPR/Cas adaptive bacterial immunity system. *Nat. Commun.* **2012**, *3*, doi:10.1038/ncomms1937.

56. Cady, K.C.; Bondy-Denomy, J.; Heussler, G.E.; Davidson, A.R.; O'Toole, G.A. The CRISPR/Cas Adaptive Immune System of *Pseudomonas aeruginosa* Mediates Resistance to Naturally Occurring and Engineered Phages. *J. Bacteriol.* **2012**, *194*, 5728–5738.

57. Lopez-Sanchez, M.J.; Sauvage, E.; Da Cunha, V.; Clermont, D.; Ratsima Hariniaina, E.; Gonzalez-Zorn, B.; Poyart, C.; Rosinski-Chupin, I.; Glaser, P. The highly dynamic CRISPR1 system of *Streptococcus agalactiae* controls the diversity of its mobilome. *Mol. Microbiol.* **2012**, *85*, 1057–1071.

58. Erdmann, S.; Garrett, R.A. Selective and hyperactive uptake of foreign DNA by adaptive immune systems of an archaeon via two distinct mechanisms. *Mol. Microbiol.* **2012**, *85* 1044–1056.

59. Sapranauskas, R.; Gasiunas, G.; Fremaux, C.; Barrangou, R.; Horvath, P.; Siksnys, V. The *Streptococcus thermophilus* CRISPR/Cas system provides immunity in *Escherichia coli*. *Nucleic Acids Res.* **2011**, *39*, 9275–9282.

60. Hatoum-Aslan, A.; Maniv, I.; Marraffini, L.A. Mature clustered, regularly interspaced, short palindromic repeats RNA (crRNA) length is measured by a ruler mechanism anchored at the precursor processing site. *Proc. Natl. Acad. Sci. USA* **2011**, *108*, 21218–21222.

61. Wiedenheft, B.; Zhou, K.; Jinek, M.; Coyle, S.M.; Ma, W.; Doudna, J.A. Structural basis for DNase activity of a conserved protein implicated in CRISPR-mediated genome defense. *Structure* **2009**, *17*, 904–912.

62. Han, D.; Lehmann, K.; Krauss, G. SSO1450—A CAS1 protein from *Sulfolobus solfataricus* P2 with high affinity for RNA and DNA. *FEBS Lett.* **2009**, *583*, 1928–1932.

63. Babu, M.; Beloglazova, N.; Flick, R.; Graham, C.; Skarina, T.; Nocek, B.; Gagarinova, A.; Pogoutse, O.; Brown, G.; Binkowski, A.; *et al.* A dual function of the CRISPR-Cas system in bacterial antivirus immunity and DNA repair. *Mol. Microbiol.* **2011**, *79*, 484–502.

64. Beloglazova, N.; Brown, G.; Zimmerman, M.D.; Proudfoot, M.; Makarova, K.S.; Kudritska, M.; Kochinyan, S.; Wang, S.; Chruszcz, M.; Minor, W.; *et al.* A novel family of sequence-specific endoribonucleases associated with the clustered regularly interspaced short palindromic repeats. *J. Biol. Chem.* **2008**, *283*, 20361–20371.

65. Samai, P.; Smith, P.; Shuman, S. Structure of a CRISPR-associated protein Cas2 from *Desulfovibrio vulgaris. Acta Crystallogr. F Struct. Biol. Cryst. Commun.* **2010**, *66*, 1552–1556.

66. Nam, K.H.; Ding, F.; Haitjema, C.; Huang, Q.; Delisa, M.P.; Ke, A. Double-stranded Endonuclease Activity in B. halodurans Clustered Regularly Interspaced Short Palindromic Repeats (CRISPR)-associated Cas2 Protein. *J. Biol. Chem.* **2012**, Epub ahead of print.

67. Plagens, A.; Tjaden, B.; Hagemann, A.; Randau, L.; Hensel, R. Characterization of the CRISPR/Cas subtype I-A system of the hyperthermophilic crenarchaeon *Thermoproteus tenax. J. Bacteriol.* **2012**, *194*, 2491–2500.

68. Goren, M.G.; Yosef, I.; Auster, O.; Qimron, U. Experimental Definition of a Clustered Regularly Interspaced Short Palindromic Duplicon in *Escherichia coli. J. Mol. Biol.* **2012**, *423*, 14–16.

69. Westra, E.R.; Brouns, S.J. The rise and fall of CRISPRs—Dynamics of spacer acquisition and loss. *Mol. Microbiol.* **2012**, *85*, 1021–1025.

70. Mojica, F.J.; Diez-Villasenor, C.; Garcia-Martinez, J.; Almendros, C. Short motif sequences determine the targets of the prokaryotic CRISPR defence system. *Microbiology* **2009**, *155*, 733–740.

71. Richter, H.; Zoephel, J.; Schermuly, J.; Maticzka, D.; Backofen, R.; Randau, L. Characterization of CRISPR RNA processing in *Clostridium thermocellum* and *Methanococcus maripaludis. Nucleic Acids Res.* **2012**, Epub ahead of print.

72. Sashital, D.G.; Jinek, M.; Doudna, J.A. An RNA-induced conformational change required for CRISPR RNA cleavage by the endoribonuclease Cse3. *Nat. Struct. Mol. Biol.* **2011**, *18*, 680–687.

73. Gesner, E.M.; Schellenberg, M.J.; Garside, E.L.; George, M.M.; MacMillan, A.M. Recognition and maturation of effector RNAs in a CRISPR interference pathway. *Nat. Struct. Mol. Biol.* **2011**, *18*, 688–692.

74. Haurwitz, R.E.; Jinek, M.; Wiedenheft, B.; Zhou, K.; Doudna, J.A. Sequence- and structure-specific RNA processing by a CRISPR endonuclease. *Science* **2010**, *329*, 1355–1358.

75. Sternberg, S.H.; Haurwitz, R.E.; Doudna, J.A. Mechanism of substrate selection by a highly specific CRISPR endoribonuclease. *RNA* **2012**, *18*, 661–672.

76. Haurwitz, R.E.; Sternberg, S.H.; Doudna, J.A. Csy4 relies on an unusual catalytic dyad to position and cleave CRISPR RNA. *EMBO J.* **2012**, *31*, 2824–2832.

77. Lintner, N.G.; Kerou, M.; Brumfield, S.K.; Graham, S.; Liu, H.; Naismith, J.H.; Sdano, M.; Peng, N.; She, Q.; Copie, V.; *et al.* Structural and Functional Characterization of an Archaeal Clustered

Regularly Interspaced Short Palindromic Repeat (CRISPR)-associated Complex for Antiviral Defense (CASCADE). *J. Biol. Chem.* **2011**, *286*, 21643–21656.

78. Sashital, D.G.; Wiedenheft, B.; Doudna, J.A. Mechanism of foreign DNA selection in a bacterial adaptive immune system. *Mol. Cell.* **2012**, *46*, 606–615.

79. Semenova, E.; Jore, M.M.; Datsenko, K.A.; Semenova, A.; Westra, E.R.; Wanner, B.; van der Oost, J.; Brouns, S.J.; Severinov, K. Interference by clustered regularly interspaced short palindromic repeat (CRISPR) RNA is governed by a seed sequence. *Proc. Natl. Acad. Sci. USA* **2011**, *108*, 10098–10103.

80. Westra, E.R.; van Erp, P.B.; Kunne, T.; Wong, S.P.; Staals, R.H.; Seegers, C.L.; Bollen, S.; Jore, M.M.; Semenova, E.; Severinov, K.; *et al.* CRISPR Immunity Relies on the Consecutive Binding and Degradation of Negatively Supercoiled Invader DNA by Cascade and Cas3. *Mol. Cell* **2012**, *46*, 595–605.

81. Howard, J.A.; Delmas, S.; Ivančić-Baće, I.; Bolt, E.L. Helicase dissociation and annealing of RNA-DNA hybrids by *Escherichia coli* Cas3 protein. *Biochem. J.* **2011**, *439*, 85–95.

82. Sinkunas, T.; Gasiunas, G.; Fremaux, C.; Barrangou, R.; Horvath, P.; Siksnys, V. Cas3 is a single-stranded DNA nuclease and ATP-dependent helicase in the CRISPR/Cas immune system. *EMBO J.* **2011**, *30*, 1335–1342.

83. Beloglazova, N.; Petit, P.; Flick, R.; Brown, G.; Savchenko, A.; Yakunin, A.F. Structure and activity of the Cas3 HD nuclease MJ0384, an effector enzyme of the CRISPR interference. *EMBO J.* **2011**, *30*, 4616–4627.

84. Deltcheva, E.; Chylinski, K.; Sharma, C.M.; Gonzales, K.; Chao, Y.; Pirzada, Z.A.; Eckert, M.R.; Vogel, J.; Charpentier, E. CRISPR RNA maturation by *trans*-encoded small RNA and host factor RNase III. *Nature* **2011**, *471*, 602–607.

85. Jinek, M.; Chylinski, K.; Fonfara, I.; Hauer, M.; Doudna, J.A.; Charpentier, E. A Programmable Dual-RNA-Guided DNA Endonuclease in Adaptive Bacterial Immunity. *Science* **2012**, *337*, 816–821.

86. Carte, J.; Wang, R.; Li, H.; Terns, R.M.; Terns, M.P. Cas6 is an endoribonuclease that generates guide RNAs for invader defense in prokaryotes. *Genes Dev.* **2008**, *22*, 3489–3496.

87. Carte, J.; Pfister, N.T.; Compton, M.M.; Terns, R.M.; Terns, M.P. Binding and cleavage of CRISPR RNA by Cas6. *RNA* **2010**, *16*, 2181–2188.

88. Wang, R.; Preamplume, G.; Terns, M.P.; Terns, R.M.; Li, H. Interaction of the Cas6 riboendonuclease with CRISPR RNAs: Recognition and cleavage. *Structure* **2011**, *19*, 257–264.

89. Hale, C.R.; Majumdar, S.; Elmore, J.; Pfister, N.; Compton, M.; Olson, S.; Resch, A.M.; Glover, C.V., 3rd.; Graveley, B.R.; Terns, R.M.; *et al.* Essential features and rational design of CRISPR RNAs that function with the Cas RAMP module complex to cleave RNAs. *Mol. Cell* **2012**, *45*, 292–302.

90. Hale, C.R.; Zhao, P.; Olson, S.; Duff, M.O.; Graveley, B.R.; Wells, L.; Terns, R.M.; Terns, M.P. RNA-guided RNA cleavage by a CRISPR RNA-Cas protein complex. *Cell* **2009**, *139*, 945–956.

91. Hale, C.; Kleppe, K.; Terns, R.M.; Terns, M.P. Prokaryotic silencing (psi)RNAs in *Pyrococcus furiosus*. *RNA* **2008**, *14*, 2572–2579.

92. Zhang, J.; Rouillon, C.; Kerou, M.; Reeks, J.; Brugger, K.; Graham, S.; Reimann, J.; Cannone, G.; Liu, H.; Albers, S.V.; *et al.* Structure and mechanism of the CMR complex for CRISPR-mediated antiviral immunity. *Mol. Cell* **2012**, *45*, 303–313.

93. Marraffini, L.A.; Sontheimer, E.J. Self versus non-self discrimination during CRISPR RNA-directed immunity. *Nature* **2010**, *463*, 568–571.

94. Young, J.C.; Dill, B.D.; Pan, C.; Hettich, R.L.; Banfield, J.F.; Shah, M.; Fremaux, C.; Horvath, P.; Barrangou, R.; VerBerkmoes, N.C. Phage-induced expression of CRISPR-associated proteins is revealed by shotgun proteomics in *Streptococcus thermophilus*. *PLoS One* **2012**, *7*, e38077.

95. Dorman, C.J. H-NS, the genome sentinel. *Nat. Rev. Microbiol.* **2007**, *5*, 157–161.

96. Westra, E.R.; Pul, Ü.; Heidrich, N.; Jore, M.M.; Lundgren, M.; Stratmann, T.; Wurm, R.; Raine, A.; Mescher, M.; van Heereveld, L.; *et al.* H-NS-mediated repression of CRISPR-based immunity in *Escherichia coli* K12 can be relieved by the transcription activator LeuO. *Mol. Microbiol.* **2010**, *77*, 1380–1393.

97. Lucchini, S.; Rowley, G.; Goldberg, M.D.; Hurd, D.; Harrison, M.; Hinton, J.C. H-NS mediates the silencing of laterally acquired genes in bacteria. *PLoS Pathog.* **2006**, *2*, e81.

98. Navarre, W.W.; Porwollik, S.; Wang, Y.; McClelland, M.; Rosen, H.; Libby, S.J.; Fang, F.C. Selective silencing of foreign DNA with low GC content by the H-NS protein in *Salmonella*. *Science* **2006**, *313*, 236–238.

99. Mojica, F.J.; Díez-Villaseñor, C. The on-off switch of CRISPR immunity against phages in *Escherichia coli*. *Mol. Microbiol.* **2010**, *77*, 1341–1345.

100. Medina-Aparicio, L.; Rebollar-Flores, J.E.; Gallego-Hernández, A.L.; Vázquez, A.; Olvera, L.; Gutiérrez-Ríos, R.M.; Calva, E.; Hernández-Lucas, I. The CRISPR/Cas Immune System Is an Operon Regulated by LeuO, H-NS, and Leucine-Responsive Regulatory Protein in *Salmonella enterica* Serovar Typhi. *J. Bacteriol.* **2011**, *193*, 2396–2407.

101. Hernández-Lucas, I.; Gallego-Hernández, A.L.; Encarnación, S.; Fernández-Mora, M.; Martínez-Batallar, A.G.; Salgado, H.; Oropeza, R.; Calva, E. The LysR-type transcriptional regulator LeuO controls expression of several genes in *Salmonella enterica* serovar Typhi. *J. Bacteriol.* **2008**, *190*, 1658–1670.

102. Majumder, A.; Fang, M.; Tsai, K.J.; Ueguchi, C.; Mizuno, T.; Wu, H.Y. LeuO expression in response to starvation for branched-chain amino acids. *J. Biol. Chem.* **2001**, *276*, 19046–19051.

103. Drahos, D.J.; Hendrix, R.W. Effect of bacteriophage lambda infection on synthesis of *groE* protein and other *Escherichia coli* proteins. *J. Bacteriol.* **1982**, *149*, 1050–1063.

104. MacRitchie, D.M.; Buelow, D.R.; Price, N.L.; Raivio, T.L. Two-component signaling and gram negative envelope stress response systems. *Adv. Exp. Med. Biol.* **2008**, *631*, 80–110.

105. Perez-Rodriguez, R.; Haitjema, C.; Huang, Q.; Nam, K.H.; Bernardis, S.; Ke, A.; DeLisa, M.P. Envelope stress is a trigger of CRISPR RNA-mediated DNA silencing in *Escherichia coli*. *Mol. Microbiol.* **2011**, *79*, 584–599.

106. Baranova, N.; Nikaido, H. The *baeSR* two-component regulatory system activates transcription of the *yegMNOB* (*mdtABCD*) transporter gene cluster in *Escherichia coli* and increases its resistance to novobiocin and deoxycholate. *J. Bacteriol.* **2002**, *184*, 4168–4176.

107. Yosef, I.; Goren, M.G.; Kiro, R.; Edgar, R.; Qimron, U. High-temperature protein G is essential for activity of the *Escherichia coli* clustered regularly interspaced short palindromic repeats (CRISPR)/Cas system. *Proc. Natl. Acad. Sci. USA* **2011**, *108*, 20136–20141.

108. Chen, C.C.; Ghole, M.; Majumder, A.; Wang, Z.; Chandana, S.; Wu, H.Y. LeuO-mediated transcriptional derepression. *J. Biol. Chem.* **2003**, *278*, 38094–38103.

109. Viswanathan, P.; Murphy, K.; Julien, B.; Garza, A.G.; Kroos, L. Regulation of *dev*, an operon that includes genes essential for *Myxococcus xanthus* development and CRISPR-associated genes and repeats. *J. Bacteriol.* **2007**, *189*, 3738–3750.

110. Shinkai, A.; Kira, S.; Nakagawa, N.; Kashihara, A.; Kuramitsu, S.; Yokoyama, S. Transcription activation mediated by a cyclic AMP receptor protein from *Thermus thermophilus* HB8. *J. Bacteriol.* **2007**, *189*, 3891–3901.

111. Deng, L.; Kenchappa, C.S.; Peng, X.; She, Q.; Garrett, R.A. Modulation of CRISPR locus transcription by the repeat-binding protein Cbp1 in *Sulfolobus*. *Nucleic Acids Res.* **2012**, *40*, 2470–2480.

112. Peng, X.; Brugger, K.; Shen, B.; Chen, L.; She, Q.; Garrett, R.A. Genus-specific protein binding to the large clusters of DNA repeats (short regularly spaced repeats) present in *Sulfolobus* genomes. *J. Bacteriol.* **2003**, *185*, 2410–2417.

Permissions

The contributors of this book come from diverse backgrounds, making this book a truly international effort. This book will bring forth new frontiers with its revolutionizing research information and detailed analysis of the nascent developments around the world.

We would like to thank all the contributing authors for lending their expertise to make the book truly unique. They have played a crucial role in the development of this book. Without their invaluable contributions this book wouldn't have been possible. They have made vital efforts to compile up to date information on the varied aspects of this subject to make this book a valuable addition to the collection of many professionals and students.

This book was conceptualized with the vision of imparting up-to-date information and advanced data in this field. To ensure the same, a matchless editorial board was set up. Every individual on the board went through rigorous rounds of assessment to prove their worth. After which they invested a large part of their time researching and compiling the most relevant data for our readers.

The editorial board has been involved in producing this book since its inception. They have spent rigorous hours researching and exploring the diverse topics which have resulted in the successful publishing of this book. They have passed on their knowledge of decades through this book. To expedite this challenging task, the publisher supported the team at every step. A small team of assistant editors was also appointed to further simplify the editing procedure and attain best results for the readers.

Apart from the editorial board, the designing team has also invested a significant amount of their time in understanding the subject and creating the most relevant covers. They scrutinized every image to scout for the most suitable representation of the subject and create an appropriate cover for the book.

The publishing team has been an ardent support to the editorial, designing and production team. Their endless efforts to recruit the best for this project, has resulted in the accomplishment of this book. They are a veteran in the field of academics and their pool of knowledge is as vast as their experience in printing. Their expertise and guidance has proved useful at every step. Their uncompromising quality standards have made this book an exceptional effort. Their encouragement from time to time has been an inspiration for everyone.

The publisher and the editorial board hope that this book will prove to be a valuable piece of knowledge for researchers, students, practitioners and scholars across the globe.

List of Contributors

Hongduo Bao, Hui Zhang, Yan Zhou, Lili Zhang and Ran Wang
Institute of Food Safety, Jiangsu Academy of Agricultural Sciences, State Key Laboratory Cultivation Base of MOST- Jiangsu Key Laboratory of Food Quality and Safety, Nanjing 210014, China

Pengyu Zhang
Ginling College, Nanjing Normal University, Nanjing 210097, China

Erin Noble, Michelle M. Spiering and Stephen J. Benkovic
Pennsylvania State University, Department of Chemistry, 414 Wartik Laboratory, University Park, PA 16802, USA

Md Zahidul Islam Pranjol
Institute of Clinical and Biomedical Science, University of Exeter Medical School, Exeter, Devon EX1 2LU, UK

Amin Hajitou
Phage Therapy Group, Department of Medicine, Burlington Danes Building, Imperial College London, Hammersmith Hospital, Du Cane Road, London W12 0NN, UK

Lindsey B. Coulter and Rodney E. Rohde
Clinical Laboratory Science Program, Texas State University, 601 University Drive, San Marcos, TX 78666, USA

Robert J. C. McLean and Gary M. Aron
Department of Biology, Texas State University, San Marcos, 601 University Drive, TX 78666, USA

David S. Morris and Peter E. Prevelige, Jr
Department of Microbiology, University of Alabama at Birmingham, 845 19th Street S, BBRB 414, Birmingham, AL 35294, USA

Liliana Costa, Ângela Cunha and Adelaide Almeida
Department of Biology and CESAM, University of Aveiro, Aveiro 3810-193, Portugal

Maria Amparo F. Faustino and Maria Graça P. M. S. Neves
Department of Chemistry and QOPNA, University of Aveiro, Aveiro 3810-193, Portugal

Jennifer Mahony
Department of Microbiology, University College Cork, Western Road, Cork, Ireland

Douwe van Sinderen
Department of Microbiology, University College Cork, Western Road, Cork, Ireland
Alimentary Pharmabiotic Centre, Biosciences Institute, University College Cork, Western Road, Cork, Ireland

Britt Koskella and Sean Meaden
BioSciences, University of Exeter, Cornwall Campus, Tremough, TR10 9EZ, UK

Sujoy Chatterjee and Eli Rothenberg
Department of Biochemistry and Molecular Pharmacology, NYU Medical School, 550 First Avenue, New York, NY 10016, USA

Marie Deghorain and Laurence van Melderen
Laboratoire de Génétique et Physiologie Bactérienne, Faculté de Sciences, IBMM, Université Libre de Bruxelles (ULB), Gosselies B-6141, Belgium

Pieter Moons, David Faster and Abram Aertsen
Laboratory of Food Microbiology, Department of Microbial and Molecular Systems (M2S), Faculty of Bioscience Engineering, Katholieke Universiteit Leuven, Kasteelpark Arenberg 22, Leuven 3001, Belgium

Stephanie D. Friedman and Fred J. Genthner
US Environmental Protection Agency, Gulf Ecology Division, 1 Sabine Island Drive, Gulf Breeze, FL, 32561, USA

Wyatt C. Snellgrove
William Carey University College of Osteopathic Medicine, 498 Tuscan Avenue, Hattiesburg, MS 39401, USA

Corinna Richter, James T. Chang and Peter C. Fineran
Department of Microbiology and Immunology, University of Otago, PO Box 56, Dunedin 9054, New Zealand

Index

A

Adenovirus, 36, 41, 46-47, 49, 80-81, 90, 94

Adsorption Rate, 157

Amino Acid, 39, 66, 92, 112, 137, 165, 179, 201-203, 205-209, 211-217, 219, 235

Astrovirus, 216, 221

B

Bacillus Subtilis, 116-117, 119, 127, 129, 177

Bacterial Communities, 132-133, 137, 139-140, 142, 146, 149

Bacterial Isolate, 141

Bacterial Phenotype, 134, 139

Bacterial Population, 132, 135

Bacteriophages, 1-3, 13-14, 16-18, 42, 49, 58-59, 72, 74-75, 78, 80-81, 85-88, 90-91, 93, 96, 98, 102-104, 107-108, 110-111, 130, 132-134, 139, 143, 147, 149-150, 153, 169, 182-188, 190, 198-202, 208, 219, 223

Binding Site, 98, 123, 128, 192, 233, 235

Biofilm, 51-59, 170, 183, 185, 189-200

C

Capsid, 42-43, 60-61, 65, 71-74, 87, 89, 93-94, 96-97, 119, 121, 124, 129, 139, 153, 171-173, 176-177, 180, 201, 206-208, 215-216, 219

Cas Protein, 223, 226, 233-234, 242

Cascade, 39, 125, 222, 227-232, 239-240, 242

Catalytic Domain, 208, 213, 215

Cell Length, 157

Cell Surface, 39, 42, 57, 61, 118-119, 125, 130, 151, 153-158, 161-166, 168, 223

Chloramphenicol Resistance, 191, 194, 196

Chromosome, 32, 152, 170-171, 174, 177, 184, 229, 235, 238

Codon, 71, 203, 205-206, 208-209, 213

Coevolution, 132-138, 140-141, 146, 148, 200

Cophenetic Correlation, 219

D

Dairy Phages, 116-117, 121, 126-127, 147, 186

Dengue, 81, 101, 109, 215, 221

Dilution Series, 141, 144-145, 194-195

E

Endoribonuclease, 230, 232-233, 241-242

Enterovirus, 81, 94, 96, 110, 221

Escherichia Coli, 2, 16-18, 31, 51-53, 59, 73, 96, 113, 116, 119, 128, 131, 137, 149, 151-152, 154, 165-168, 189, 197-200, 217, 223, 237-244

F

Food-borne Pathogen, 2, 139, 190

G

Genetic Distance, 134, 211

Genogroup, 201-209, 211-215, 217-219

Genome Length, 205, 215-216

Gram-positive Bacteria, 119, 176, 183, 185

H

Hepatitis, 76-77, 105, 107, 113, 215, 220

Horizontal Gene Transfer, 170, 174, 222-223, 238

Horizontal Transfer, 134, 169, 174-176, 179, 181, 239

Host Cell, 13, 42, 95, 108, 119-120, 129, 134, 147, 151-154, 159, 162, 172

Host Range, 1-2, 122-123, 125, 132-142, 144-145, 147, 149-150, 165, 181, 188, 190

L

Lactococcal Phage, 119-121, 123-124, 127, 129-131

Lactococcus Lactis, 116-117, 119, 127-128, 130-131, 148

Lambda Phage, 100, 108, 151, 155

Leviviridae, 201-204, 206, 208-209, 211-213, 215, 217, 219-220

Lipoteichoic Acid, 119, 123

Lysogenic Conversion, 174-175, 182, 184, 189-191

Lytic Cycle, 134, 174, 196

M

Male-specific Coliphage, 201, 203

Mammalian Viruses, 75, 78, 80-81, 85, 87, 89-91, 93-94, 98-99, 101-104, 130

Manual Alignment, 211, 213, 215-218

Molecular Biology, 151-152, 168, 202

Molecular Mechanism, 48, 50, 175, 184

Morphological Families, 171, 176, 179-180

Morphology, 3, 64-68, 117, 121-122, 176-177, 180, 182, 186

Mosaicism, 169, 177, 179-181

N

Nucleic Acid, 77, 89, 96, 99, 103-104, 134, 222, 224, 227

Nucleotide Sequence, 66, 166, 183, 187, 199, 202, 204, 206, 216, 218, 220, 237

P

Phage Genome, 43, 97, 119, 139, 174, 191

Phage Therapy, 16, 34, 52, 57-58, 132, 139, 149, 181, 190, 198

Phage-host Interaction, 124, 126-127

Phosphorylation, 208, 235

Photosensitizer, 75-76, 78, 80-81, 98, 106, 113

Phylogeny, 123, 197, 203, 226

Plasmids, 61, 134, 137, 174, 197, 222-227, 230

Priming Acquisition, 228-229

Proflavine, 80, 85-86, 108

Protospacer, 228-230, 233-234

R

Recombinant Strain, 213, 221

Recombination Analysis, 211, 218, 221

Recombination Event, 201-203, 215, 219

Replicase Gene, 202, 205-206, 208-215, 219-220

Replicase Nucleotide, 206, 211, 218

Resistance Development, 104, 108, 189-190, 192-193

Ribonucleoprotein, 222, 227, 230

S

Siphophage, 118

Siphoviridae, 117-119, 121, 124, 126, 129, 169, 171-173, 176-178, 180, 186

Spacer Acquisition, 226-227, 229, 240-241

Staphylococci, 169-172, 175-176, 178-182, 184-186, 188, 238

Staphylococcus, 52, 55, 58-59, 98, 105, 114, 139, 153, 169-170, 180, 182-188, 225

T

Transduction, 14, 41, 43-46, 49, 139, 149, 174, 181, 199, 223

V

Viral Photoinactivation, 75, 77, 87, 92, 103

Virulence, 59, 139, 145, 153, 169-170, 172-177, 180-181, 183-185, 190-191, 196-198, 223, 238